Discrete Geometric Analysis

CONTEMPORARY MATHEMATICS

347

Discrete Geometric Analysis

Proceedings of the
First JAMS Symposium on
Discrete Geometric Analysis

December 12–20, 2002
Sendai, Japan

Motoko Kotani
Tomoyuki Shirai
Toshikazu Sunada
Editors

American Mathematical Society
Providence, Rhode Island

2000 *Mathematics Subject Classification.* Primary 20-06, 60-06, 37-06, 52-06;
Secondary 52Cxx, 37Axx.

Library of Congress Cataloging-in-Publication Data

Discrete geometric analysis: the first JAMS conference, discrete geometric analysis, December 12–20, 2002, Sendai, Japan /Motoko Kotani, Tomoyuki Shirai, Toshikazu Sunada, editors.

p. cm. – (Contemporary mathematics, ISSN 0271-4132 ; 347)

Includes bibliographical references.

ISBN 0-8218-3351-0 (alk. paper)

1. Geometry, Differential–Congresses. 2. Mathematical analysis–Congresses. I. Kotani, Motoko, 1960- II. Shirai, Tomoyuki. III. Sunada, T. (Toshikazu), 1948– IV. Contemporary mathematics (American Mathematical Society) ; v. 347.

QA641.D47 ~~2004~~ 2005

516.3′6–dc22   2004041141

∞ The paper used in this book is acid-free and falls within the guidelines
established to ensure permanence and durability.
Visit the AMS home page at http://www.ams.org/

10 9 8 7 6 5 4 3 2 1   09 08 07 06 05 04

Contents

Foreword

The idea of this JAMS symposium was brought about a few years ago by the board members of our Foundation, the Japan Association for Mathematical Sciences, which was founded in 1984 primarily for the purpose of supporting mathematical research and education.

It was inspired by the spirit of a series of meetings and gatherings of pure mathematicians which began in 1956 and continued with generous financing by Mr. Toyosaburo Taniguchi. After 1974, the annual meetings were expanded into broader fields of basic research and named the Taniguchi International Symposia, supported by the Taniguchi Foundation. In any event, Taniguchi's generosity was unique in those days because it was directed to help pure mathematicians who, in his words, could hardly get enough financial support from government or from industry. He once told us that he was deeply moved by the sight of pure mathematicians talking in earnest to each other about difficult subject matters that he could hardly understand.

Mr. Taniguchi was a highly respected businessman. Though his profession was rather far from the disciplines of advanced mathematics, he was unusually close to our heart in understanding the true needs of pure mathematicians; that is, their desire to meet each other, to get acquainted with each other, and to discuss ideas and thoughts person to person. Pure mathematicians don't need big machinery, spacious laboratories, nor a large staff for expensive experiments. Their desire is to meet each other and exchange ideas. They love to chat face-to-face in search of new ideas. Taniguchi wanted to help mathematicians do that. He was happier to see that invited mathematicians to his symposia were amicably talking to each other, rather than to see them spending too much time in formal lectures. He once told us that researchers should first become mutual friends in order to readily exchange new ideas, even raw and uncooked ones.

In 1999, after the 42nd and the last pure math Taniguchi International Symposia since 1974 (which followed the pre-series of about 20 annual meetings supported by Taniguchi during those post-WWII years in Japan), the final memorial symposium was held in homage to Mr. Taniguchi who had died in 1994. With that, the Taniguchi Foundation was officially closed.

Mr. Taniguchi and his Symposia have gone, but they are dearly remembered by many Japanese mathematicians as well as by many of the invitees from abroad. It is our sincere wish at the Japan Association for Mathematical Sciences to restore something similar to Taniguchi's ideal, even though our means and scale are clearly below his. This JAMS symposium was indeed the first trial after the Taniguchi's, which was brought to reality with much thanks to the planning and organization of Professor Kotani and her colleagues at the Tohoku University.

Heisuke Hironaka

Preface

This volume of proceedings is an outcome of the symposium on Discrete Geometric Analysis held at Matsusima, Japan 12th-16th December, 2002, and the workshop at Tohoku Uniersity, Japan, 18th-20th December, 2002, under the auspices of the Japan Association for Mathematical Sciences (JAMS). The members of the organizing committee were Motoko Kotani, Koji Fujiwara, Toshikazu Sunada, and Hajime Urakawa.

According to the policy of JAMS, as explained in the Foreword by Professor Heisuke Hironaka, the Chair of the Foundation, the JAMS symposium was organized in a small and closed form and participants were mainly young researchers who are actively working in the field. The workshop that followed was open to the public so that many mathematicians took part in the activity. Such a style of meetings made possible the opportunity to effectively exchange their ideas and research findings.

As the title of the conference suggests, the contributed papers are concerned with analysis and geometry of discrete objects; for instance, graphs and discrete groups. The most recent developments in this field are accounted in each paper.

On behalf of the organizers, we wish to express our sincere thanks to JAMS for providing generous support to hold this conference and to all who made the conference and this volume possible.

Feb. 2004

Motoko Kotani
Tomoyuki Shirai
Toshikazu Sunada
Editors

List of Participants

Georgios Alexopoulos
Universite Paris-Sud

Thierry Coulhon
Universite de Cergy-Pontoise

Vadim Kaimanovich
IRMAR, Universite Rennes1

Alex Lubotzky
Hebrew University of Jerusalem

Nicolas Monod
The University of Chicago

Guoliang Yu
Vanderbilt University

Andrzej Zuk
The University of Chicago

Koji Fujiwara
Tohoku University

Yusuke Higuchi
Showa University

Hiroyasu Izeki
Tohoku University

Masaki Izumi
Kyoto University

Tsuyoshi Kato
Kyoto University

Atsushi Katsuda
Okayama University

Sadayoshi Kojima
Tokyo Institute of Technology

Motoko Kotani
Tohoku University

Takashi Kumagai
Kyoto University

Hirofumi Nagoshi
Kyoto University

Shu Nakamura
University of Tokyo

Tomoyuki Shirai
Kanazawa University

Toshikazu Sunada
Tohoku University

Hajime Urakawa
Tohoku University

Contemporary Mathematics
Volume **347**, 2004

On the asymptotic behavior of convolution powers and heat kernels on Lie groups

Noël Lohoué and Georgios Alexopoulos

ABSTRACT. We prove Gaussian estimates for the convolution powers of a continuous density with compact support on a connected semisimple Lie group G. We also prove similar estimates for the heat kernel of a left invariant sub-Laplacian on G. The density and the sub-Laplacian need not be symmetric. When G is a compact extension of an abelian Lie group, then we obtain an asymptotic expansion.

1. Introduction

Let G be a connected noncompact semisimple Lie group with finite center and let e be the identity element and $\mathfrak{g}$ the Lie algebra of G. We identify $\mathfrak{g}$ with the left invariant vector fields on G.

Let us fix an Iwasawa decomposition $G = KAN$ of G. K is a maximal compact subgroup of G, $A \cong \mathbb{R}^d$ is abelian and N is simply connected nilpotent. $S = NA$ is a solvable Lie group. Every $g \in G$ can be written in a unique way as $g = \kappa a n$, with $\kappa \in K$, $a \in A$, $n \in N$; we set $\mathcal{H}(g) = a$. Let $\mathfrak{k}$, $\mathfrak{a}$ and $\mathfrak{n}$ be the Lie algebras of K, A and N respectively. We have $\mathfrak{g} = \mathfrak{k} \oplus \mathfrak{a} \oplus \mathfrak{n}$. Let $\mathfrak{a}^+$ be the corresponding positive Weyl chamber and let us denote by Σ the set of positive restricted roots and by Σ_0 the set of indivisible ones. Note that we have the Cartan decomposition $G = K \exp \overline{\mathfrak{a}^+} K$.

For every root $\zeta \in \Sigma$ let us denote by $\mathfrak{g}_\zeta$ the corresponding root space. Then $\mathfrak{n} = \sum_{\zeta \in \Sigma} \mathfrak{g}_\zeta$. Let m_ζ denote the multiplicity of the root $\zeta \in \Sigma$ and let $\rho = (1/2) \sum_{\zeta \in \Sigma} m_\zeta \zeta$. Note that if G is complex, then $\Sigma_0 = \Sigma$ and $m_\zeta = 2$, for all $\zeta \in \Sigma$. Let also $\mathfrak{a}^*$ be the dual of $\mathfrak{a}$ and $\mathfrak{a}^*_{\mathbb{C}}$ its complexification.

A function f is called spherical if $f(\kappa_1 x \kappa_2) = f(x)$, $\kappa_1, \kappa_2 \in K$, $x \in G$. The elementary spherical function associated to $\lambda \in \mathfrak{a}^*_{\mathbb{C}}$, is given by

$$\varphi_\lambda(x) = \int_K e^{(i\lambda - \varrho)\mathcal{H}(x\kappa)} d\kappa.$$

2000 *Mathematics Subject Classification.* Primary 22E30; Secondary 43A80, 43A90, 60B99.

Key words and phrases. Semisimple Lie groups, density, convolution, sub-Laplacian, Harnack inequality, heat kernel, homogenization.

Let U be a compact neighborhood of the identity element e of G, let $U^n = \{x_1 x_2 ... x_n : x_1, x_2, ..., x_n \in U\}$ and let $|x| = \inf\{n : x \in U^n\}$.

The goal of this article is to study the asymptotic behavior of the convolution powers of densities and heat kernels on G. The diagonal behavior has been studied in [**B1, L2**]. In this article, we are mainly interested in the off-diagonal behavior.

1.1. Densities. Let us fix a function ϕ (density) with the following properties:

1. $\phi \geq 0$ and $\int \varphi(x)dx = 1$,
2. ϕ is continuous ,
3. $\phi(e) > 0$ and $\text{supp}(\phi) = \overline{\{x \in G : \phi(x) > 0\}}$ is compact.

Given $f, h \in L^1$, let us set $f * h(x) = \int f(y)h(y^{-1}x)dy$, $x \in G$ and let us denote by $\phi^{*k} = \phi * \phi * \cdots * \phi$ the kth convolution power of the density ϕ.

THEOREM 1.1.1. *Let $C > 0$. Then, there are $c, \mu > 0$ such that*

$$\phi^{*k}(x) \leq ck^{-(d+2|\Sigma_0|)/2} e^{-\mu k} \varphi_0(x) \exp\left(-\frac{|x|^2}{ck}\right), \qquad k \in \mathbb{N},\ x \in G, \tag{1.1.1}$$

$$\phi^{*k}(x) \geq \frac{1}{c} k^{-(d+2|\Sigma_0|)/2} e^{-\mu k} \varphi_0(x), \qquad k \in \mathbb{N}, \quad |x| \leq C\sqrt{k}. \tag{1.1.2}$$

Let us observe that if $\Phi : L^2 \to L^2$ is defined by $\Phi(f)(x) = \int f(xy)\phi(y)dy$, then $e^{-\mu}$ is equal to the spectral radius of Φ. In particular, if the density ϕ is also symmetric, i.e. if $\phi(x) = \phi(x^{-1})$, $x \in G$, then $e^{-\mu} = \|\Phi\|_{2\to 2}$. We point out that, if ϕ symmetric, then we have the following weaker but more general estimate.

$$\phi^{*k}(x) \leq ce^{-\mu k} \exp\left(-\frac{|x|^2}{ck}\right), \qquad k \in \mathbb{N},\ x \in G,$$

This estimate can be proved, for example, by considering the convolution operator $\Phi' = e^{\mu}\Phi$ which is symmetric and satisfies $\|\Phi\|_{2\to 2} = 1$ and by arguing as in [**A3**].

We can also obtain estimates by making use of the property of conservation of support for the Abel transform. This idea has been exploited, in the context of symmetric spaces, in [**An2, An3**]. As we shall show, following this approach, we can obtain the following estimate.

$$\phi^{*k}(x) \leq c\varphi_0(x) e^{-\mu k} \exp\left(-\frac{|x|^2}{ck}\right), \qquad x \in G,\ k \in \mathbb{N}. \tag{1.1.3}$$

1.2. Heat kernels. Let us fix $E_0, E_1, ..., E_n$ left invariant vector fields on G and let us assume that the vector fields $E_1, ..., E_n$ satisfy Hörmander's condition, i.e. their successive Lie brackets $[E_{i_1}, [E_{i_2}, [..., [E_{i_{k-1}}, E_{i_k}]...]]$, $1 \leq i_1, ..., i_k \leq n$ generate the tangent space $T_x G$ at very point $x \in G$.

Let us consider the sub-Laplacian $L = -(E_1^2 + ... + E_n^2) + E_0$ and let us denote by $P_t(x, y)$ the associated heat kernel, i.e. the fundamental solution of the heat equation $((\partial/\partial t) + L)\, u = 0$.

THEOREM 1.2.1. *Let $C > 0$. Then, there are $c, \mu > 0$ such that*

$$P_t(x, e) \leq ct^{-(d+2|\Sigma_0|)/2} e^{-\mu t} \varphi_0(x) \exp\left(-\frac{|x|^2}{ct}\right), \qquad t \geq 1,\ x \in G, \tag{1.2.1}$$

$$P_t(x,e) \geq \frac{1}{c} t^{-(d+2|\Sigma_0|)/2} e^{-\mu t} \varphi_0(x), \qquad t \geq 1, \quad |x| \leq C\sqrt{t}. \tag{1.2.2}$$

If the sub-Laplacian L is symmetric, i.e. if $E_0 = 0$, then the above result was proved in [**LA**]. Note that then μ is the spectral gap, i.e. L has spectral decomposition on L^2, $L = \int_\mu^\infty \lambda dE_\lambda$. We point out again that if L is symmetric, then we have the following weaker but more general stimate.

$$P_t(x,e) \leq e^{-\mu t} \exp\left(-\frac{|x|^2}{ct}\right), \qquad t \geq 1,\ x \in G.$$

It can be proved by making use of the property of finite propagation speed for the operator $\cos t\sqrt{L-\mu}$ and by arguing, for example, as in [**V, p. 238**].

Also, as in the case of a density, by exploiting the property of conservation of support for the Abel transform we can obtain the following estimate.

$$P_t(x,e) \leq c\varphi_0(x) e^{-\mu t} \exp\left(-\frac{|x|^2}{ct}\right), \qquad x \in G,\ t \geq 1. \tag{1.2.3}$$

1.3. Compact extensions of abelian Lie groups. As a byproduct we obtain an asymptotic expansion, when the Lie group G is a compact extension of an abelian Lie group and the density ϕ or the sub-Laplacian L is centered.

More precisely, let us assume that G is a connected Lie group which has a closed analytic abelian subgroup $A \lhd G$ such that G/A is compact. By taking the quotient G/C with the maximal torus C of A, if necessary, we can assume that A is simply connected and hence that $A \cong \mathbb{R}^d$ and $G = AK$, with K a compact subgroup.

Let $G_1 = [A,K]K$ and let $\mathfrak{g}_1$ denote the Lie algebra of G_1. Then, $G/G_1 \cong \mathbb{R}^p$. Let $\mathfrak{g}_1$ denote the Lie algebra of G_1.

Let the density ϕ be as in section 1.1 and let π denote the projection $\pi : G \to G/G_1 \cong \mathbb{R}^p$. We say that ϕ is centered, if $\int \pi(x)_i \phi(x)dx = 0$, $1 \leq i \leq p$, where y_i is the i-th coordinate of $y = (y_1, ..., y_p) \in \mathbb{R}^p$.

We can associate with ϕ a so called, homogenized operator L_H (cf. [**A2**]). This is a second order elliptic differential operator with constant coefficients on A.

If ϕ is centered, then L_H does not have first order terms. Let $P_t^H(x,y)$ be the heat kernel of L_H and let us extend $P_t^H(x,y)$ to G by setting $P_t^H(xz,yw) = P_t^H(x,y)$, $x,y \in A$, $z,w \in K$.

Let us now assume that ϕ is centered and, in order to simplify the exposition, that $\phi \in C^{d+1}$ (the interested reader will observe that this last condition can be relaxed and replaced by a condition similar to the one we have in the case of $\mathbb{R}^d$, (cf. [**Fe**])). We have the following.

THEOREM 1.3.1. *There are functions $C_{m,n}(z)$, $m \in \mathbb{N}^d$, on K, such that for all $x \in A$, $z \in K$ and $k \in \mathbb{N}$*

(1.3.1)

$$\phi^{*k}(xz) \quad = \quad P_k^H(x,0) \ + \sum_{\substack{1\leq|m|\leq N\\|n|=2\nu\\|m|\geq 2\nu}} C_{m,n}(z) k^\nu \frac{\partial^{m+n}}{\partial x^{m+n}} P_k^H(x,0) \ + \ O(k^{-(d+N)/2}).$$

Let L be a left invariant sub-Laplacian on G as in section 1.2. We say that L is centered, if $E_0 \in \mathfrak{g}_1$. As before, we can associate with L a homogenized sub-Laplacian L_H (cf. [**A1**]). Again, if L is centered then L_H has no first order terms.

Let us assume that L is centered, let $P_t^H(x,y)$ be the heat kernel of L_H and let us extend $P_t^H(x,y)$ to G as before. Then, we have the following.

THEOREM 1.3.2. *There are functions $\psi^j(z)$, $\psi^{*j}(z)$, $1 \le j \le d$ and $C_{m,n}(z,w)$, $m,n \in \mathbb{N}^d$, on K, such that for all $x \in A$, $z,w \in K$ and $t \ge 1$*

(1.3.2)

$$P_t(xz,w) = P_t^H(x,0) + \sum_{1\le j\le d} \psi^j(z)\frac{\partial}{\partial x_j}P_t^H(x,0) - \sum_{1\le j\le d}\psi^{*j}(w)\frac{\partial}{\partial x_j}P_t^H(x,0)$$
$$+ \sum_{\substack{2\le|m|\le N\\|n|=2\nu\\|m|\ge\nu}} C_{m,n}(z,w)t^\nu \frac{\partial^{m+n}}{\partial x^{m+n}}P_t^H(x,0) + O(t^{-(d+N)/2}).$$

1.4. Notations. We fix Haar measures dg, dn, da and dk on G, N, A and K respectively and we assume that dk-measure$(K) = 1$ and that $dg = e^{2\rho(a)}d\kappa da dn = e^{-2\rho(a)}dn da d\kappa$. Note that $e^{-2\rho(a)}$ is equal to the modular function of the group $S = NA$. When working on AK we shall denote by dx the measure $da d\kappa$.

If a function f is defined on A, then we assume that it is extended to AK, by setting $f(xz) = f(x)$, $x \in A$, $z \in K$. If f is defined on K, then we assume that it is extended to AK, by setting $f(xz) = f(z)$, $x \in A$, $z \in K$.

The Weyl group is denoted by W.

The inner product induced by the Killing form on $\mathfrak{a}$ is denoted by $\langle .,. \rangle$ and the corresponding norm by $|.|$.

In order to simplify the notation, we identify A and $\mathfrak{a}$, with $\mathbb{R}^d$, using the exponential map and hence a will stand for both $a \in A$ and the element $\log a \in \mathfrak{a}$ satisfying $a = \exp(\log a)$. We also assume that $\langle .,. \rangle$ coincides with the Euclidean inner product on $\mathbb{R}^d$.

Since we have identified A with $\mathbb{R}^d$, every $x \in AK$ can be written as $x = az$, with $a = (a_1, ..., a_d) \in \mathbb{R}^d$, $z \in K$. We set $x_i = a_i$ and $P_i(x)$, $1 \le i \le d$.

If $y = bw$, with $b = (b_1, ..., b_d) \in \mathbb{R}^d$, $w \in K$, then we set $x+y = (a_1+b_1, ..., a_d+b_d)zw$ and $\langle x,y \rangle = a_1b_1 + ... + a_db_d$.

We set $\partial_i = (\partial/\partial x_i)$, $1 \le i \le d$.

If λ is a linear form on $\mathfrak{a}$ and $g = na\kappa$, with $n \in N, a \in A, \kappa \in K$, then we set $\lambda(g) = \lambda(a) = \lambda(\log a)$. Note that then $\lambda(g) = -\lambda(\mathcal{H}(g^{-1}))$, $g \in G$. Also $\rho(g) = \rho(a)$, if $g = na\kappa$, $n \in N$, $a \in A$, $\kappa \in K$.

The spherical Fourier transform $\mathcal{F}f(\lambda)$ of a spherical function $f \in L^1(G)$ is defined by

$$\mathcal{F}f(\lambda) = \int_G f(x)\varphi_\lambda(x^{-1})dx, \quad \lambda \in \mathfrak{a}^*.$$

We have the invertion formula

$$f(x) = c\int_{\mathfrak{a}} \mathcal{F}f(\lambda)\varphi_\lambda(x)\frac{d\lambda}{|\mathbf{c}(\lambda)|^2},$$

where $\mathbf{c}(\lambda)$ is the $\mathbf{c}$-function of Harish-Chandra.

The Abel transform, is defined by

$$\mathcal{A}f(a) = e^{-\rho(a)} \int_N f(na)dn, \qquad a \in A.$$

Let us denote by $\widehat{f}$ the Euclidean Fourier transform of f. Then, we have the important relation

$$\mathcal{F}f(\lambda) = \widehat{\mathcal{A}f}(\lambda) = \int_A \mathcal{A}f(a)e^{-i\lambda(a)}da. \tag{1.4.1}$$

Note that if G is complex, then we have the inversion formula

$$f(a) = c\left(\prod_{\zeta\in\Sigma_0} \frac{1}{e^{\zeta(a)} - e^{-\zeta(a)}}\right)\left(\prod_{\zeta\in\Sigma_0} D_\zeta\right)\mathcal{A}f(a), \qquad a \in A, \tag{1.4.2}$$

where D_ζ denotes differentiation in the direction of ζ.

We set $\Phi_k(x,y) = \varphi^{*k}(x^{-1}y)$. Note that

$$\Phi_k(f)(x) = \int \Phi_k(x,y)f(y)dy = \int \varphi^{*k}(y)f(xy)dy.$$

When $n = 1$, then we shall just write $\Phi(x,y)$ instead of $\Phi_1(x,y)$. We shall also set $\Phi_0(x,y) = \delta_{x^{-1}y}$, where δ_x is the Dirac mass at x.

We set

$$\mathcal{A}\Phi_k(x,y) = e^{-\rho(x)}e^{-\rho(y)} \int_N \Phi_k(nx,y)dn.$$

Note that

$$\mathcal{A}\Phi_k(ax,ay) = \mathcal{A}\Phi_k(x,y), \quad a \in A, x,y \in AK. \tag{1.4.3}$$

We also set

$$\overline{\Phi_k}(x,y) = \int_K\int_K \Phi_k(xz,yw)dzdw,$$

$$\mathcal{A}\overline{\Phi_k}(x,y) = e^{-\rho(x)}e^{-\rho(y)} \int_N \overline{\Phi_k}(nx,y)dn.$$

Note that

$$\mathcal{A}\overline{\Phi_k}(x,y) = \int_K\int_K \mathcal{A}\Phi_k(xz,yw)dzdw. \tag{1.4.4}$$

Let us fix a spherical function σ such that $\sigma \in C^\infty$, $\sigma \geq 0$, $\mathrm{supp}(\sigma)$ is compact and $\int \sigma(x)dx = 1$ and let us consider the kernel

$$\begin{aligned}\Phi_{\sigma,k}(x,y) &= \int_G\int_G \sigma(z)\Phi_k(xz,yw)\sigma(w)dzdw \\ &= \int_G\int_G \sigma(z)\overline{\Phi_k}(xz,yw)\sigma(w)dzdw.\end{aligned}$$

Let also

$$\mathcal{A}\Phi_{\sigma,k}(x,y) = e^{-\rho(x)}e^{-\rho(y)} \int_N \Phi_{\sigma,k}(nx,y)dn.$$

Note that

$$\mathcal{A}\Phi_{\sigma,k}(x,y) = \int_A\int_A \mathcal{A}\sigma(x_1)\mathcal{A}\overline{\Phi_k}(xx_1,yy_1)\mathcal{A}\sigma(y_1)dx_1dy_1, \quad x,y \in A. \tag{1.4.5}$$

Let us set

$$\Phi_k(x) = \Phi_k(x,e), \quad \overline{\Phi_n}(x) = \overline{\Phi_k}(x,e), \quad \Phi_{\sigma,k}(x) = \Phi_{\sigma,k}(x,e).$$

Then

$$\mathcal{A}\overline{\Phi_k}(x) = \mathcal{A}\overline{\Phi_k}(x,e), \quad \mathcal{A}\Phi_{\sigma,k}(x) = \mathcal{A}\Phi_{\sigma,k}(x,e), \quad x \in A. \tag{1.4.6}$$

Also it follows from (1.4.1) and (1.4.5) that

$$\mathcal{F}\Phi_{\sigma,k}(\lambda) = \mathcal{F}\sigma(\lambda)\mathcal{F}\overline{\Phi_k}(\lambda)\mathcal{F}\sigma(-\lambda). \tag{1.4.7}$$

We define in the same way the kernels $\overline{P_t}(x,y)$, $AP_t(x,y)$, $\mathcal{A}\overline{P_t}(x,y)$ and $P_t(x)$, $\overline{P_t}(x)$. These kernels satisfy similar relations.

Given a kernel $K(x,y)$, we denote also by K the associated operator $Kf(x) = \int K(x,y)f(y)dy$. If $S(x,y)$ is another kernel then we denote by KS the convolution product

$$KS(x,y) = \int K(x,z)S(z,y)dz.$$

We set

$$\|K\|_p = \sup\{\|K(x,.)\|_p, \|K(.,y)\|_p; x,y \in G\}, \quad 1 \le p \le \infty.$$

Throughout this paper the different constants are always denoted with the same letter c. When their dependence, or independence is significant, it is clearly stated.

Finally, in order to simplify the notation, we drop the summation signs for repeated indices.

1.5. The general strategy. The first difficulty is already the fact that the kernel $\Phi_k(x,y)$ is not spherical. To remedy this let us observe that there is a compact neighorhood V of the identity element e of G and $c > 0$ such that

$$\varphi(xy) \le c\varphi^{*2}(x), \qquad x \in G, \ y \in V.$$

This implies that for every compact neighorhood U of e , there is $n \in \mathbb{N}$ and $c > 0$ such that for all $k \in \mathbb{N}$ and $x, y \in G$ and $z, w \in U$,

$$\Phi_k(x,y) \le c\Phi_{k+n}(xz, yw). \tag{1.5.1}$$

So, it is enough to prove the estimates (1.1.1) and (1.1.2) for $\overline{\Phi_k}(x)$ or for $\Phi_{\sigma,k}(x)$.

The problem now is to obtain estimates for the spherical Fourier transform $\mathcal{F}\overline{\Phi_k}(\lambda)$. In view of (1.4.1) and (1.4.2), we should study the Abel transform $\mathcal{A}\Phi_k(x,y)$.

Using technics from Homogenization theory and in particular the method of Bloch wave expansions (cf. [**BLP, JKO, Se, Z**]), we obtain an asymptotic expansion for the kernel $\mathcal{A}\Phi_k(x,y)$ and its Euclidean Fourier transform over $A \cong \mathbb{R}^d$. Theorem 1.3.1 follows from this expansion, as a particular case. By averaging this expansion over K, we obtain an expansion for the Abel transform $\mathcal{A}\overline{\Phi_k}(a)$ and its Fourier transform.

The problem now is to invert these transforms. When G is complex, this can be done by using formula (1.4.2). In the general case we must repeat some of the work of [**AJ**]. The situation here is more complicated though because we have some extra polynomial terms in the expansion. We obtain the desired estimates by exploiting the action of the Weyl group which implies a certain number of cancelations. These calculations have been carried out in [**LA**] and we shall refer to that paper for the omitted details.

In order to prove Theorems 1.2.1 amd 1.2.2, we follow a similar approach.

2. Densities

2.1. Conjugation with e^ρ and projections on AK and K. Let us consider the kernel

$$\Phi_k^\rho(x,y) = e^{-\rho(x)}\Phi_k(x,y)e^{\rho(y)}, \qquad x,y \in G,\ k \in \mathbb{N}$$

and let us denote also by Φ_k^ρ the associated operator

$$\Phi_k^\rho f(x) = \int_G \Phi_k^\rho(x,y)f(y)dy.$$

We want to consider the projection $\Phi_{AK,k}^\rho$ of the operator Φ_k^ρ on AK. To this end, let us fix a function $f \in C^\infty$, such that $f(ng) = f(g)$, $g \in G, n \in N$. Then

$$\begin{aligned}
\int \Phi_k^\rho(x,y)f(y)dy =&e^{-\rho(x)} \int \Phi_k(x,y)e^{\rho(y)}f(y)dy \\
=&e^{-\rho(x)} \int_N\int_A\int_K \Phi_k(x,na\kappa)e^{\rho(a)}f(a\kappa)e^{-2\rho(a)}dnda d\kappa \qquad (2.1.1)\\
=&e^{-\rho(x)} \int_N\int_A\int_K \Phi_k(x,nak)e^{-\rho(a)}f(a\kappa)dnda d\kappa.
\end{aligned}$$

This shows that the kernel of the operator $\Phi_{AK,k}^\rho$ with respect to the measure $dad\kappa$ is given by

$$\begin{aligned}
\Phi_{AK,k}^\rho(x,y) =&e^{-\rho(x)}e^{-\rho(y)} \int_N \Phi_k(x,ny)dn \\
=&e^{-\rho(x)}e^{-\rho(y)} \int_N \Phi_k(nx,y)dn \\
=&\mathcal{A}\Phi_k(x,y), \quad x,y \in AK.
\end{aligned}$$

Note that this kernel is left invariant with respect to A, i.e.

$$\Phi_{AK,k}^\rho(ax,ay) = \Phi_{AK,k}^\rho(x,y), \qquad a \in A,\ x,y \in AK.$$

Let $\Phi_{K,k}^\rho$ be the projection of the operator $\Phi_{AK,k}^\rho$ on K. Arguing as before, we can see that the kernel of $\Phi_{K,k}^\rho$ with respect to the measure $d\kappa$ is given by

$$\Phi_{K,k}^\rho(z,w) = \int_A \Phi_{AK,k}^\rho(z,aw)da, \qquad z,w \in K.$$

Let us denote by $\Phi_{AK,k}^{\rho\ *}$ and $\Phi_{K,k}^{\rho\ *}$ the adjoints of $\Phi_{AK,k}^\rho$ and $\Phi_{K,k}^\rho$ with respect to the measures $dad\kappa$ and dk respectively. Then

$$\Phi_{AK,k}^{\rho\ *}(x,y) = \Phi_{AK,k}^\rho(y,x),$$

$$\Phi_{K,k}^{\rho\ *}(z,w) = \Phi_{K,k}^\rho(w,z).$$

Since the kernel $\Phi_K^\rho(z,w)$ is positive, the corresponding operator Φ_K^ρ has a simple and isolated largest eigenvalue $e^{-\mu}$, $\mu > 0$ and a corresponding eigenfunction p satisfying $p(z) > 0$, $z \in K$ (cf. [**KLS**]). Since $e^{-\mu}$ is also eigenvalue of $\Phi_K^{\rho\ *}$, let us consider a corresponding eigenfunction p^* also satisfying $p^*(z) > 0$, $z \in K$. Let us also assume that

$$\int_K p(z)p^*(z)dz = 1.$$

We extend p^*, p to AK by setting $p(a\kappa) = p(\kappa)$, $p^*(a\kappa) = p^*(\kappa)$ $a \in A$, $\kappa \in K$.

Let us consider the operator

$$\mathcal{K}_k = e^{\mu k}\frac{1}{p}\Phi^\rho_{AK,k}p.$$

We have

$$\begin{aligned}\mathcal{K}_k f(x) =& e^{\mu k}\int_{AK}\frac{1}{p(x)}\Phi^\rho_{AK,k}(x, a\kappa)p(\kappa)f(a\kappa)dad\kappa\\ =& e^{\mu k}\int_{AK}\frac{1}{p(x)}\Phi^\rho_{AK,k}(x, a\kappa)\frac{1}{p^*(\kappa)}f(a\kappa)p^*(\kappa)p(\kappa)dad\kappa.\end{aligned}$$

So the kernel of $\mathcal{K}_k$ with respect to the measure $p^*pdad\kappa$ is equal to

$$\mathcal{K}_k(x, y) = \frac{1}{p(x)}\Phi^\rho_{AK,k}(x, y)\frac{1}{p^*(y)}. \tag{2.1.2}$$

Note that the adjoint $\mathcal{K}^*_k$ of $\mathcal{K}_k$ with respect to the measure $p^*pdad\kappa$, has kernel $\mathcal{K}^*_k(x, y) = \mathcal{K}_k(y, x)$. Also,

$$\begin{aligned}&\int_{AK}\mathcal{K}_k(x, y)p^*(y)p(y)dy = 1,\\ &\mathcal{A}\Phi_k(x, y) = \Phi^\rho_{AK,k}(x, y) = e^{-\mu k}p(x)\mathcal{K}_k(x, y)p^*(y), \quad x, y \in AK.\end{aligned} \tag{2.1.3}$$

We set $\mathcal{K} = \mathcal{K}_1$.

2.2. Homogenization. If f is a continuous function on K, then we shall set

$$\langle f\rangle = \int_K f(z)p^*(z)p(z)dz.$$

Let $U = \{xz,\ x \in A, |x| < r, z \in K\}$. By taking r large enough we can assume that $\operatorname{supp}\mathcal{K}(x, .) \subseteq xU$, $x \in AK$.

Let us set $\partial_i = (\partial/\partial x_i)$, $1 \le i \le d$. Then, by the Taylor formula, there is $c > 0$ such that for all $f \in C^\infty(A)$, $x \in AK$, $y \in U$,

(2.2.1)

$$\begin{aligned}f(x + y) =& f(x) + \sum_{1\le i\le d} P_i(y)\partial_i f(x) + \frac{1}{2}\sum_{1\le i,j\le d} P_i(y)P_j(y)\partial_i\partial_j f(x) + F_0(y),\\ \partial_j f(x + y) =& \partial_j f(x) + \sum_{1\le i\le d} P_i(y)\partial_i\partial_j f(x) + F_j(y), \quad 1 \le j \le d,\end{aligned}$$

with $|F_j(x)| \le c\|\nabla^3 f\|_{L^\infty(x+U)}$, $0 \le j \le d$.

We set, for $z \in K$ and $1 \le i, j \le d$,

$$\begin{aligned}\beta_i(z) =& \int \mathcal{K}(z, yw)P_i(yw)p^*(w)p(w)dydw,\\ q_i =& \langle\beta_i\rangle, \quad a_i = \beta_i - \langle\beta_i\rangle,\\ a_{ij}(z) =& \int \mathcal{K}(z, yw)P_i(yw)P_j(yw))p^*(w)p(w)dydw.\end{aligned}$$

Combining (2.2.1) and the above definitions we have the following.

LEMMA 2.2.1. *Let $\psi \in C(K)$. Then, there is $c > 0$, such that for all $f \in C^\infty(A)$, $x \in A$, $\kappa \in K$ and all $1 \le j \le d$,*

$$(2.2.2) \qquad (I - \mathcal{K})f(xz) = -\frac{1}{2}a_{ij}(z)\partial_i\partial_j f(x) - a_i(z)\partial_i f(x) - q_i\partial_i f(x) + F_1(xz),$$

$$(I - \mathcal{K})(\psi\partial_j f)(xz) = \big((I - \mathcal{K})\psi\big)(z)\partial_j f(x) - \sum_{1\le i\le d} \mathcal{K}(\psi P_i)(z)\partial_i\partial_j f(x) + F_2(xz),$$

where $|F_i(xz)| \le c\|\nabla_X^3 f\|_{L^\infty(xU)}$, $i = 1, 2$.

We define the (first order) correctors $\psi^j, 1 \le j \le d$ (cf. [**A2**]) as functions (defined on K and extended to AK) satisfying

$$(2.2.3) \qquad (I - \mathcal{K})\psi^j = a_j, \quad \langle\psi^j\rangle = 0.$$

Note that the correctors ψ^j are unique.

We set

$$\beta_{ij}(z) = \int \mathcal{K}(z, yw)\psi^j(w)P_i(y)p^*(w)p(w)dydw, \quad z \in K,\ 1 \le i, j \le d.$$

Combining (2.2.1), (2.2.2) and (2.2.3) we have that

$$(2.2.4) \quad (I - \mathcal{K})\big(f + \sum_{1\le j\le d} \psi^j\partial_j f\big)(xz) =$$
$$- \sum_{1\le j\le d} q_j\partial_j f(x) - \sum_{1\le i,j\le d} \big(\frac{1}{2}a_{ij}(z) + \beta_{ij}(z)\big)\partial_i\partial_j f(x) + F(xz),$$

where $|F(xz)| \le c\|\nabla_X^3 f\|_{L^\infty(xU)}$.

The homogenised operator $\mathcal{D}_H$ associated to $\mathcal{K}$ is defined as the operator

$$\mathcal{D}_H = - \sum_{1\le i,j\le n_1} q_{ij}\partial_i\partial_j - \sum_{1\le j\le d} q_j\partial_j$$

where

$$q_{ij} = \left\langle \frac{1}{2}a_{ij} + \beta_{ij}\right\rangle, \quad 1 \le i, j \le d.$$

The operator $\mathcal{D}_H$ is elliptic. The proof of this fact is the same as the one given in a similar context in [**A2**] and we shall omit it.

We define the (second order) correctors $\psi^{ij}, 1 \le i, j \le d$ as functions satisfying

$$(I - \mathcal{K})\psi^{ij} = \frac{1}{2}a_{ij} + \beta_{ij} - q_{ij}, \qquad \langle\psi^{ij}\rangle = 0.$$

The following lemma is a consequence of (2.2.4) and the above definitions.

LEMMA 2.2.2. *There is $c > 0$, such that for all $f \in C^\infty(A)$,*

$$(2.2.5) \qquad (I - \mathcal{K})\left(f + \sum_{1\le j\le d} \psi^j\partial_j f + \sum_{1\le i,j\le d} \psi^{ij}\partial_i\partial_j f\right) = \mathcal{D}_H f + F$$

with the function F satisfying $|F(xz)| \le c\|\nabla_X^3 f(x)\|_{L^\infty(xU)}$, $x \in A$, $z \in K$.

2.3. Estimates for the kernel $\mathcal{K}_k$. We start with the observation that there is a compact neighbohood U of e and $c > 0$ such that

$$\mathcal{K}_k(x, y) \geq c, \quad x, y \in U.$$

From this observation, and by adapting certain arguments from [**VSC**] (cf. [**LA**] for the details), we can prove the following estimate concerning the uniform decay of the kernel $\mathcal{K}_k(x, y)$.

THEOREM 2.3.1. *There is $c > 0$ such that for all $x, y \in AK$ and $k \in \mathbb{N}$*

$$|\mathcal{K}_k(x, y)| \leq ck^{-d/2}. \tag{2.3.1}$$

As we shall show in section 2.6, we have

$$q_j = \langle \beta_j \rangle = 0, \quad 1 \leq j \leq d.$$

So in what follows we shall make this assumption. Then, making use of (2.3.1) and (2.2.5) and by working in the same way as in [**A2**], we can prove the following Gaussian estimate.

THEOREM 2.3.2. *There is $c > 0$ such that for all $x, y \in A$, $z, w \in K$ and $k \in \mathbb{N}$*

$$\mathcal{K}_k(xz, yw) \leq c\, k^{-d/2} \exp\left(-\frac{|x-y|^2}{ck}\right). \tag{2.3.2}$$

We can also prove similar estimates for the space and time differences. More precisely, given $x_0 \in A$ and $z_0 \in K$, let us set $\partial_{x_0}\mathcal{K}_k(xz, yw) = \mathcal{K}_k(xx_0z, yw) - \mathcal{K}_k(xz, yw)$ and $\partial_{z_0}\mathcal{K}_k(xz, yw) = \mathcal{K}_k(xzz_0, yw) - \mathcal{K}_k(xz, yw)$, for $x, y \in A$, $z, w \in K$. Let us also set $\partial_1\mathcal{K}_k = \mathcal{K}_{k+1} - \mathcal{K}_k$. Then, arguing as in [**A2**], we can prove the following.

THEOREM 2.3.3. *Let $\ell, m \in \mathbb{N}$ and $z_0 \in K$, $x_1, ... x_m \in A$. Then there is $c > 0$ such that for all $x, y \in A$, $z, w \in K$ and $k \in \mathbb{N}$*

$$\begin{aligned} \partial^\ell \partial_{z_0} \mathcal{K}_k(xz, yw) &\leq c\, k^{-(d+1+2\ell)/2} \exp\left(-\frac{|x-y|^2}{ck}\right), \\ \partial^\ell \partial_{x_1} ... \partial_{x_m} \mathcal{K}_k(xz, yw) &\leq c\, k^{-(d+m+2\ell)/2} \exp\left(-\frac{|x-y|^2}{ck}\right). \end{aligned} \tag{2.3.3}$$

Let $P_t^H(x, y)$ be the heat kernel of the homogenised operator $\mathcal{D}_H$ and let us extend it to to AK by setting $P_t^H(xz, yw) = P_t^H(x, y)$, $x, y \in A$, $z, w \in K$. Then, arguing again as in [**A2**], we can prove the following Berry-Esseen estimate.

THEOREM 2.3.4. *For all $x_0 \in A$ there is $c > 0$ such that for all $x, y \in A$, $z, w \in K$ and $k \in \mathbb{N}$*

$$\begin{aligned} |\mathcal{K}_k(xz, yw) - P_k^H(x, y)| &\leq ck^{-(d+1)/2}, \\ |\partial_{x_0}\mathcal{K}_k(xz, yw) - \partial_{x_0} P_k^H(x, y)| &\leq ck^{-(d+2)/2}. \end{aligned} \tag{2.3.4}$$

We point out that if $|\Sigma_0| = 1$, then (2.3.4) suffices for the proof of Theorem 1.2.1 and we do not need the Bloch wave expansion below.

2.4. Proof of the estimate (1.1.3). Let us set

$$V_r = \{x \in A : |x| \leq r\}, \qquad U_r = KV_rK \ .$$

Then, the conservation of supports property of the Abel transform says that if f is a smooth spherical function then $\operatorname{supp} f \subseteq U_r$ iff $\operatorname{supp} \mathcal{A}f \subseteq V_r$ (cf. for example [**An2**]).

Let us fix a smooth function τ on A, such that $|\nabla\tau(y)| \leq 1$, $y \in A$, $\tau(y) = 0$, $|y| \leq 1$ and $\tau(y) = 1$, $|y| \geq 2$. Let also $C > 0$ and set $\tau_r(y) = \tau(C|y|/r)$, $r \geq 1$.

Let us set

$$\overline{\mathcal{K}_k}(x) = \int_K \int_K p(z)K_k(xz, w)p^*(w)dzdw, \quad x \in A.$$

Then, by (2.1.3),

$$\begin{aligned}\mathcal{A}\overline{\Phi_k}(x) =& e^{-\mu k}\overline{\mathcal{K}_k}(x)\\ \mathcal{A}\Phi_{\sigma,k}(x) =& e^{-\mu k}\int_A\int_A \sigma(y^{-1})\overline{\mathcal{K}_k}(yxh)\sigma(h)dydh.\end{aligned}$$

Let

$$\begin{aligned}F_{r,k}(x) =& \tau_r(x)\overline{\mathcal{K}_k}(x),\\ F_{\sigma,r,k}(yx) =& \int_A\int_A \mathcal{A}\sigma(y^{-1})\ F_{r,k}(x)(yxh)\ \mathcal{A}\sigma(h)dydh.\end{aligned}$$

Then

$$\widehat{F_{\sigma,r,k}}(\xi) = \int_A F_{\sigma,r,k}(y)e^{-i\langle\xi,y\rangle}dy = \widehat{\sigma}(\xi)\widehat{F_{r,k}}(\xi)\widehat{\sigma}(-\xi).$$

Also, by (2.3.2),

$$|\widehat{F_{r,k}}(\xi)| \leq c\exp\left(-\frac{r^2}{ck}\right), \quad \xi \in A.$$

Let us recall that σ is a compactly supported smooth function. So, by chosing the constant C large enough, there is $c > 0$ such that

$$\mathcal{A}\Phi_{\sigma,k}(x) = e^{-\mu k}F_{\sigma,r,k}(x), \quad x \notin V_{r/2}, r \geq c.$$

By the conservation of the supports property of the Abel transform, this implies that for all $x \notin U_{r/2}$, $r \geq c$

$$\begin{aligned}\Phi_{\sigma,k}(x) =& ce^{-\mu k}\int_A \widehat{F_{\sigma,r,k}}(\xi)\varphi_\xi(x)\frac{d\xi}{|\mathbf{c}(\xi)|^2}\\ \leq& ce^{-\mu k}\int_A |\widehat{F_{\sigma,r,k}}(\xi)||\varphi_\xi(x)|\frac{d\xi}{|\mathbf{c}(\xi)|^2}\\ \leq& ce^{-\mu k}\int_A |\widehat{F_{r,k}}(\xi)||\widehat{\sigma}(\xi)||\widehat{\sigma}(-\xi)|\varphi_0(x)\frac{d\xi}{|\mathbf{c}(\xi)|^2}\\ \leq& ce^{-\mu k}\varphi_0(x)\exp\left(-\frac{r^2}{ck}\right)\int_A |\widehat{\sigma}(\xi)||\widehat{\sigma}(-\xi)|\frac{d\xi}{|\mathbf{c}(\xi)|^2}\\ \leq& ce^{-\mu k}\varphi_0(x)\exp\left(-\frac{r^2}{ck}\right).\end{aligned}$$

The estimate (1.1.3) follows by taking $r = |x|$.

2.5. Bloch wave expansions (cf. [**BLP, JKO, Se**]). Given $\xi \in A$, let us consider the kernel

$$(2.5.1)\qquad \begin{aligned} G_k(\xi, z, w) &= \int_A \mathcal{K}_k(z, yw) e^{i\langle \xi, y\rangle} dy \\ &= \int_A \mathcal{K}_k(xz, w) e^{-i\langle \xi, x\rangle} dx, \qquad z, w \in K, \quad \xi \in A. \end{aligned}$$

Let us also set $G(\xi, z, w) = G_1(\xi, z, w)$. Then, we have

$$(2.5.2)\qquad \mathcal{K}_k(xz, w) = (2\pi)^{-n} \int_A e^{i\langle \xi, x\rangle} G_k(\xi, z, w) d\xi, \qquad z, w \in K, \quad x \in A.$$

Let us denote also by $G_k(\xi)$ the operator

$$G_k(\xi)(f)(z) = \int_K G_k(\xi, z, w) f(w) p^*(w) p(w) dw$$

and let us observe that $\{G_k(\xi),\ \xi \in A\}$ is an analytic family of compact operators.

Since the kernel of $G_k(\xi)$ is defined as a Fourier transform of $\mathcal{K}_k$, we have that

$$(2.5.3)\qquad \|G_k(\xi)\|_{2\to 2} < 1, \quad \xi \neq 0 \quad \text{and} \quad \|G_k(\xi)\|_{2\to 2} \to 0 \ (|\xi| \to \infty).$$

Let us denote by $e^{-\lambda_0(\xi)k}$ the eigenvalue with largest absolute value of the operator $G_k(\xi)$, whenever $G_k(\xi)$ admits such an eigenvalue and let $\phi_0(\xi, z)$ be a corresponding eigenfunction.

The operator $G_k(0)$ is positive. So, it admits an eigenvalue with largest absolute value, which is isolated and simple (cf. [**KLS**]). By perturbation theory, (cf. [**DS, Se**]), the same holds for the operators $G_k(\xi)$, $|\xi| \le r_0$, for some small $r_0 > 0$. The eigenvalue $e^{-\lambda_0(\xi)k}$ depends analytically on ξ, for $|\xi| \le r_0$ and we can also assume that the same holds for the corresponding eigenfunctions $\phi_0(\xi, z)$. Note also that we have $\lambda_0(0) = 1$ and that we can take $\phi_0(0, z) = 1$.

By (2.5.3), there are $c, c' > 0$ such that

$$(2.5.4)\qquad \|G_k(\xi)\|_{2\to 2} < ce^{-c'k}, \quad |\xi| \ge r_0.$$

Let us consider eigenvalues $e^{-\lambda_0^*(\xi)k}$ with corresponding eigenfunctions $\phi_0^*(\xi, z)$ for the adjoint operator $G_k^*(\xi)$ of $G_k(\xi)$. Note that $\lambda_0^*(\xi) = \overline{\lambda_0(\xi)}$, $\lambda_0^*(0) = 0$ and that we can take $\phi_0^*(0, z) = 1$. We shall assume that the eigenfunctions $\phi_0^*(\xi, z)$ depend analytically on ξ and that

$$(2.5.5)\qquad \langle \phi_0(\xi), \phi_0^*(\xi)\rangle = \int_K \phi_0(\xi, z)\overline{\phi_0^*(\xi, z)} p^*(z) p(z) dz = 1.$$

Then, the kernel $E_k(\xi, z, w)$ of the spectral projection E_k associated with the eigenvalue $e^{-\lambda_0(\xi)k}$ is given by

$$E_k(\xi, z, w) = e^{-\lambda_0(\xi)k} \phi_0(\xi, z)\overline{\phi_0^*(\xi, w)}.$$

Let $H_k(\xi) = G_k(\xi) - E_k(\xi)$. Then there are $c, c' > 0$ such that

$$(2.5.6)\qquad \|H_k(\xi)\|_{2\to 2} < ce^{-c'k}, \quad |\xi| \le r_0.$$

Let $H_t(\xi, z, w)$ denote the kernel of $H_t(\xi)$.

LEMMA 2.5.1. *There are $c, c' > 0$ such that for all $\xi \in A$, $z, w \in A$ and $k \in \mathbb{N}$*

$$\|H_k(\xi, z, w)\|_\infty \leq ce^{-c'k}. \tag{2.5.7}$$

PROOF. We have

$$|H_{k+1}(\xi, z, w)| \leq \|H_k(\xi, z, .)\|_2 \|H_1(\xi, ., w)\|_2. \tag{2.5.8}$$

Also

$$\|H_{k+1}(\xi, z, .)\|_2 \leq \|H_k^*(\xi)\|_{2\to 2}\|H_1(\xi, z, .)\|_2 \leq ce^{-c'k}\|H_1(\xi, z, .)\|_2. \tag{2.5.9}$$

The lemma follows from (2.5.8) and (2.5.9), since

$$\|H_k(\xi)\|_2 \leq \|G_k(\xi)\|_2^2 + \|E_k(\xi)\|_2^2.$$

LEMMA 2.5.2. *There are $c, c' > 0$ such that*

$$|G_k(\xi, z, w)| \leq c'e^{-ck}\|G(\xi)\|_\infty, \qquad |\xi| \geq r_0,\ z, w \in K,\ k \in \mathbb{N}. \tag{2.5.10}$$

PROOF. We have

$$|G_{k+1}(\xi, z, w)| \leq \|G_k(\xi, z, .)\|_2 \|G_1(\xi, ., w)\|_2 \leq c\|G_k(\xi, z, .)\|_2 \|G(\xi)\|_\infty \tag{2.5.11}$$

Also, by (2.5.4)

$$\|G_{k+1}(\xi)(z, .)\|_2 = \|G_k^*(\xi)\|_{2\to 2}\|G(\xi)(z, .)\|_2 \leq ce^{-c'k}, \quad \text{for } |\xi| \geq r_0. \tag{2.5.12}$$

The lemma follows from (2.5.11) and (2.5.12).

Arguing as in [**Se**], we have, for $|\xi| \leq r_0$, the expansions:

$$\begin{aligned}
\lambda_0(\xi) &= \sum_{1\leq i\leq d} b_i\xi_i + \sum_{1\leq i,j\leq d} b_{ij}\xi_i\xi_j + \sum_{|m|\geq 3} b_m\xi^m,\\
\phi_0(\xi, z) &= 1 + \sum_{1\leq j\leq d} c_j(z)\xi_j + \sum_{1\leq i,j\leq d} c_{ij}(z)\xi_i\xi_j - \sum_{|m|\geq 3} c_m(z)\xi^m,\\
\phi_0^*(\xi, z) &= 1 + \sum_{1\leq j\leq d} c_j^*(z)\xi_j + \sum_{1\leq i,j\leq d} c_{ij}^*(z)\xi_i\xi_j + \sum_{|m|\geq 3} c_m^*(z)\xi^m,
\end{aligned} \tag{2.5.13}$$

where $m = (m_1, ..., m_d)$, $m_i \in \mathbb{N}$, $\xi^m = \xi_1^{m_1}...\xi_d^{m_d}$ and $|m| = m_1 + ... + m_d$.

LEMMA 2.5.3. *The eigenvalue $\lambda_0(\xi)$ is invariant with respect to the Weyl group W, i.e. $\lambda_0(w \cdot \xi) = \lambda_0(\xi)$, $w \in W$.*

PROOF. Since $\overline{\Phi_k}(x)$ is spherical, its spherical Fourier transform is W-invariant, i.e.

$$\mathcal{F}\overline{\Phi_k}(w \cdot \xi) = \mathcal{F}\overline{\Phi_k}(\xi), \quad \xi \in A,\ w \in W.$$

Let us set

$$\overline{G_k}(\xi) = \int_K \int_K p(z)G_k(\xi, z, w)p^*(w)dzdw.$$

Then, by (2.1.3) and (2.5.1), $\mathcal{F}\overline{\Phi_k}(\xi) = e^{-\mu k}\overline{G_k}(\xi)$.

Let us also set

$$\overline{H_k}(\xi) = \int_K \int_K p(z) H_k(\xi, z, w) p^*(w) dz dw$$

$$C_0(\xi) = \int_K p(z) \phi_0(\xi, z) dz$$

$$C_0^*(\xi) = \int_K p^*(z) \overline{\phi_0^*(\xi, z)} dz.$$

Then

$$\overline{H_k}(\xi) = O\left(e^{-c'k}\right),$$

$$C_0(\xi) \to C_0(0)\ (\xi \to 0), \quad C_0(0) = \int_K p(z) dz$$

$$C_0^*(\xi) \to C_0(0)\ (\xi \to 0), \quad C_0^*(0) = \int_K p^*(z) dz.$$

Let $w \in W$. Then

$$\begin{aligned} 1 = & \frac{\mathcal{F}\overline{\Phi_k}(\xi)}{\mathcal{F}\overline{\Phi_k}(w \cdot \xi)} = \frac{\overline{G_k}(\xi)}{\overline{G_k}(w \cdot \xi)} \\ = & \frac{C_0(\xi) C_0^*(\xi) e^{-\lambda_0(\xi) k} + \overline{H_k}(\xi)}{C_0(w \cdot \xi) C_0^*(w \cdot \xi) e^{-\lambda_0(w \cdot \xi) k} + \overline{H_k}(w \cdot \xi)} \\ = & \frac{C_0(\xi) C_0^*(\xi) e^{-\lambda_0(\xi) k} + O\left(e^{-c'k}\right)}{C_0(w \cdot \xi) C_0^*(w \cdot \xi) e^{-\lambda_0(w \cdot \xi) k} + O\left(e^{-c'k}\right)}. \end{aligned}$$

If we take $|\xi|$ small and let $k \to \infty$, we get that $\lambda_0(\xi) = \lambda_0(w \cdot \xi)$.

COROLLARY 2.5.4. *i) We have* $b_i = 0,\ 1 \le i \le d$.
ii) The quadratic form $b_{ij}\xi_i\xi_j$ *is invariant with respect to the Weyl group.*

Note that we also have $b_m = 0,\ |m| = 3$.

2.6. The coefficients b_{ij}, c_j and c_{ij}, $1 \le i, j \le d$. Let us set $G(\xi) = G_1(\xi)$. We have the identity

$$G(\xi) \phi_0(\xi, z) = e^{-\lambda_0(\xi)} \phi_0(\xi, z).$$

By (2.5.13) and Corollary 2.5.4, we have

$$e^{-\lambda_0(\xi)} \phi_0(\xi, z) = 1 + c_j(z)\xi_j + \big(c_{ij}(z) - b_{ij}\big)\xi_i\xi_j + \dots \tag{2.6.1}$$

On the other hand,

$$\begin{aligned} G(\xi, z, w) = & \int_A \mathcal{K}(z, yw) e^{i\langle \xi, y \rangle} dy = \sum_{k \ge 0} \frac{1}{k!} \int_A \mathcal{K}(z, yw) \big(i \langle \xi, y \rangle\big)^k dy \\ = & \int_A \mathcal{K}(z, yw) dy + i\xi_j \int_A \mathcal{K}(z, yw) y_j dy - \frac{1}{2}\xi_i\xi_j \int_A \mathcal{K}(z, yw) y_i y_j dy + \dots \\ = & G_n(0, z, w) + i\xi_j \int_A \mathcal{K}(z, yw) y_j dy - \frac{1}{2}\xi_i\xi_j \int_A \mathcal{K}(z, yw) y_i y_j dy + \dots \end{aligned}$$

So
(2.6.2)

$$\begin{aligned}\Big(G(\xi)\phi_0(\xi)\Big)(z) = \int_K \Bigg(\bigg(G(0,z,w) + i\xi_j \int_A \mathcal{K}(z,yw)y_j dy \\ - \frac{1}{2}\xi_i\xi_j \int_A \mathcal{K}(z,yw)y_i y_j dy + \ldots\bigg) \\ \bigg(1 + c_j(w)\xi_j + c_{ij}(w)\xi_i\xi_j + \ldots\bigg)\Bigg) p^*(w)p(w)dw \\ =1 + \Big(\mathcal{K}c_j + i\beta_j\Big)\xi_j + \Big(\mathcal{K}c_{ij} + i\mathcal{K}(P_i c_j) - \frac{1}{2}a_{ij}\Big)\xi_i\xi_j + \ldots.\end{aligned}$$

Comparing the terms which are linear in ξ, in (2.6.1) and (2.6.2), we have

$$c_j = (\mathcal{K}c_j) + i\beta_j \Rightarrow (I - \mathcal{K})c_j = i\beta_j.$$

Hence

$$\langle \beta_j \rangle = q_j = 0 \tag{2.6.3}$$

and we can take $c_j = i\psi^j$.

Comparing the second order terms in ξ, in (2.6.1) and (2.6.2), we get

$$(I - \mathcal{K})c_{ij} = i\big(\mathcal{K}(P_i c_j)\big) - \frac{1}{2}a_{ij} + b_{ij} = -\big(\mathcal{K}(P_i\psi^j)\big) - \frac{1}{2}a_{ij} + b_{ij}. \tag{2.6.4}$$

Since $\langle (I - \mathcal{K})c_{ij} \rangle = 0$ we should have

$$\Big\langle -\big(\mathcal{K}(P_i\psi^i)\big) - \frac{1}{2}a_{ij} + b_{ij} \Big\rangle = 0$$

which implies that

$$b_{ij} = \Big\langle \big(\mathcal{K}(P_i\psi^j)\big) + \frac{1}{2}a_{ij} \Big\rangle = q_{ij}. \tag{2.6.5}$$

Now (2.6.4) becomes

$$(I - \mathcal{K})c_{ij} = -\big(\mathcal{K}(y_i\psi^j)\big) - \frac{1}{2}a_{ij} + q_{ij}$$

which implies that $c_{ij} = -\psi^{ij} + c$ where c is a constant.

Arguing in the same way for $\phi_0^*(\xi, z)$, we have that $c_j^* = -\psi^{*j}$, $c_{ij}^* = -\psi^{*ij} + c^*$, where ψ^{*j} and ψ^{*ij} are the first and second order correctors associated with the kernel $\mathcal{K}^*(x,y)$ and c^* is a constant. The constants c, c^* should be such that $\langle \phi_0(\xi), \phi_0^*(\xi) \rangle = 1$

2.7. The expansions for $E_k(\xi, z, w)$ and $E_{\sigma,k}(\xi)$. Let us fix a large $N \in \mathbb{N}$. Then it follows from (2.5.13) (cf. [**LA**] for the omitted details) that

$$
\begin{aligned}
E_k(\xi,z,w) =& e^{-k\lambda_0(\xi)}\phi_0(\xi,z)\overline{\phi_0^*(\xi,w)} \\
=& e^{-kq_{ij}\xi_i\xi_j}\Big(1+i\psi^j(z)\xi_j - i\psi^{*j}(w)\xi_j \\
&+ \sum_{\substack{2\le|m|\le N\\|n|=2\nu\\|m|\ge 2\nu}} c_{m,n}(z,w)\xi^{m+n}k^\nu + c_N(k,z,w,\xi)\Big),
\end{aligned} \tag{2.7.1}
$$

where

$$
c_N(k,z,w,\xi) = \sum_{1\le\ell\le N_1} c_{N,\ell}(k,z,w,\xi)e^{k\theta_\ell(\xi)}, \tag{2.7.2}
$$

with

$$
\theta_\ell(\xi)\le c|\xi|^3, \quad |c_{N,\ell}(k,z,w,\xi)|\le c|\xi|^{4\nu+m}k^\nu, \quad 2\nu+m\ge N.
$$

Let σ be as in Section 1.4. Since $\mathcal{F}\sigma$ is W-invariant, we have the expansion

$$
\mathcal{F}\sigma(\xi) = s_0 + s_{ij}\xi_i\xi_j + \sum_{|m|\ge 4} s_m\xi^m \tag{2.7.3}
$$

Let us set

$$
\begin{aligned}
\overline{E_k}(\xi) =& \int_K\int_K p(z)E_k(\xi,z,w)p^*(w)dzdw \\
E_{\sigma,k}(\xi) =& \mathcal{F}\sigma(\xi)\overline{E_k}(\xi)\mathcal{F}\sigma(-\xi).
\end{aligned} \tag{2.7.4}
$$

Then, it follows from (2.7.1) and (2.7.3) that

$$
E_{\sigma,k}(\xi) = e^{-kq_{ij}\xi_i\xi_j}\Big(C_{\sigma,0} + \sum_{\substack{1\le|m|\le N\\|n|=2\nu\\|m|\ge 2\nu}} C_{\sigma,m,n}\xi^{m+n}k^\nu + C_N(\sigma,\xi)\Big), \tag{2.7.5}
$$

where the analogue to (2.7.2) still holds.

2.8. An expansion for $\mathcal{K}_k(x,y)$ and the proof of Theorem 1.3.1. If $\phi\in C^{d+1}$, then $\mathcal{K}_k\in C^{d+1}$ and hence, by (2.5.10),

$$
\int_{|\xi|\ge r_0}|G_k(\xi,z,w)|d\xi \le ce^{-c'k}\int_{|\xi|\ge r_0}\|G(\xi)\|_\infty d\xi \le ce^{-c'k}. \tag{2.8.1}
$$

Also , by (2.5.7)

$$
\int_{|\xi|\le r_0}|H_k(\xi,z,w)|d\xi \le ce^{-c'k}. \tag{2.8.2}
$$

It follows from (2.5.3) (2.8.1) and (2.8.2) that

$$
\mathcal{K}_k(xz,w) = (2\pi)^{-n}\int_A e^{i\langle\xi,x\rangle}E_k(\xi,z,w)d\xi + O(e^{-c'k}). \tag{2.8.3}
$$

Now we can use the expansion (2.7.1). Let us observe that

$$\int_{|\xi|\le r_0} e^{i\langle\xi,x\rangle} e^{-tq_{ij}\xi_i\xi_j}\xi^{m+n}d\xi = \int e^{i\langle\xi,x\rangle} e^{-kq_{ij}\xi_i\xi_j}\xi^{m+n}d\xi + O(e^{-ck})$$
$$= i^{|m|+|n|}\frac{\partial^{m+n}}{\partial x^{m+n}}P_k^H(x,0) + O(e^{-ct}),$$
$$\int_{|\xi|\le r_0} e^{-tq_{ij}\xi_i\xi_j}|c_N(k,z,w,\xi)|d\xi \le ck^{-(N+d)/2}.$$

So, we have

(2.8.4)
$$\mathcal{K}_k(xz,w) = P_k^H(x,0) + \sum_{1\le j\le d}\psi^j(z)\frac{\partial}{\partial x_j}P_k^H(x,0) - \sum_{1\le j\le d}\psi^{*j}(w)\frac{\partial}{\partial x_j}P_k^H(x,0)$$
$$+ \sum_{\substack{2\le|m|\le N\\ |n|=2\nu\\ |m|\ge 2\nu}} C_{m,n}(z,w)k^{\nu}\frac{\partial^{m+n}}{\partial x^{m+n}}P_k^H(x,0) + O(k^{-(d+N)/2}).$$

If $G = AK$, then we can assume that $\Phi_k(x,y) = \mathcal{K}_k(x,y)$ and Theorem 1.3.1, follows from the above expansion. The only change is that we should replace the subindex $|m| \ge 2\nu$ with $|m| \ge \nu$.

2.9. Proof of Theorem 1.1.1. Let us introduce some notations. Let us set

$$\overline{G_k}(\xi) = \int_K\int_K p(z)G_k(\xi,z,w)p^*(w)dzdw, \qquad G_{\sigma,k}(\xi) = \mathcal{F}\sigma(\xi)\overline{G_k}(\xi)\mathcal{F}\sigma(-\xi).$$

Let us also define $\overline{H_k}(\xi)$, $H_{\sigma,k}(\xi)$, $\overline{E_k}(\xi)$, $E_{\sigma,k}(\xi)$ in the same way. Then

$$\mathcal{F}\overline{\Phi_k}(\xi) = e^{-\mu k}\overline{G_n}(\xi), \quad \mathcal{F}\Phi_{\sigma,k}(\xi) = e^{-\mu k}G_{\sigma,k}(\xi)$$

Hence

$$\Phi_{\sigma,k}(x) = e^{-\mu k}\int_A G_{\sigma,k}(\xi)\varphi_\xi(x)\frac{d\xi}{|\mathbf{c}(\xi)|^2}.$$

Since, $|\mathbf{c}(\xi)|^{-2}$ grows polynomially we have

$$\int_{|\xi|\ge r_0}|G_{\sigma,k}(\xi)||\varphi_\xi(x)|\frac{d\xi}{|\mathbf{c}(\xi)|^2} \le ce^{-c'k}\varphi_0(x)\int_{|\xi|\ge r_0}|G_{\sigma,1}(\xi)|\frac{d\xi}{|\mathbf{c}(\xi)|^2}$$
$$\le ce^{-c'k}\varphi_0(x).$$

Also, by (2.5.7)

$$\int_{|\xi|\le r_0}|H_{\sigma,k}(\xi)||\varphi_\xi(x)|\frac{d\xi}{|\mathbf{c}(\xi)|^2} \le ce^{-c'k}\varphi_0(x)\int_{|\xi|\le r_0}H_{\sigma,k}(\xi)|\frac{d\xi}{|\mathbf{c}(\xi)|^2}$$
$$\le ce^{-c'k}\varphi_0(x).$$

So

(2.9.1) $$\Phi_{\sigma,k}(x) = e^{-\mu k}\int_{|\xi|\le r_0} E_{\sigma,k}(\xi)\varphi_\xi(x)\frac{d\xi}{|\mathbf{c}(\xi)|^2} + e^{-\mu k}\varphi_0(x)O(e^{-c'k}).$$

Let us now use the expansion (2.7.5). As in the previous section, we observe that

$$\begin{aligned}\int_{|\xi|\le r_0} e^{-kq_{ij}}|C_N(\sigma,\xi)||\varphi_\xi(x)|\frac{d\xi}{|\mathbf{c}(\xi)|^2} &\le \varphi_0(x)O(k^{-(N+D)/2}),\\ \int_{|\xi|\ge r_0} e^{-kq_{ij}\xi_i\xi_j}\xi^{m+n}\varphi_\xi(x)\frac{d\xi}{|\mathbf{c}(\xi)|^2} &= \varphi_0(x)O(e^{-ck}).\end{aligned} \tag{2.9.2}$$

Let us set

$$\begin{aligned}F_{0,\sigma,k}(\xi) =&C_{\sigma,0}e^{-kq_{ij}\xi_i\xi_j},\\ F_{1,\sigma,k}(\xi) =&e^{-kq_{ij}\xi_i\xi_j}\sum_{\substack{1\le|m|\le N\\ |n|=2\nu\\ |m|\ge 2\nu}} C_{\sigma,m,n}\xi^{m+n}k^\nu,\\ F_{\sigma,k} =&F_{0,\sigma,k}+F_{1,\sigma,k}.\end{aligned}$$

Let us also denote by $S_{\sigma,k}$, $S_{0,\sigma,k}$, $S_{1,\sigma,k}$ and $A_{\sigma,k}$, $A_{0,\sigma,k}$, $A_{1,\sigma,k}$ the inverse spherical and Euclidean Fourier tramsforms of $F_{\sigma,k}$, $F_{0,\sigma,k}$, $F_{1,\sigma,k}$ respectively. Then, it follows from (2.7.5), (2.9.1) and (2.9.2)

$$\Phi_{\sigma,k}(x) = S_{\sigma,k}(x) + \varphi_0(x)e^{-\mu k}O(k^{-(N+d)/2}). \tag{2.9.3}$$

Note also that $\mathcal{A}S_{i,\sigma,k} = A_{i,\sigma,k}$, $i = 0, 1$.

2.9.1. Proof of Theorem 1.1.1, in the general case. Let us recall that we have identified $\mathfrak{a}$ and A with $\mathbb{R}^d$ and the inner product $\langle\cdot,\cdot\rangle$ induced by the killing form with the Euclidean inner product on $\mathbb{R}^d$. Let us set $\omega(x) = \min\{\zeta(x),\ \ \zeta\in\Sigma_0\}$, $x \in A$. Note that $\omega(x) \ge 0$ iff $x = \exp H$ with $H \in \overline{\mathfrak{a}^+}$.

By the local Harnack inequality (1.5.1), it is enough to prove that $\Phi_{\sigma,k}(x)$ satisfies (1.1.1) and (1.1.2) for those $x \in A$ that satisfy $\omega(x) > c_0$, for some large $c_0 > 0$. We have the following.

LEMMA 2.9.1. *There is $c_0 > 0$ and $c > 0$ such that for all $k \in \mathbb{N}$ and all $\omega(x) \ge c_0$*

$$|S_{1,\sigma,k}(x)| \le \frac{c}{\sqrt{k}}k^{-d/2}k^{-|\Sigma_0|}e^{-\mu k}\varphi_0(x)e^{-\frac{|x|^2}{ct}}, \tag{2.9.4}$$

$$S_{0,\sigma,k}(x) \le ck^{-|\Sigma_0|}e^{-\mu k}\varphi_0(x)\left\{\prod_{\zeta\in\Sigma_0}\left(1+\frac{\zeta(x)}{k}\right)^{((m_\zeta+m_{2\zeta})/2)-1}\right\}P_k^H(x,0), \tag{2.9.5}$$

$$S_{0,\sigma,k}(x) \ge \frac{1}{c}k^{-|\Sigma_0|}e^{-\mu t}\varphi_0(x)\left\{\prod_{\zeta\in\Sigma_0}\left(1+\frac{\zeta(x)}{k}\right)^{((m_\zeta+m_{2\zeta})/2)-1}\right\}P_k^H(x,0). \tag{2.9.6}$$

The proof of (2.9.5) and (2.9.6) is contained in [**AJ**]. For the proof of (2.9.4) we refer to [**LA**].

Let us fix a large $N \in \mathbb{N}$ and let $N' > N$, $N' \in \mathbb{N}$. Then by (2.9.3)

$$|\Phi_{\sigma,k}(x) - S_{\sigma,k}(x)| \le ck^{-(N+d)/2}e^{-\mu k}\varphi_0(x). \tag{2.9.7}$$

Also by (1.1.3), and Lemma 2.9.1, for all $\omega(x) \ge c_0$,

$$|\Phi_{\sigma,k}(x) - S_{\sigma,k}(x)| \le ce^{-\mu k}\varphi_0(x)e^{-\frac{|x|^2}{ck}}. \tag{2.9.8}$$

Let $\varepsilon \in (0,1)$. Then by interpolating (2.9.7) and (2.9.8) we have

$$|\Phi_{\sigma,k}(x) - S_{\sigma,k}(x)| \leq ck^{-(1-\varepsilon)(N+d)/2} e^{-\mu k} \varphi_0(x) e^{-\varepsilon \frac{|x|^2}{ck}}. \tag{2.9.9}$$

Let us observe that for all $c > 0$ there is $c' > 0$ suct that

$$\frac{|x|}{\sqrt{t}} e^{-\varepsilon \frac{|x|^2}{ct}} \leq c' e^{-\varepsilon \frac{|x|^2}{c't}}. \tag{2.9.10}$$

It follows from Lemma 2.9.1, (2.9.9) and (2.9.10) and by chosing N sufficiently large, that $\Phi_{\sigma,k}(x)$ satisfies (1.1.1) and (1.1.2) for $x \in A$, $\omega(x) > c_0$. Theorem 1.1.1 follows.

2.9.2. Proof of Theorem 1.1.1, when G is complex. The procf of Theorem 1.1.1, when G is complex, is relatively easy. More precisely, by (2.8.4) there are constants $C_{\sigma,0}$, $C_{\sigma,m,n}$ such that

$$A_{\sigma,k}(x) = C_{\sigma,0} P_k^H(x,0) + \sum_{\substack{1 \leq |m| \leq N \\ |n| = 2\nu \\ |m| \geq 2|n|}} C_{\sigma,m,n} t^\nu \frac{\partial^{m+n}}{\partial x^{m+n}} P_k^H(x,0) \tag{2.9.11}$$

Since $A_{\sigma,k}$ is equal to the Abel transform of $S_{\sigma,k}$, using formula (1.4.2), we can obtain an explicit expession for $S_{\sigma,k}$. The lower estimate (1.1.2) follows from this expression and (2.9.3). The upper Gaussian estimate (1.1.1), follows by interpolating with the estimate (1.1.3), as before. We point out that a number of terms in this explicit expression for $S_{\sigma,k}$ vanish, because the expression should be invariant with respect to the action of the Weyl group W. We refer the reader to [**LA**], for the details.

Finally note that, if the density ϕ is spherical, then the above argument gives a full asymptotic expansion for the convolution powers ϕ^{*k}. Thus, we can obtain a generalisation for the classical Edgeworth expansion (cf. [**Fe**]). It would be interesting to extend this result, to densities which are not necessarily spherical.

3. Heat kernels

3.1. The sub-Laplacian L. Let us identify the elements $X \in \mathfrak{n} + \mathfrak{a}$ with the left invariant vector fields on $S \times K$ satisfying $X(e) = X$ and the elements $X \in \mathfrak{k}$ with the right invariant vector fields on $S \times K$ satisfying $X(e) = X$. Let us fix $\{X_q, ..., X_{d+1}\}$, $\{X_d, ..., X_1\}$ and $\{X_0, ..., X_{-r}\}$ bases of $\mathfrak{n}$, $\mathfrak{a}$ and $\mathfrak{k}$, respectively. Since we have identified A with $\mathbb{R}^d$, we assume that $X_i(e) = \partial/\partial x_i$, $1 \leq i \leq d$.

We want to express the left invariant sub-Laplacian $L = -(E_1^2 + ... + E_p^2) + E_0$ in terms of the vector fields $\{X_q, ..., X_{-r}\}$. To this this end, let us observe that

$$s\kappa \exp(tE) = s \exp(t \operatorname{Ad} \kappa(E))\kappa, \quad E \in \mathfrak{g}, s \in S, \ \kappa \in K. \tag{3.1.1}$$

Also, by the Baker-Campbell-Hausdorff formula,

$$\exp(t(X+Y)) = \exp tX \exp tY \exp O(t^2). \tag{3.1.2}$$

It follows from (3.1.1) and (3.1.2) that every $E \in \mathfrak{g}$ can be written as

$$E(sk) = \Big(\operatorname{Ad} \kappa(E)\Big)(s\kappa) = a_q(\kappa) X_q + ... + a_{-r}(\kappa) X_{-r}, \qquad s \in S, \ \kappa \in K. \tag{3.1.3}$$

Now, let us fix $f, h \in C_0^\infty(G)$. Then

$$\begin{aligned}\int_G f(x)Lh(x)dx &= \int_G (E_if)(x)(E_ih)(x)dx + \int_G f(x)(E_0h)(x)dx \\ &= \int_N\int_A\int_K (X_if)(na\kappa)a_{ij}(\kappa)(X_ih)(na\kappa)e^{-2\rho(a)}dndad\kappa \\ &+ \int_N\int_A\int_K f(na\kappa)a_j(\kappa)(X_jh)(na\kappa)e^{-2\rho(a)}dndad\kappa \\ &= -\int_N\int_A\int_K f(na\kappa)\big(X_ia_{ij}(\kappa)X_jh\big)(na\kappa)e^{-2\rho(a)}dndad\kappa \\ &+ \int_N\int_A\int_K f(x)a_{ij}(\kappa)(X_jh)(na\kappa)(2X_i\rho(a))e^{-2\rho(a)}dndad\kappa \\ &+ \int_N\int_A\int_K f(na\kappa)a_j(\kappa)(X_jh)(na\kappa)e^{-2\rho(a)}dndad\kappa.\end{aligned}$$

So, we have the expression

$$L = -X_ia_{ij}X_j + (a_j + 2a_{ij}(X_i\rho))\,X_j. \tag{3.1.4}$$

Note that $a_{ij} = a_{ji}$.

3.2. The heat kernel $P_t(x, y)$. As we already mentioned, the heat kernel $P_t(x, y)$ of L is defined as the fundamental solution of the heat equation $((\partial/\partial t) + L)\,u = 0$. It is also the kernel (with respect to the measure dg) of the heat semigroup $P_t = e^{-tL}$, $t > 0$. Since L is left invariant, we have $P_t(x, y) = P_t(y^{-1}x, e)$, $x, y \in G$.

Let $L^* = -(E_1^2 + \dots + E_p^2) - E_0$. Then $\big((\partial/\partial t) + L\big)P_t(., y) = 0$, $\big((\partial/\partial t) + L^*\big)P_t(x, .) = 0$. Hence, by the local Harnack inequallity (cf. [**VSC**]), for every compact $U \subseteq G$, there is $c > 0$ such that

$$P_t(x, y) \le cP_{t+1}(xz, yw), \qquad x, y \in G,\ z, w \in K,\ t \ge 1.$$

It follows, that there is $c > 0$ such that

$$\frac{1}{c}P_t(x, y) \le \overline{P}_{t+1}(x, y) \le cP_{t+2}(x, y), \quad x, y \in G,\ t \ge 1, \tag{3.2.1}$$

where

$$\overline{P}_t(x, y) = \int_K\int_K P_t(xz, yw)dzdw.$$

3.3. Conjugation with e^ρ and projection on AK. Let us consider the operator $L^\rho = e^{-\rho}Le^\rho$. By (3.1.4),

$$L^\rho = -\,X_ia_{ij}X_j + a_jX_j + a_j(X_j\rho) - (X_ia_{ij})(X_j\rho) + a_{ij}(X_i\rho)(X_j\rho).$$

Let $P_t^\rho(x, y)$ denote the heat kernel associated with L^ρ. Then

$$P_t^\rho(x, y) = e^{-\rho(x)}P_t(x, y)e^{\rho(y)}.$$

Let us consider the operator

$$\begin{aligned}L_{AK}^\rho = &-\sum_{-r\le i,j\le d} X_ia_{ij}X_j + \sum_{-r\le j\le d} a_jX_j \\ &+ \sum_{1\le j\le d} a_j(X_j\rho) - \sum_{-r\le i,j\le d} \big((X_ia_{ij})(X_j\rho) - a_{ij}(X_i\rho)(X_j\rho)\big).\end{aligned}$$

Let $f \in C^\infty$, such that $f(ng) = f(g)$, $g \in G, n \in N$. Then, $L^\rho f = L^\rho_{AK} f$. This shows that L^ρ_{AK} is the projection of L^ρ on $AK = N \backslash G$.

Let $P^\rho_{AK,t}(x,y)$ denote the heat kernel of L^ρ_{AK}, with respect to the measure $dad\kappa$. Arguing as in section 2.1, we can see that if the function f is as before, then

$$\int P^\rho_t(x,y)f(y)dg = e^{-\rho(x)} \int_N \int_A \int_K P_t(nx, ak)e^{-\rho(a)} f(a\kappa) dn da d\kappa.$$

It follows that

$$\begin{aligned} P^\rho_{AK,t}(x,y) =& e^{-\rho(x)} e^{-\rho(y)} \int_N P_t(nx,y)dn \\ =& \mathcal{A}P_t(x,y), \qquad x,y \in AK. \end{aligned}$$

3.4. The operator $\mathcal{D}$. Let us consider functions $p^*, p \in C^\infty(K)$ such that $p^*(\kappa) > 0$, $p(\kappa) > 0$, $z \in K$ and let us extend them to AK by setting $p(a\kappa) = p(\kappa)$, $p^*(a\kappa) = p^*(\kappa)$ $a \in A$, $\kappa \in K$. Let also $\mu \in \mathbb{R}$. Then, we can easily verify that

$$\begin{aligned} p^*(L^\rho_{AK} - \mu)p = & - X_i(p^* a_{ij} p X_j) \\ & + [(X_i p^*)a_{ij}p - p^* a_{ji}(X_i p) + p^* a_j p]X_j + p^*(L^\rho_{AK} p - \mu p), \end{aligned}$$

$$X_j[(X_i p^*)a_{ij}p - p^* a_{ji}(X_i p) + p^* a_j p] = -p(L^{\rho *}_{AK} - \mu)p^* + p^*(L^\rho_{AK} - \mu)p.$$

Let us now assume that μ is the lowest eigenvalue of the projection L^ρ_K of L^ρ_{AK} on K and let p and p^* be corresponding eigenfunctions, i.e.

$$L^\rho_K p - \mu p = 0, \qquad L^{\rho *}_K p^* - \mu p^* = 0$$

and that

$$\int_K p^*(\kappa)p(\kappa)d\kappa = 1.$$

Let us set, for simplification,

$$\begin{aligned} \mathcal{D} =& p^*(L^\rho_{AK} - \mu)p, \\ \alpha_{ij} =& p^* a_{ij} p, \\ \beta_j =& (X_i p^*)a_{ij}p - p^* a_{ji}(X_i p) + p^* a_j p, \\ q_j =& \int_K \beta_j(\kappa)d\kappa, \quad \text{for } 1 \le j \le d \ \text{ and } \ q_j = 0, \quad \text{for } -r \le j \le 0, \\ \alpha_j =& \beta_j - q_j. \end{aligned}$$

Then

$$\begin{aligned} \mathcal{D} = -X_i \alpha_{ij} X_j + \alpha_j X_j + q_j X_j, \qquad X_j \alpha_j = 0 \\ \frac{1}{p}(L^\rho_{AK} - \mu)p = \frac{1}{p^* p}\mathcal{D}. \end{aligned}$$

Let us denote by T_t the semigroup of operators with kernel $P^\rho_{AK,t}(x,y)$ (with respect to the measure $dad\kappa$ on AK). Of course, the infinitessimal generator of T_t is equal to L^ρ_{AK}.

Let us consider the semigroup $K_t = (1/p)e^{\mu t} T_t p$. Then, K_t has as infinitessimal generator the operator

$$\frac{1}{p}(L^\rho_{AK} - \mu)p = \frac{1}{p^* p}\mathcal{D}.$$

The kernel of K_t with respect to the measure $p^*pdad\kappa$ it is given by

$$\mathcal{K}_t(x,y) = \frac{1}{p(x)} e^{\mu t} P^{\rho}_{AK,t}(x,y) \frac{1}{p^*(y)}.$$

It follows that

$$\mathcal{A}P_t(x,y) = P^{\rho}_{AK,t}(x,y) = e^{-\mu t} p(x)\mathcal{K}_t(x,y)p^*(y), \quad x,y \in AK. \tag{3.4.1}$$

The formal adoint $\mathcal{D}^*$ of $\mathcal{D}$ with respect to the measure $dad\kappa$ is given by

$$\mathcal{D}^* = -X_i\alpha_{ij}X_j - \alpha_j X_j - q_j X_j.$$

Also the formal adoint of $(1/p^*p)\mathcal{D}$ with respect to the measure $p^*pdad\kappa$ is equal to $(1/p^*p)\mathcal{D}^*$.

3.5. Homogenization. If f is a continuous function on K, then we shall set

$$\langle f \rangle = \int_K f(z)dz.$$

Let us fix a function $f \in C^\infty(A)$ and let us extend f to AK by setting $f(x\kappa) = f(x), x \in A, \kappa \in K$. Then,

$$\mathcal{D}f = -\alpha_{ij}X_iX_jf - (X_i a_{ij})\,X_jf + \alpha_j X_j f. \tag{3.5.1}$$

We define the (first order) correctors $\psi^j, 1 \le j \le d$ (cf. [**A1, BLP, JKO**]) as solutions of the problem

$$\mathcal{D}\psi^j = -\alpha_j + X_i\alpha_{ij}, \quad \langle \psi^j \rangle = 0. \tag{3.5.2}$$

Note that $\langle \alpha_j \rangle = 0$, $1 \le j \le d$, that $\langle X_i\alpha_{ij} \rangle = 0$ and that $\mathcal{D}^*u = 0$ iff u =const. So, the correctors ψ^j are well defined.

Combining (3.5.1) and (3.5.2), we have that there is $c > 0$ independent of f such that
(3.5.3)

$$\mathcal{D}\Big(f + \psi^j X_j f\Big) = -\Big(\alpha_{ij} - \alpha_i\psi^j - q_i\psi^j + \alpha_{\ell i}X_\ell\psi^j + X_\ell(\alpha_{\ell i}\psi^j)\Big)X_iX_jf + q_jX_jf + F$$

with the function F satisfying $|F(x\kappa)| \le c|\nabla^3 f(x)|$, $x \in A, \kappa \in K$. This expression motivates the following definition.

We define the homogenised sub-Laplacian $\mathcal{D}_H$ associated with $\mathcal{D}$ by

$$\mathcal{D}_H = -\sum_{1\le i,j\le d} q_{ij}X_iX_j + \sum_{1\le j\le d} q_jX_j, \tag{3.5.4}$$

where

$$q_{ij} = \left\langle \alpha_{ij} - \alpha_i\psi^j + \sum_{-r\le \ell\le 0} \alpha_{i\ell}X_\ell\psi^j \right\rangle, \quad 1 \le i,j \le d.$$

The operator $\mathcal{D}_H$ is elliptic. The proof of this fact is the same as the one given in a similar context in [**A1**] and we shall omit it.

We define the (second order) correctors $\psi^{ij}, 1 \le i,j \le d$ as solutions of the problem

$$\mathcal{D}\psi^{ij} = \alpha_{ij} - \alpha_i\psi^j - q_i\psi^j + \alpha_{i\ell}X_\ell\psi^j + X_\ell(\alpha_{\ell i}\psi^j) - q_{ij}, \quad \langle \psi^{ij} \rangle = 0. \tag{3.5.5}$$

Combining (3.5.3), (3.5.4) and (3.5.5), we have the following expression which explains the relation between the operators $\mathcal{D}$ and $\mathcal{D}_H$.

$$\mathcal{D}\Big(f+\psi^j X_j f + \psi^{ij} X_i X_j f\Big) = \mathcal{D}_H f + F \tag{3.5.6}$$

where the function F satisfies $|F(x\kappa)| \le c\left(|\nabla^3 f(x)| + |\nabla^4 f(x)|\right), x \in A, \kappa \in K$.

Now, let us consider the corrector ψ^t defined by

$$\mathcal{D}_K \psi^t = 1 - p^* p, \quad \langle \psi^t \rangle = 0.$$

Let also $u(t,x) \in C^\infty(\mathbb{R} \times A)$ and let us extend u to $\mathbb{R} \times AK$ by setting $u(t, x\kappa) = u(t,x), x \in A, \kappa \in K, t \in \mathbb{R}$. Then

$$\Big(\frac{\partial}{\partial t} + \frac{1}{p^*p}\mathcal{D}\Big)\Big(u+\psi^j X_j u + \psi^t \frac{\partial}{\partial t} u + \psi^{ij} X_i X_j u\Big) = \frac{1}{p^*p}\big(\frac{\partial}{\partial t} + \mathcal{D}_H\big)u + U, \tag{3.5.7}$$

where the function U satisfies

$$\begin{aligned} |U(t,x\kappa)| \le c\big(|\nabla^3 u(t,x)| + |\nabla^4 u(t,x)| + |\nabla \frac{\partial}{\partial t} u(t,x)| \\ + |\nabla^2 \frac{\partial}{\partial t} u(t,x)| + |\frac{\partial^2}{\partial t^2} u(t,x)|\big), \qquad x \in A, \kappa \in K. \end{aligned}$$

3.6. Estimates for $\mathcal{K}_t$. As in section 2.3, we observe that there is a compact neighbohood U of e and $c > 0$ such that

$$\mathcal{K}_t(x,y) \ge c, \quad x, y \in U.$$

From this observation, and by adapting certain arguments from [**VSC**] (cf. [**LA**] for the details), we can prove the following.

THEOREM 3.6.1. *There is $c > 0$ such that for all $x, y \in AK$ and $t \ge 1$*

$$|\mathcal{K}_t(x,y)| \le ct^{-d/2}. \tag{3.6.1}$$

As we shall show in section 3.8, we have

$$q_j = \langle a_j \rangle = 0, \quad 1 \le j \le d.$$

So, in what follows we shall make this assumption. Then, making use of (3.6.1) and (3.5.7) and working in the same way as in [**A1**], we can prove the following estimates.

THEOREM 3.6.2. *There is a constant $c > 0$ such that*

$$K_t(xz, yw) \le c\, t^{-d/2} \exp\left(-\frac{|x-y|^2}{ct}\right), \quad t \ge 1,\ x, y \in A,\ z, w \in K. \tag{3.6.2}$$

Concerning the spacial derivatives, we have the following.

THEOREM 3.6.3. *For all $\ell \in \mathbb{N}$, $\varepsilon > 0$, there is $c > 0$ such that for all $1 \leq i_2, ..., i_\ell \leq d$ $-r \leq i \leq d$ and all $x, y \in AK$ and $t \geq 1$*

$$\tag{3.6.3} \begin{aligned} |X_i \mathcal{K}_t(x,y)| &\leq ct^{-1/2}\mathcal{K}_{(1+\varepsilon)t}(x,y), \\ |X_{i_1}....X_{i_\ell}\mathcal{K}_t(x,y)| &\leq ct^{-\ell/2}\mathcal{K}_{(1+\varepsilon)t}(x,y). \end{aligned}$$

Let $P_t^H(x,y)$ be the heat kernel of the homogenised operator $\mathcal{D}_H$ and let us extend it to to AK by setting $P_t^H(xz, yw) = P_t^H(x,y)$, $x, y \in A$, $z, w \in K$. Then, arguing again as in [**A1**], we can prove the following.

THEOREM 3.6.4. *There is $c > 0$ such that for all $-r \leq i \leq d$, $x, y \in AK$ and $t \geq 1$*

$$\tag{3.6.4} \begin{aligned} |\mathcal{K}_t(x,y) - P_t^H(x,y)| &\leq ct^{-(d+1)/2}, \\ |X_i\mathcal{K}_t(x,y) - X_iP_t^H(x,y)| &\leq ct^{-(d+2)/2}. \end{aligned}$$

We point out that if $|\Sigma_0| = 1$, then (3.6.4) suffices for the proof of Theorem 1.2.1 and we do not need the Bloch wave expansion below.

3.7. Proof of the estimate (1.2.3). The proof of (1.2.3) is similar to the proof of (1.1.3) given in section 2.4. Note that by (3.6.3), we do not need to consider the auxiliary function σ, but we can work only with $\overline{P_t}(x)$.

3.8. Bloch wave expansions (cf. [**JKO, BLP, Se, Z**]). Given $\xi \in A$, let us consider the operator

$$\begin{aligned} \mathcal{D}(\xi) = e^{-i\langle\xi,x\rangle}\mathcal{D}e^{i\langle\xi,x\rangle} =& - X_i\alpha_{ij}X_j - i\xi_i\alpha_{ij}X_j - i\xi_j\,(X_i\alpha_{ij}) - i\xi_j\alpha_{ij}X_i + \alpha_{ij}\xi_i\xi_j \\ &+ \alpha_iX_i + q_iX_i + i\alpha_i\xi_i + iq_i\xi_i. \end{aligned}$$

The heat kernel of the operator $(1/p^*p)\mathcal{D}(\xi)$ is given by

$$e^{-i\langle\xi,x\rangle}\mathcal{K}_t(x,y)e^{i\langle\xi,y\rangle}.$$

Let $G_t(\xi, x, y)$ be the heat kernel of the projection of $(1/p^*p)\mathcal{D}(\xi)$ (also denoted by $(1/p^*p)\mathcal{D}(\xi)$) on the compact space K. Then, we have

$$\tag{3.8.1} \begin{aligned} G_t(\xi, z, w) &= \int_A \mathcal{K}_t(z, yw)e^{i\langle\xi,y\rangle}dy \\ &= \int_A \mathcal{K}_t(xz, w)e^{-i\langle\xi,x\rangle}dx, \qquad z, w \in K, \ \ \xi \in A. \end{aligned}$$

Hence

$$\tag{3.8.2} \mathcal{K}_t(xz, w) = (2\pi)^{-n}\int_A e^{i\langle\xi,x\rangle}G_t(\xi, z, w)d\xi, \qquad z, w \in K, \ x, y \in A.$$

Thus we have the analytic family of compact operators

$$G_t(\xi)(f)(z) = \int_K G_t(\xi, z, w)f(w)p^*(w)p(w)dw.$$

Now let us observe that $\lambda(\xi)$ is an eigenvalue for $(1/p^*p)\mathcal{D}(\xi)$ with corresponding eigenfunction $\phi(\xi, z)$ iff $e^{-\lambda(\xi)t}$ is an eigenvalue with corresponding eigenfunction $\phi(\xi, z)$ for the operator $G_t(\xi)$. So arguing as in section 2.5, we have that there is $r_0 > 0$ such that for every $|\xi| \leq r_0$, the operator $(1/p^*p)\mathcal{D}(\xi)$ admits a simple and isolated eigenvalue $\lambda_0(\xi)$ with largest absolute value which depends analytically on ξ. We consider $\phi_0(\xi,z)$ and $\phi_0^*(\xi,z)$ corresponding eigenfunctions for $(1/p^*p)\mathcal{D}(\xi)$ and $(1/p^*p)\mathcal{D}^*(\xi)$ respectively, satisfying

$$\int_K \phi_0(\xi,z)\overline{\phi_0^*(\xi,z)}p^*(z)p(z)dz = 1, \qquad \phi_0(0,z) = \phi_0^*(0,z) = 1,\ z \in K.$$

As in section 2.5, we have, for $|\xi| \leq r_0$,

$$\tag{3.8.3} \begin{aligned} \lambda_0(\xi) &= \sum_{1\leq i\leq d} b_i\xi_i + \sum_{1\leq i,j\leq d} b_{ij}\xi_i\xi_j + \sum_{|m|\geq 3} b_m\xi^m, \\ \phi_0(\xi,x) &= 1 + \sum_{1\leq j\leq d} c_j(z)\xi_j + \sum_{1\leq i,j\leq d} c_{ij}(z)\xi_i\xi_j + \sum_{|m|\geq 3} c_m(x)\xi^m, \\ \phi_0^*(\xi,x) &= 1 + \sum_{1\leq j\leq d} c_j^*(z)\xi_j + \sum_{1\leq i,j\leq d} c_{ij}^*(z)\xi_i\xi_j + \sum_{|m|\geq 3} c_m^*(x)\xi^m, \end{aligned}$$

where $m = (m_1,...,m_d)$, $m_i \in \mathbb{N}$, $\xi^m = \xi_1^{m_1}...\xi_d^{m_d}$ and $|m| = m_1 + ... + m_d$.

As before, we set

$$\begin{aligned} E_t(\xi,z,w) &= e^{-\lambda_0(\xi)t}\phi_0(\xi,z)\overline{\phi_0^*(\xi,w)}, \\ H_t(\xi) &= G_t(\xi) - E_t(\xi). \end{aligned}$$

Arguing in the same way as in section 2.5, we can prove that there is $c, c' > 0$ such that

$$\tag{3.8.4} \begin{aligned} \|H_t(\xi)\|_\infty &\leq ce^{-c't}, \quad t \geq 1, |\xi| \leq r_0, \\ \|G_t(\xi)\|_\infty &\leq ce^{-c't}\|G_1(\xi)\|_\infty, \quad t \geq 1, |\xi| \geq r_0. \end{aligned}$$

Note that by (3.6.3), for all $N \in \mathbb{N}$, there is $c > 0$ such that

$$\tag{3.8.5} \|G_1(\xi)\|_\infty \leq \frac{c}{(1+|\xi|)^N}, \quad \xi \in A.$$

By arguing as in Lemma 2.5.3, we have that the eigenvalue $\lambda_0(\xi)$ is invariant with respect to the Weyl group W, i.e. $\lambda_0(w\cdot\xi) = \lambda_0(\xi)$, $w \in W$. This implies that $b_i = 0$, $1 \leq i \leq d$, $b_m = 0$, $|m| = 3$ and that the quadratic form $b_{ij}\xi_i\xi_j$ is invariant with respect to the Weyl group.

We argue now as in section 2.6. By replacing $\phi_0(\xi,z)$ and $\lambda_0(\xi)$ by their respective expansions in the identity

$$\frac{1}{p^*p}\mathcal{D}(\xi)\phi_0(\xi,z) = \lambda_0(\xi)\phi_0(\xi,z)$$

and by comparing the first order terms in ξ we have that

$$\mathcal{D}c_j = iX_\lambda\alpha_{\lambda j} - i\alpha_j - iq_j,$$

which implies that

$$q_j = 0, \quad 1 \leq j \leq d$$

and that we may take $c_j = i\psi^j$.

Also, by comparing the terms of second order in ξ we have

$$\mathcal{D}c_{ij} = p^*pb_{ij} - a_{ij} - \alpha_{i\mu}X_\mu\psi^j - X_\lambda(\alpha_{\lambda i}\psi^j) + \alpha_i\psi^j,$$

which implies that

$$b_{ij} = \int_K b_{ij} p^*(z) p(z) dz = \langle a_{ij} + \alpha_{i\mu} X_\mu \psi^j - \alpha_i \psi^j \rangle = q_{ij},$$
$$c_{ij} = -\, \psi^{ij} + c,$$

where c is a constant.

3.9. Proof of Theorems 1.2.1 and 1.3.2. The proof of Theorems 1.2.1 and 1.3.2 is similar to the proof of Theorems 1.1.1 and 1.3.1, respectively. Note that by (3.8.5) we do not need to consider the auxiliary function σ in the proof of Theorem 1.2.1, but we can work only with $\overline{P_t}(x)$.

References

[A1] G. K. Alexopoulos, *Sub-Laplacians with drift on Lie groups of polynomial volume growth*, Memoirs of the AMS **155** (2002), no. 739.

[A2] G. K. Alexopoulos, *Centered densities on Lie groups of polynomial volume growth*, Probab. Th. Rel. Fields **124** (2002), 112-150.

[A3] G. K. Alexopoulos, *On the mean distance of random walks on groups*, Bull. Sci. Math. **111** (1987), 189-199.

[An1] J.-P. Anker, *The spherical Fourier transform for rapidly decreasing functions. A simple proof of a caracterisation due to Harish-Chandra, Helgason, Trompi and Varadarajan.*, J. Funct. Anal. **96** (1991), 331-349.

[An2] J.-P. Anker, *L_p Fourier multipliers on Riemannian symmetric spaces of the noncompact type*, Ann. of Math. **132** (1990), 597-628.

[An3] J.-P. Anker, *Sharp estimates for some functions of the Laplacian on noncompact symmetric spaces*, Duke Math. J. **65** (1992), 257-297.

[AJ] J.-P. Anker and L. Ji, *Heat kernel and Green function estimates on noncompact symmetric spaces*, Geom. Funct. Anal. **9** (1999), 1035–1091.

[BLP] A. Bensoussan, J. L. Lions and G. Papanicolaou, *Asymptotic Analysis of Periodic Structures*, North Holland Publ., 1978.

[B1] Ph. Bougerol, *Théorème centrale limite local sur certains groupes de Lie*, Ann. Sci. Ec. Norm. Sup. 4ème série **14** (1981), 403-432.

[B2] Ph. Bougerol, *Comportement asymptotique des puissances de convolution d'une probabilité sur un espace symétrique*, Astérisque **74** (1980), 29-45.

[DS] N. Dunford and J. T. Schwarz, *Linear operators*, vol. I, Interscience, New York, 1953.

[Fe] W.Feller, *An introduction to the Theory of Probability and its Applications*, vol. II, Wiley, 1971.

[Ga] R. Gangoli, *Asymptotic behavior of spectra of compact quotients of certain symmetric spaces*, Acta Math. **12** (1968), 151-192.

[JKO] V. V. Jikov, S. M. Kozlov and O. A. Oleinik, *Homogeneization of Differential Operators and Integral Functionals*, Springer-Verlag, 1994.

[KS] M. Kotani and T. Sunada, *Albanese maps and off diagonal long time asymptotics for the heat kernel*, Comm. Math. Phys. **209** (2000), 633-670.

[KLS] M. A. Krasnosel'skij, J. A. Lifshits and A. V. Sobolev, *Positive Linear Systems: The method of positive operators*, 1989.

[L1] N. Lohoué, *Estimations L^p des coefficients de représentations et opérateurs de convolution*, Adv. Math. **38** (1980), 178-221.

[L2] N. Lohoué, *Inégalités de Sobolev pour les sous-laplaciens de certains groupes unimodulaires*, Geom. and Funct. Anal. **4** (1992), 394-420.

[LA] N. Lohoué and G. Alexopoulos, *On the large time behavior of heat kernels on Lie groups*, Duke Math. J. **120** (2003), 311-351.

[LR2] N. Lohoué and Th. Rychener, *Some function spaces on symmetric spaces related to convolution operators*, prépublication, Orsay (1982).

[O] P. Ostellari, *Comportement global du noyau de la chaleur associé à certains sous-laplaciens sur les groupes de Lie semi-simples*, J. Funct. Anal. **199** (2003), 521-534.

[Se] E. V. Sevost'janova, *An assymptotic expansion of the solution of a second order elliptic equation with periodic coefficients*, Math. USSR Sbornik **43** (1982), 181-198.

[V] N. Th. Varopoulos, *Long range estimates for Markov chains.*, Bull. Sci. Math. **109** (1985), 225-252.

[VSC] N. Th. Varopoulos, L. Saloff-Coste and Th. Coulhon, *Analysis and Geometry on Groups*, Cambridge Tracts in Mathematics, 1993.

[W] G. Warner, *Harmonic Analysis on semisimple Lie groups*, vol. 1 and 2, Sringer, 1972.

[Z] V. V. Zhikov, *A spectral approach to the asymptotic problems of diffusion*, Diff. Ouravnenia **25** (1989), no. 1, 44-55; English trasl., Differ. Equations **265** (1989), no. 1, 33-39.

UNIVERSITÉ DE PARIS-SUD, MATHÉMATIQUES, BÂT. 425, 91405 ORSAY CEDEX, FRANCE

Contemporary Mathematics
Volume **347**, 2004

Some spectral and geometric properties for infinite graphs

Yusuke Higuchi and Tomoyuki Shirai

Dedicated to Professor Tokuzo Shiga on his sixtieth birthday

ABSTRACT. For infinite graphs we discuss some topics on spectral and geometric properties, for example, full spectral property, isoperimetric constants, graph-operations, bipartiteness and so on. We also collect some problems around these topics.

CONTENTS

1. Introduction

For finite or infinite graphs, there are many kinds of researches on the relationship between geometric and spectral properties. Some of them clarify the similarities between finite (infinite) graphs and compact (non-compact) manifolds; others clarify the differences between them. The present paper is mainly concerned with spectral and geometric properties for infinite graphs from the latter point of view. There are three main topics, which are discussed in Sections 2, 3 and 4.

In Section 2 we treat laplacians on infinite, abelian covering graphs of finite graphs. The spectrum of our laplacian is a closed subset of the interval $[0, 2]$ for every graph and we say that a graph has "full spectrum property" if the laplacian on it has the whole interval $[0, 2]$ as its spectrum. We give sufficient conditions for an abelian covering graph to have full spectrum property. This property can also

2000 *Mathematics Subject Classification.* Primary 47B39, 58J50; Secondary 05C50, 05C99.
Key words and phrases. Spectral geometry, discrete laplacian, abelian covering graph, full spectrum property, para-line graph, isoperimetric constant.

be considered in the manifold setting if one replaces $[0,2]$ by $[0,\infty)$, however, our technique used in the proof cannot be directly applied to the manifold case since it heavily depends on graph structure.

In Section 3 we introduce a kind of curvature for planar graphs and give a sufficient condition for both the bottom of the spectrum and the isoperimetric constant to be away from 0. In particular, we determine the isoperimetric constants explicitly for a special class of planar graphs that are called (d,f)-regular planar graphs. We also consider the upper bound of the spectrum. It is unique to graphs simply because a discrete laplacian is a bounded operator while a Laplace-Beltrami operator on a manifold is usually unbounded. We show that the upper bound of the spectrum of a laplacian is closely related to a geometric property, so-called the bipartiteness.

In Section 4 we discuss about three graph-operations: line graphs, subdivisions and para-line graphs. There might be no analogous objects for manifolds unless one considers singular manifolds. We show how the spectra change under those graph-operations, and as an application of these results we determine the spectrum of an infinite Sierpiński lattice graph.

In Section 5, we collect some problems around topics which are discussed in the present paper.

The rest of this section is devoted to the explanation of our setting which is used throughout the present paper. Let $G=(V(G),E(G))$ be a connected, locally finite graph, where $V(G)$ is the set of its vertices and $E(G)$ is the set of its unoriented edges. A graph G may have self-loops and multiple edges. Considering each edge in $E(G)$ to have two orientations, we introduce the set of all oriented edges; we denote it by $A(G)$. For an edge $e\in A(G)$, the origin vertex and the terminal one of e are denoted by $o(e)$ and $t(e)$, respectively. The inverse edge of e is denoted by $\bar{e}$. We sometimes write $[e]$ for the unoriented edge which is obtained from $e\in A(G)$ (or $\overline{e}$) by forgetting its orientation.

Let $p: A(G)\to(0,1]$ be a transition probability such that

$$\sum_{e\in A_x(G)} p(e)=1,$$

where $A_x(G)=\{e\in A(G)\mid o(e)=x\}$. We assume that p is reversible, that is, there exists a positive valued function $m: V(G)\to(0,\infty)$ such that

$$m(o(e))p(e)=m(t(e))p(\bar{e}) \ \ (=: m_A(e)) \tag{1.1}$$

for every oriented edge $e\in A(G)$. The function m is called a reversible measure for p and it is unique, if it exists, up to a multiple constant. We set

$$C^0(G,\mathbf{C})=\{f : V(G)\to\mathbf{C}\}$$

as the space of all functions on G and denote by $\ell^2(G)$ or $\ell^2(G,m)$ the space of all square summable functions on $V(G)$ with respect to the inner product $\langle f_1,f_2\rangle=\sum_{x\in V(G)} f_1(x)\overline{f_2(x)}m(x)$. Define the discrete laplacian $\Delta_G : \ell^2(G)\to\ell^2(G)$ by

$$\Delta_G f(x)=\sum_{e\in A_x(G)} p(e)f(t(e))-f(x).$$

It becomes a self-adjoint operator on $\ell^2(G)$ because of the reversibility (1.1). We set

$$C^1(G,\mathbf{C})=\{\omega : A(G)\to\mathbf{C}\mid \omega(\overline{e})=-\omega(e) \text{ for } e\in A(G)\}$$

as the space of all 1-forms on G and define the coboundary operator $d : C^0(G, \mathbf{C}) \to C^1(G, \mathbf{C})$ by

$$df(e) = f(t(e)) - f(o(e))$$

for every $e \in A(G)$. We introduce an inner product on $C^1(G, \mathbf{C})$ by

$$\langle\langle \omega_1, \omega_2 \rangle\rangle = \frac{1}{2} \sum_{e \in A(G)} \omega_1(e)\overline{\omega_2(e)} m_A(e).$$

Then the dual δ of d is defined by the equality $\langle\langle df, \omega \rangle\rangle = \langle f, \delta\omega \rangle$ and expressed by

$$\delta\omega(x) = -\sum_{e \in A_x} p(e)\omega(e).$$

It is easy to check that $\Delta_G = -\delta d$. In Section 3 we use a quadratic form $\mathcal{E}$ defined by

$$\mathcal{E}(f, f) = \langle -\Delta_G f, f \rangle = \langle\langle df, df \rangle\rangle. \tag{1.2}$$

In this paper, we often take $(\deg_G o(e))^{-1}$ as $p(e)$ and $\deg_G x$ as $m(x)$, which are associated with the simple random walk on G. Here $\deg_G x = \# A_x(G)$ and it is called the degree of x. In this case, they satisfy (1.1) and the laplacian is expressed by

$$\Delta_G f(x) = (\deg_G x)^{-1} \sum_{e \in A_x(G)} f(t(e)) - f(x).$$

Our main concern of this paper is to study the relationship between the spectral properties of Δ_G and the geometric properties of G.

Finally let us recall the notion of bipartiteness, which plays an important role throughout this paper. We call a graph G bipartite if G has no closed path of odd length. In other words, the vertex set $V(G)$ can be partitioned into two disjoint sets V_1 and V_2 in such a way that every edge in $E(G)$ connects a vertex in V_1 with a vertex in V_2.

Acknowledgments. The authors would like to express their sincere gratitude to JAMS for its hospitality in the JAMS Symposium on Discrete Geometric Analysis 2002. The first named author is partially supported by JSPS under the Grant-in-Aid for Scientific Research No.14204010 and No.13340030; the second named author is partially supported by JSPS under the Grant-in-Aid for Scientific Research No.13740057.

2. Full spectrum property for maximal abelian covering graphs

After giving some preliminaries around abelian covering structure of graphs, we discuss some conditions for abelian covering graphs to have full spectrum property in Section 2.2.

2.1. Abelian covering graphs and twisted operators. For a finite connected graph M, we define the chain groups with coefficients in $\mathbf{Z}$ by

$$C_0(M, \mathbf{Z}) = \{ \sum_{x \in V(M)} a_x x \mid a_x \in \mathbf{Z} \}, \quad C_1(M, \mathbf{Z}) = \{ \sum_{e \in A(M)} a_e e \mid a_e \in \mathbf{Z} \} / \sim,$$

where $\sim$ is the relation $\bar{e} = -e$ for $e \in A(M)$. The boundary map $\partial : C_1(M, \mathbf{Z}) \to C_0(M, \mathbf{Z})$ is defined by $\partial(e) = t(e) - o(e)$. The 1-homology group $H_1(M, \mathbf{Z})$ is equal to $\ker\partial$. We also define the cochain groups with coefficients in an abelian group $\mathcal{A}$ by

$$\begin{aligned} C^0(M,\mathcal{A}) &= \{f : V(M) \to \mathcal{A}\}, \\ C^1(M,\mathcal{A}) &= \{\omega : A(M) \to \mathcal{A} \mid \omega(\bar{e}) = -\omega(e) \text{ for } e \in A(M)\}. \end{aligned}$$

The coboundary map $d : C^0(M,\mathcal{A}) \to C^1(M,\mathcal{A})$ is the same one as was defined in Section 1 and the 1-cohomology group is defined by $H^1(M,\mathcal{A}) = C^1(M,\mathcal{A})/\mathrm{Im}d$. It is known as the discrete Kodaira-Hodge theorem that $H^1(M,\mathbf{R})$ is identified with the space $\{\omega \in C^1(M,\mathbf{R}) \mid \delta\omega = 0\}$ of harmonic 1-forms. In this identification, the integral 1-cohomology group $H^1(M,\mathbf{Z})$ corresponds to the lattice subgroup

$$\{\omega \in H^1(M,\mathbf{R}) \mid \int_c \omega \in \mathbf{Z} \text{ for every closed path } c \text{ in } M\},$$

where $\int_c \omega = \sum_{i=1}^n \omega(e_i)$ for $c = (e_1, e_2, \dots, e_n)$. Here a path $p = (e_1, e_2, \dots, e_n)$ of length n is a sequence of oriented edges with $t(e_i) = o(e_{i+1})$ for $i = 1, \dots, n-1$; a closed path is a path such that $t(e_n) = o(e_1)$.

We recall that a map $\gamma : V(G) \to V(G)$ is an automorphism if it is a bijection satisfying the following: two vertices x_1 and x_2 are adjacent in G if and only if the vertices $\gamma(x_1)$ and $\gamma(x_2)$ are adjacent in G. See also Section 4.3. Now we assume that an abelian group Γ acts freely on $G = (V(G), E(G))$ as an automorphism group. The quotient $M = \Gamma\backslash G$ naturally has a graph structure and we assume $M = (V(M), E(M))$ is finite. The canonical morphism $\pi : G \to M$ is a covering map in the sense that

(1) π is surjective as a map between the sets of vertices,
(2) the map $\pi|_{A_x(G)} : A_x(G) \to A_{\pi(x)}(M)$ is bijective for every $x \in V(G)$.

Hence G can be regarded as an abelian covering graph of M with covering transformation group Γ. In particular, when the transformation group is the 1-homology group $H_1(M,\mathbf{Z})$ of M, the corresponding covering graph of M is said to be the maximal abelian covering graph of M and is denoted by M^{ab}. The covering graph M^{ab} is maximal in the sense that for any abelian covering graph G of M with covering transformation group Γ there exists a covering map $M^{ab} \to G$ which factorizes the covering map $M^{ab} \to M$. The covering transformation group Γ is isomorphic to the quotient group $H_1(M,\mathbf{Z})/\Gamma_1$, where Γ_1 is the covering transformation group of the covering map $M^{ab} \to G$.

The group $\widehat{H}_1(M,\mathbf{Z})$ of unitary characters of the 1-homology group $H_1(M,\mathbf{Z})$ is identified with the Jacobian torus $J(M) = H^1(M,\mathbf{R})/H^1(M,\mathbf{Z})$ in the following way. For a given $\omega \in H^1(M,\mathbf{R})$, define the unitary character χ_ω by

$$\chi_\omega(\sigma) = \exp(2\pi\sqrt{-1}\int_c \omega),$$

where σ is represented by a closed path c, that is a $\mathbf{Z}$-linear combination of closed paths in M. The correspondence $\omega \mapsto \chi_\omega$ gives rise to an isomorphism of the Jacobian torus onto the group of unitary characters of $H_1(M,\mathbf{Z})$.

We denote by $\widehat{\Gamma}$ the group of unitary characters of Γ. The surjective homomorphism $\eta : H_1(M,\mathbf{Z}) \to \Gamma$ gives rise to an injective homomorphism $\widehat{\eta} : \widehat{\Gamma} \to \widehat{H}_1(M,\mathbf{Z}) \cong J(M)$. We identify $\widehat{\Gamma}$ with its image in $J(M)$. Then $\chi_\omega \in \widehat{\Gamma}$ if and only if $\int_c \omega \in \mathbf{Z}$ for every closed path c in M with $\eta(c) = 1$.

Now we assume that the probability p and the reversible measure m on G are invariant under the Γ-action. In other words, p and m are defined as the lift of a transition probability p_M and m_M on M, that is, $p(e) = p_M(\pi(e))$ for any $e \in A(G)$ and $m(x) = m_M(\pi(x))$ for any $x \in V(G)$, respectively. For any $\chi \in \widehat{\Gamma}$, we put

$$\ell^2_\chi = \{f : V(G) \to \mathbf{C} \mid f(\sigma x) = \chi(\sigma) f(x) \text{ for any } \sigma \in \Gamma\}$$

with the inner product $\langle f, g\rangle_\chi = \sum_{x\in\mathcal{D}} f(x)\overline{g(x)}m(x)$, where $\mathcal{D}$ is a fundamental set of $V(G)$ for Γ. Since Δ_G commutes with the Γ-action, we can restrict it to ℓ^2_χ. We call the restriction $\Delta_G|_{\ell^2_\chi}$ the twisted laplacian and denote it by $\Delta_{G,\chi}$. It follows from the theory of direct integral that

$$Spec(-\Delta_G) = \bigcup_{\chi\in\widehat{\Gamma}} Spec(-\Delta_{G,\chi}), \tag{2.1}$$

where $Spec(-\Delta_G)$ is the spectrum set of $-\Delta_G$ on $\ell^2(G)$. For any $\chi \in \widehat{H}_1(M, \mathbf{Z})$, one can find $\theta \in H^1(M, \mathbf{R})$ such that $\chi = \chi_\theta$. Moreover, it is known that $\Delta_{G,\chi}$: $\ell^2_\chi \to \ell^2_\chi$ is unitarily equivalent to $L_{\theta,M} : \ell^2(M) \to \ell^2(M)$ which is defined by

$$L_{\theta,M} f(x) = \sum_{e\in A_x(M)} p(e)\exp(2\pi\sqrt{-1}\theta(e)) f(t(e)) - f(x). \tag{2.2}$$

Consequently, we can rewrite (2.1) in terms of θ as

$$Spec(-\Delta_G) = \bigcup_{\theta\in H^1(M,\mathbf{R})\,;\,\chi_\theta\in\widehat{\Gamma}} Spec(-L_{\theta,M}). \tag{2.3}$$

In particular, if G is the maximal abelian covering graph M^{ab} of M, we have

$$Spec(-\Delta_{M^{ab}}) = \bigcup_{\theta\in H^1(M,\mathbf{R})} Spec(-L_{\theta,M}). \tag{2.4}$$

2.2. Full spectrum property. By the definition of a discrete laplacian, it holds that $Spec(-\Delta_G)$ is a closed subset of $[0, 2]$.

DEFINITION 2.1 (Full Spectrum Property). We say that an infinite graph G has Full Spectrum Property (FSP) if it holds that $Spec(-\Delta_G) = [0, 2]$ for the laplacian with respect to the simple random walk: for all $e \in A(G)$, $p(e) = (\deg_G o(e))^{-1}$.

Hereafter in Section 2.2, we discuss FSP under the assumption that graphs have abelian covering structure.

The d-dimensional integer lattice $\mathbf{Z}^d$ is the maximal abelian covering graph of a d-bouquet graph for every natural number d, and it is known that $Spec(-\Delta_{\mathbf{Z}^d}) = [0, 2]$. Then we can say that the infinite graph $\mathbf{Z}^d$ has FSP. On the other hand, the triangular lattice is an abelian covering graph of a 3-bouquet graph but not the maximal one. It does not have FSP since its spectrum set is $[0, 3/2]$. More generally, it follows from Remark 3.12 that a non-bipartite abelian covering graph of a finite graph does not have FSP since it turns out to be "essentially non-bipartite". It should be noted that the maximal abelian covering graph of a finite graph is always bipartite.

PROPOSITION 2.2 (cf.[**10**]). *The maximal abelian covering graph M^{ab} of a finite graph M is bipartite. So the spectrum set $Spec(-\Delta_{M^{ab}})$ is symmetric with respect to* 1.

Thus it might be natural to ask whether infinite bipartite covering graphs have FSP. However, the following example shows that it is not always the case.

EXAMPLE 2.3. Let M be a finite graph such that $V(M) = \{0, 1\}$ and $A(M) = \{e_i, \overline{e}_i, i = 1, 2, 3\}$ with $o(e_i) = 0$ and $t(e_i) = 1, i = 1, 2, 3$. The 1-homology group of M is isomorphic to $\mathbf{Z}^2$ and its maximal abelian covering graph is the hexagonal lattice, which has FSP. Take $c_1 = e_1 + \overline{e}_2, c_2 = e_2 + \overline{e}_3$ as a basis of $H_1(M, \mathbf{Z})$ and let Γ_1 be the subgroup of $H_1(M, \mathbf{Z})$ generated by c_1. The quotient graph M^{ab} by Γ_1 is isomorphic to the abelian covering graph of M with covering transformation group $\Gamma \cong H_1(M, \mathbf{Z})/\Gamma_1$, which is the graph G such that $V(G) = \mathbf{Z}^1$ and $A(G) = \{e_n, \overline{e}_n, e'_m, \overline{e}'_m, \ n \in \mathbf{Z}, m \in 2\mathbf{Z}\}$, where $o(e_n) = o(e'_n) = n$ and $t(e_n) = t(e'_n) = n + 1$. It is easy to compute the spectrum of $-\Delta_G$ by using (2.3); it is $[0, 2/3] \cup [4/3, 2]$, not equal to $[0, 2]$. Then G does not have FSP. Consequently, G does not necessarily have FSP even if it is an infinite bipartite covering.

In [**10**], the following problem is raised as a conjecture:

CONJECTURE 2.4. *Assume that G is an infinite connected graph such that G is isomorphic to the maximal abelian covering graph M^{ab} of a finite connected graph M. Then G has FSP.*

We have some partial affirmative answers toward this conjecture:

PROPOSITION 2.5 ([**10**]). *Assume G is an infinite connected graph and G is isomorphic to the maximal abelian covering graph M^{ab} of a finite connected graph M. Then G has FSP if one of the following conditions holds:*
(1) $\deg_G x \in 2\mathbf{N}$ for any vertex $x \in V(G)$.
(2) M is a $(2d+1)$-regular graph having a 2ℓ-factor such that $\lceil (d+1)/2 \rceil \leq \ell \leq d$, where $d \geq 1$ and $\lceil x \rceil$ is the minimum integer which is equal or greater than x.

In the statement above, the terminology "k-factor" appears. Although this is a well-known notion in graph theory (cf.[**1, 2**]), here we give the definition:

DEFINITION 2.6 (k-factor). Let $G = (V(G), E(G))$ be a finite graph. A graph $H = (V(H), E(H))$ is called a spanning subgraph or a *factor* of G if $V(H) = V(G)$ and $E(H) \subset E(G)$. In addition, a factor H is called a *k-factor* if $\deg_H x = k$ for every vertex $x \in V(H)$.

Here we give a proof of Proposition 2.5 after recalling a key notion and property which is used in the proof:

DEFINITION 2.7 (Euler circuit). A closed path $c = (e_1, e_2, \ldots, e_n)$ in a finite graph M is called an *Euler circuit* if $n = \#A(M)/2$ and, for each $e \in A(M)$, there exists i $(1 \leq i \leq n)$ such that e_i or $\overline{e}_i$ equals to e.

REMARK 2.8. It is well-known as a classical and famous fact in graph theory (cf.[**1, 2**]) that a finite graph M has an Euler circuit if and only if the degree of each vertex is even.

PROOF OF PROPOSITION 2.5. Firstly we assume the condition (1). Then the quotient graph M also satisfies $\deg_M x \in 2\mathbf{N}$ for any vertex $x \in V(M)$, and M has an Euler circuit $c = (e_1, e_2, \ldots, e_n)$. Then we put a 1-form $\theta \in C^1(M, \mathbf{R})$ as $\theta(e_i) = \alpha \in \mathbf{R}$ and $\theta(\overline{e}_i) = -\alpha$ for every $i = 1, 2, \ldots, n$. It is easy to see that $\{e_i, \overline{e}_i\}_{i=1}^n = A(M)$ and

$$\#\{\, e \in A_x(M) \mid \theta(e) = \alpha \,\} = \#\{\, e \in A_x(M) \mid \theta(e) = -\alpha \,\} = (\deg_M x)/2$$

for every vertex $x \in V(M)$. For every α, the 1-form θ defined as above is harmonic, that is, $\theta \in H^1(M, \mathbf{R})$. Denote by $\mathbf{1}$ a constant function on $V(M)$ such that $\mathbf{1}(x) = 1$. Then

$$\begin{aligned} -L_{\theta,M}\mathbf{1}(x) &= -(\deg_M x)^{-1} \sum_{e \in A_x(M)} \exp\left(2\pi\sqrt{-1}\theta(e)\right) \mathbf{1}(t(e)) + \mathbf{1}(x) \\ &= 1 - (\deg_M x)^{-1} \cdot (\deg_M x)/2 \cdot \left(\exp\left(2\pi\sqrt{-1}\alpha\right) + \exp\left(-2\pi\sqrt{-1}\alpha\right) \right) \\ &= (1 - \cos 2\pi\alpha)\mathbf{1}(x). \end{aligned}$$

This implies that the quantity $1 - \cos 2\pi\alpha$ is an eigenvalue of $-L_{\theta,M}$. By (2.4),

$$\bigcup_{\alpha \in \mathbf{R}} (1 - \cos 2\pi\alpha) \subset \bigcup_{\theta \in H^1(M,\mathbf{R})} Spec(-L_{\theta,M}) = Spec(-\Delta_G),$$

thus we get the conclusion.

Secondly we assume the condition (2). Then each component M_j of a 2ℓ-factor has an Euler circuit $c_j = (e_{j1}, e_{j2}, \dots, e_{jn_j})$ in M_j, and we set a 1-form $\theta \in C^1(M, \mathbf{R})$ as $\theta(e_{ji}) = -\theta(\overline{e}_{ji}) = \alpha$ for every i, j, and $\theta(\cdot) = 0$ for other edges. It is easy to see that

$$\#\{ e \in A_x(M) \mid \theta(e) = \alpha \} = \#\{ e \in A_x(M) \mid \theta(e) = -\alpha \} = \ell$$

for every vertex $x \in V(M)$. As was in the proof under the condition (1),

$$\begin{aligned} -L_{\theta,M}\mathbf{1}(x) &= -(2d+1)^{-1} \sum_{e \in A_x(M)} \exp\left(2\pi\sqrt{-1}\theta(e)\right) \mathbf{1}(t(e)) + \mathbf{1}(x) \\ &= \frac{2\ell}{2d+1}(1 - \cos 2\pi\alpha)\mathbf{1}(x). \end{aligned}$$

Thus we have $[\,0, 4\ell/(2d+1)\,] \subset Spec(-\Delta_G)$, where $4\ell/(2d+1) > 1$. The conclusion follows from Proposition 2.2 and the proof is completed. □

We have obtained some affirmative partial results for this conjecture, however, we also found an example which shows that Conjecture 2.4 needs an additional assumption.

EXAMPLE 2.9. Let $M_{m,n}$ be a finite graph which looks like a "magnifier glass":

$$\begin{aligned} V(M_{m,n}) &= \{x_0, x_1, \dots, x_{m-1}, x_m = y_1, y_2, \dots, y_n\}, \\ E(M_{m,n}) &= \{x_0x_1, x_1x_2 \dots, x_{m-1}x_m\} \cup \{y_1y_2, y_2y_3, \dots, y_{n-1}y_n, y_ny_1\}. \end{aligned}$$

In other words, $M_{m,n}$ is isomorphic to $P_m + C_n$, where the symbol $+$ signifies

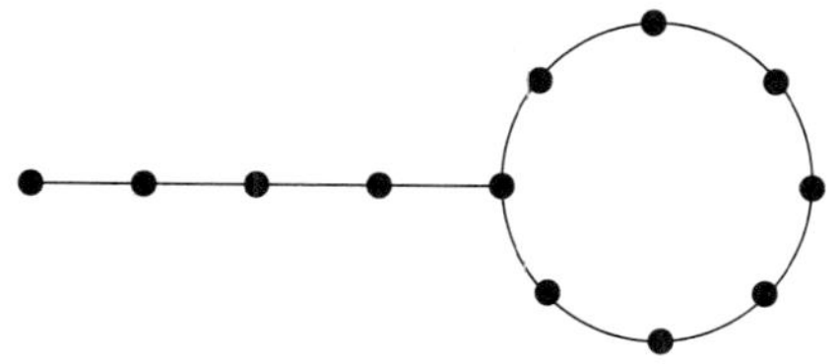

FIGURE 1. $M_{m,n}$ ($m = 4$ and $n = 8$).

that an endpoint of the path P_m with length m and a vertex in the cycle C_n are identified. Let $G_{m,n}$ be the maximal abelian covering graph $M^{cb}_{m,n}$ of $M_{m,n}$. Then the graph $G_{m,n}$ is an infinite path to which finite paths are attached periodically. Direct calculation shows that the characteristic equation of the twisted operator $L_{\theta,M_{m,n}}$ with $\int_{C_n} \theta = \alpha \in \mathbf{R}$ is equivalent to

$$T_{m+n}(\lambda) = (2\cos 2\pi\alpha - T_n(\lambda)) \cdot T_m(\lambda),$$

where $T_n(\lambda) = \cos(n \arccos \lambda)$ is the Tchebyshev polynomial of the first kind. It leads us to the following: let k be the greatest common divisor $\gcd(m,n)$ of m and n, and let $N = m + n + 1 - k$. Then the spectrum set of $-\Delta_{G_{m,n}}$ is given by the form

$$Spec(-\Delta_{G_{m,n}}) = \bigcup_{i=1}^{N} [a_i, b_i],$$

where $0 = a_1 < b_1 < a_2 \leq b_2 < \cdots < a_{N-1} \leq b_{N-1} < a_N < b_N = 2$. In particular, $G_{m,n}$ does not have FSP. Moreover, when n/k is odd, then $-\Delta_{G_{m,n}}$ has no eigenvalue and $a_i < b_i$ for all $1 \leq i \leq N$; when n/k is even, then $-\Delta_{G_{m,n}}$ has k eigenvalues and there exist $1 < i_1 < i_2 < \cdots < i_k < N$ only for which $a_{i_j} = b_{i_j} (1 \leq j \leq k)$.

We showed that Conjecture 2.4 is true if M is an even-regular graph or an odd-regular graph having some properties. Then this conjecture would be true at least for regular graphs though we do not show at this stage (see also Problem 5.1). On the other hand, by Example 2.9, we now know Conjecture 2.4 itself is not true for all maximal covering graphs. However, we strongly feel that such a case is rare. For example, we believe that Conjecture 2.4 is true if we assume M has no vertex whose degree is 1. We close this section with proposing the following problem.

PROBLEM 2.10. Characterize all finite graphs whose maximal abelian covering graphs do not have FSP.

3. Spectral and geometric constants

In Section 3.1, we give the isoperimetric constants for a special class of hyperbolic graphs explicitly. In Section 3.2, we introduce a kind of curvature and discuss its properties. In Section 3.3, we show the relationship between non-bipartiteness and spectral/geometric constants.

In Section 3.1 we assume that $p(e) = (\deg_G o(e))^{-1}$ and $m(x) = \deg_G x$, so the quantities, such as $\mathcal{E}(f, f)$ or Δ_G, which appear in this section are correspond to the simple random walk. In Section 3.3 we do not assume it.

3.1. Isoperimetric constant for a (d, f)-regular planar graph. In this subsection, we consider the isoperimetric constant for a (d, f)-regular planar graph.

The isoperimetric constant is defined as

$$h(G) = \inf\{\frac{|E(\partial_v K)|}{Area(K)} \mid K \text{ is a finite subgraph of } G\},$$

where $Area(K) = \sum_{x \in V(K)} \deg_G x$ and $E(\partial_v K) = \{xy \in E(G) \mid x \in V(K), y \in V(G) \setminus V(K)\}$. Another equivalent definition is also useful. It is easy to see that

$$h(G) = \inf\{\frac{\mathcal{E}(1_K, 1_K)}{\|1_K\|^2} \mid K \text{ is a finite subgraph of } G\}, \tag{3.1}$$

where 1_K is the indicator function of $V(K)$ and $\mathcal{E}(f,f)$ was defined in (1.2). We have the following well-known inequality between the isoperimetric constant $h(G)$ and the lower bound $\lambda_0(G)$ of the spectrum of $-\Delta_G$:

$$1-\sqrt{1-h^2(G)} \le \lambda_0(G) \le h(G) < 1. \tag{3.2}$$

The second inequality follows immediately from the variational formulas (3.1) and (3.3).

If G is planar, there is another kind of isoperimetric constant which is a dual notion of $h(G)$: for an embedding φ of G in the plane $\mathbf{R}^2$, one can define

$$h^*(G) = h^*(G,\varphi) = \inf\{ \frac{|E(\partial_f K)|}{|F(K)|} \mid K \text{ is a finite subgraph of } G \},$$

where $F(K)$ is the set of all faces K and $E(\partial_f K) = \{ xy \in E(K) \cap E(F') \mid F' \in F(G)\backslash F(K) \}$.

Both of them are discrete analogues of the isoperimetric constant for a manifold: each denominator corresponds to the "area" of K, and each numerator to the "length" of the boundary of K. We remark that $h(G)$ is a graph invariant while $h^*(G)$ may depend on the way of embedding in the plane.

Now we introduce a special class of graphs. A graph G is said to be a (d,f)-*regular planar graph* if G is a planar graph satisfying the following:

(1) G is d-regular in the ordinary sense, that is, $\deg_G x = d$ for every vertex $x \in V(G)$ and $d \ge 3$.
(2) G admits an embedding in the plane $\mathbf{R}^2$ in such a way that every face F is an f-gon with $f \ge 3$ and the number of vertices contained in each compact subset in $\mathbf{R}^2$ is finite.

Here we call such an embedding a (d,f)-regular realization of G. For any pair (d,f) with $H(d,f) := 4-(d-2)(f-2) \le 0$ there exists an infinite (d,f)-regular planar graph. For example, the hexagonal lattice, the square lattice $\mathbf{Z}^2$ and the triangular lattice are $(3,6)$, $(4,4)$, $(6,3)$-regular planar graphs, respectively. They correspond to the case $H(d,f)=0$ and their spectra are explicitly known. As far as we know, the spectrum set of Δ_G for the other "hyperbolic" cases, $H(d,f)<0$, has not yet been computed explicitly. Rather we can compute the isoperimetric constant explicitly.

THEOREM 3.1 ([**7**],[**10**]). *Let $d \ge 3$ and $f \ge 3$ with $H(d,f) \le 0$ and G be a (d,f)-regular planar graph. Then we have*

$$h(G) = \frac{d-2}{d}\sqrt{\frac{H(d,f)}{H(d,f)-4}}$$

and for a (d,f)-regular realization of G, we have

$$h^*(G) = (f-2)\sqrt{\frac{H(d,f)}{H(d,f)-4}}.$$

REMARK 3.2. The case of $f=\infty$ may be understood as a d-regular tree T_d. In this case, it is well-known that $h(T_d) = (d-2)/d$ and then the expression of $h(G)$ is still valid by considering its value as the limit as f tends to ∞.

3.2. Combinatorial curvature. In this subsection we always assume G to be a planar graph embedded in the plane. The quantity $H(d,f)$ plays a role like a curvature for a (d,f)-regular planar graph G in the previous subsection. For general planar graphs embedded in the plane, one can discuss another kind of curvature which is called in [**8**] a combinatorial curvature defined as follows: for each vertex x,

$$\kappa_G(x) = 1 - \frac{\deg_G x}{2} + \sum_{i=1}^{\deg_G x} \frac{1}{d(F_i)},$$

where $F_1, F_2, \cdots, F_{\deg_G x}$ are the faces around a vertex x and $d(F)$ is equal to f if a face F is an f-gon.

REMARK 3.3. For a (d,f)-regular planar graph G, $\kappa_G(x) = H(d,f)/2f$ for all $x \in V(G)$.

While we can compute the isoperimetric constants for (d,f)-regular planar graphs, it is almost impossible to determine the isoperimetric constants for general planar graphs. Here, instead of giving the explicit values, we give a sufficient condition for the positivity of the isoperimetric constant in terms of the combinatorial curvature.

THEOREM 3.4. *If $\kappa_G(x) < 0$ for every vertex $x \in V(G)$, $h^*(G) > 0$.*

The combinatorial curvature κ_G, by definition, takes values in $\mathbf{Q}$. An interesting feature of the combinatorial curvature is that the negative values of κ_G do not accumulate to 0 while the positive values may do.

THEOREM 3.5. *If $\kappa_G(x) < 0$, then $\kappa_G(x) \leq -1/1806$.*

For details, one can refer to [**8**].

REMARK 3.6. A similar type of curvature $\kappa_G^*(\cdot)$ was introduced by M. Gromov [**6**]. This is a dual notion of the combinatorial curvature discussed above in the sense that it is defined on the set of faces while the above is on the set of vertices. In fact, for a (d,f)-regular planar graph G, $\kappa_G^*(F) = H(d,f)/2d$ for all $F \in F(G)$, where $F(G)$ is the set of faces in an infinite planar graph G. In [**8**] we also showed that $h(G) > 0$ if $\kappa_G^*(F) < 0$ for every face $F \in F(G)$.

3.3. Non-bipartiteness and Dirichlet form. Now we recall the well-known variational formula for the lower and upper bound for $-\Delta_G$, say $\lambda_0(G)$ and $\lambda_\infty(G)$, respectively:

$$\lambda_0(G) = \inf_{f \in \ell^2(G)\setminus\{0\}} \frac{\mathcal{E}(f,f)}{\|f\|^2}, \quad \lambda_\infty(G) = \sup_{f \in \ell^2(G)\setminus\{0\}} \frac{\mathcal{E}(f,f)}{\|f\|^2}. \tag{3.3}$$

Keeping this formula in mind, we generalize these constants by

$$\alpha_0(G,\mathcal{F}) = \inf_{f \in \mathcal{F}\setminus\{0\}} \frac{\mathcal{E}(f,f)}{\|f\|^2}, \quad \alpha_\infty(G,\mathcal{F}) = \sup_{f \in \mathcal{F}\setminus\{0\}} \frac{\mathcal{E}(f,f)}{\|f\|^2}, \tag{3.4}$$

where $\mathcal{F}$ is a nonempty subset of $\ell^2(G)$ containing at least one element other than 0.

EXAMPLE 3.7. When $\mathcal{F} = \ell^2(G)$, $\alpha_0(G,\mathcal{F})$ and $\alpha_\infty(G,\mathcal{F})$ are, of course, the lower and upper bound of the spectrum of $-\Delta_G$, respectively. Moreover, if we take the set $\{f \in \ell^2(G)\,;\, |f(x)| \in \{0,1\}, \forall x \in V(G)\}$ as $\mathcal{F}$, then it follows from the variational formula (3.1) that $\alpha_0(G,\mathcal{F})$ is nothing but the isoperimetric constant $h(G)$. Here $\alpha_\infty(G,\mathcal{F})$ is the corresponding upper bound, which may be also interesting quantity. See Problem 5.6.

The quantity $\alpha_\infty(G,\mathcal{F})$ is closely related to the bipartiteness of a graph G.

PROPOSITION 3.8. *Suppose $\mathcal{F}$ is closed under U_A for any subset $A \in V(G)$, where U_A is the unitary operator defined as $U_A f = f \cdot 1_{A^c} - f \cdot 1_A$. Then*

$$\alpha_0(G,\mathcal{F}) + \alpha_\infty(G,\mathcal{F}) \le 2.$$

The equality holds if G is bipartite.

PROOF. Here we simply write $\alpha(G)$ for $\alpha(G,\mathcal{F})$ and set $P = I + \Delta_G$. By (3.4) and the positivity preserving property of P, we obtain, for any $f \in \mathcal{F}$,

$$\begin{aligned} \alpha_0(G)\|f\|^2 + \mathcal{E}(f,f) &\le \mathcal{E}(|f|,|f|) + \mathcal{E}(f,f) \\ &= 2(\mathcal{E}(f_+,f_+) + \mathcal{E}(f_-,f_-)) \\ &= 2\Big(\langle (I-P)f_+, f_+\rangle + \langle (I-P)f_-, f_-\rangle\Big) \\ &\le 2\|f\|^2, \end{aligned} \tag{3.5}$$

where $f_+ = \max(f,0)$ and $f_- = \max(-f,0)$. We can choose a sequence of $f_n \in \mathcal{F}$ such that $\mathcal{E}(f_n,f_n)/\|f_n\|^2 \to \alpha_\infty(G)$. Consequently, putting $f = f_n$ in (3.5) and letting $n \to \infty$, we obtain

$$\alpha_0(G) + \alpha_\infty(G) \le 2. \tag{3.6}$$

Suppose G is bipartite. Then there exists a bipartition V_1, V_2 such that $V(G) = V_1 \sqcup V_2$ as was mentioned in Section 1. Define $\widetilde{f}$ by $\widetilde{f} = f \cdot 1_{V_1} - f \cdot 1_{V_2}$ for $f \in \mathcal{F}$. It is obvious from the assumption that $\widetilde{f} \in \mathcal{F}$ and it follows easily from the second equality in (3.5) that $\mathcal{E}(f,f) + \mathcal{E}(\widetilde{f},\widetilde{f}) = 2\|f\|^2$. Then, in the same way as in the above, we obtain $\alpha_0(G) + \alpha_\infty(G) \ge 2$, which together with (3.6) implies $\alpha_0(G) + \alpha_\infty(G) = 2$. □

For finite graphs, it is well-known that the maximal eigenvalue for any reversible transition probability p is equal to 2 if and only if a graph is bipartite. Also for infinite graphs, one can see from Proposition 3.8 that $\lambda_0(G) + \lambda_\infty(G) = 2$ if a graph G is bipartite. On the other hand, the converse does not hold in general. In fact, let $\mathbf{Z}^1 = (V(\mathbf{Z}^1), E(\mathbf{Z}^1))$ be the ordinary one dimensional lattice and $G = (V(G), E(G))$ a graph obtained from $\mathbf{Z}^1$ by adding an edge, that is, $V(G) = V(\mathbf{Z}^1)$ and $E(G) = E(\mathbf{Z}^1) \cup \{[e_0]\}$ with $o(e_0) = -1$ and $t(e_0) = 1$. Then, although G becomes non-bipartite due to the odd cycle with length 3 consisting of three vertices $\{-1,0,1\}$, $Spec(-\Delta_G)$ is still equal to $[0,2]$ since the essential spectrum does not change under such a local perturbation.

However, under some reasonable assumption, the converse assertion in the above holds. Before stating the theorem, we introduce a notion of a graph G being uniformly non-bipartite.

DEFINITION 3.9. Let N be a positive integer. An infinite graph G is called an N-non-bipartite graph if the following condition holds: for every vertex $x \in V(G)$, there exists an odd cycle C_{2n+1} going through x with $n \le N$.

THEOREM 3.10. *Let G be an N-non-bipartite graph. Assume that the reversible transition probability p is uniformly bounded below by p_0. Then it holds that*

$$\lambda_0(G) + \lambda_\infty(G) \le 2 - \epsilon(N, p_0),$$

where $\epsilon(N, p_0)$ depends only on N and p_0, and $0 < \epsilon(N, p_0) < 1$.

A SKETCH OF THE PROOF. In the discussion below, for simplicity, we restrict ourselves to a special class of N-non-bipartite graphs: G is a planar graph embedded in the plane and every face F of G is bounded by an odd cycle C_{2n+1} with $n \le N$. A (d, f)-regular planar graph for an odd $f(\le 2N+1)$ is a typical example belonging to this class. The discussion for the general N-non-bipartite graphs can be seen in [**12**]. See also Remark 3.12.

A key point is that the estimate of the quadratic form $\mathcal{E}(f, f)$ is reduced to that of local quadratic forms. In particular, under the above restriction for graphs, G admits a face decomposition and we only have to estimate the local quadratic forms on faces. Define a local quadratic form $\mathcal{E}_F(f, f)$ associated with a face $F \in F(G)$ by

$$\mathcal{E}_F(f, f) = \frac{1}{4} \sum_{e \in A(F)} |df(e)|^2 m_A(e).$$

Here we identify a face F with a cycle of length $d(F)$ as a finite subgraph of G. If we decompose G into the faces $F(G)$, we have

$$\mathcal{E}(f, f) = \frac{1}{2} \sum_{e \in A(G)} |df(e)|^2 m_A(e) = \sum_{F \in F(G)} \mathcal{E}_F(f, f) \tag{3.7}$$

since every edge $e \in A(G)$ is counted twice in the face decomposition.

Now let $C_F(x) = \sum_{e \in A_x(F)} p(e)$ and $m_F(x) = (1/2)C_F(x)m(x)1_{V(F)}(x)$. Then the operator associated with the form $\mathcal{E}_F$ on $\ell^2(G, m_F)$ is given by the laplacian $-\Delta_F$ (restricted on F) corresponding to the reversible transition probability (restricted on F)

$$p_F(e) = C_F(o(e))^{-1} p(e) 1_{A(F)}(e).$$

In other words,

$$\mathcal{E}_F(f, f) = \langle -\Delta_F f, f \rangle_{m_F},$$

where $\Delta_F = I - P_F$. Note that

$$\|f\|^2_{m_F} = \frac{1}{2} \sum_{e \in A(F)} |f(o(e))|^2 m_A(e).$$

and

$$\|f\|^2 = \sum_{F \in F(G)} \|f\|^2_{m_F}. \tag{3.8}$$

The local quadratic form $\mathcal{E}_F$ is always bounded by $2\|f\|^2_{m_F}$ for any $F \in F(G)$. Moreover, since every face F in G is identified with an odd cycle, we can obtain a stronger (uniform) estimate. Let p be a reversible transition probability on a cycle C_{2n+1} which is bounded below by a positive constant p_0, that is, $p(e) \ge p_0$ for any $e \in A(G)$. We denote the maximal eigenvalues of the laplacian on C_{2n+1} corresponding to a transition probability p by $\lambda_\infty(p, n)$. Since $\lambda_\infty(p, n)$ is continuous with respect to p and the set of all reversible measures bounded below by $p_0 > 0$ is closed, the maximum of $\lambda_\infty(p, n)$ attains at some reversible probability

p bounded below. From the remark after Proposition 3.8, $\lambda_\infty(p,n)$ is never equal to 2 since C_{2n+1} is not bipartite unless p takes values 0 or 1. Hence $\lambda_\infty(p,n)$ is bounded above uniformly in p by a constant strictly less than 2 which depends only on n and p_0. Hence we obtain the following uniform estimate for the local quadratic forms:

$$\mathcal{E}_F(f,f) \le (2-\delta)\|f\|^2_{m_F}, \quad \forall F \in F(G) \tag{3.9}$$

for some $\delta = \delta(N,p_0) > 0$ depending only on N and p_0.

It follows from (3.7), (3.8) and (3.9) that

$$\mathcal{E}(f,f) \le (2-\delta)\|f\|^2 \tag{3.10}$$

for any $f \in \ell^2(G,m)$. It should be noted that the estimate (3.10) is still valid whenever G is N-non-bipartite and the transition probability is uniformly bounded below, although δ may vary in $(0,1)$.

Let $\lambda_0(G)$ be the bottom of the spectrum of $-\Delta_G$. Then it is known that there exists a positive function h such that $-\Delta_G h = \lambda_0(G)h$ (cf.[**4**]). Define a new reversible probability p_h by harmonic transform of p as

$$p_h(e) = \frac{1}{1-\lambda_0(G)}\frac{h(t(e))p(e)}{h(o(e))}$$

with a reversible measure $m_h = mh^2$. Here $1-\lambda_0(G)$ is not equal to 0 by the inequality in (3.2).

REMARK 3.11. If the original transition probability is uniformly bounded below, then so is the harmonic transform p_h though the bound p_0 may be changed. It is an easy consequence of a discrete version of the Harnack inequality: for a positive function h such that $\Delta_G h \le 0$ it holds that

$$p(e)h(t(e)) \le h(o(e)) \le \frac{1}{p(\overline{e})}h(t(e))$$

for every $e \in A(G)$.

Consider the Dirichlet form

$$\mathcal{E}_h(f,f) = \langle -\Delta_h f, f\rangle_{m_h}$$

on $\ell^2(G,m_h)$, where $\Delta_h = I - P_h$. Then it is easy to see that

$$\mathcal{E}(f,f) = \lambda_0(G)\|Uf\|^2_{m_h} + (1-\lambda_0(G))\mathcal{E}_h(Uf,Uf)$$

for any $f \in \ell^2(G,m)$, where $U : \ell^2(G,m) \to \ell^2(G,m_h)$ is a unitary operator defined by $Uf = f/h$. Suppose G is N-non-bipartite. By Remark 3.11, one can apply the uniform estimate (3.10) for $\mathcal{E}_h$ to obtain

$$\begin{aligned}\mathcal{E}(f,f) &\le \lambda_0(G)\|Uf\|^2_{m_h} + (1-\lambda_0(G))(2-\delta)\|Uf\|^2_{m_h} \\ &= \{\lambda_0(G) + (1-\lambda_0(G))(2-\delta)\}\|f\|^2\end{aligned}$$

for any $f \in \ell^2(G,m)$. Hence by the variational formula (3.3) we get

$$\lambda_\infty(G) \le \lambda_0(G) + (1-\lambda_0(G))(2-\delta).$$

Consequently, we obtain

$$\lambda_0(G) + \lambda_\infty(G) \le 2 - (1-\lambda_0(G))\delta.$$

□

REMARK 3.12. A generalization of Theorem 3.10 is obtained in [**12**]. We call a graph G satisfying the following condition an "essentially non-bipartite graph": there exists a positive integer N such that, for every edge $e \in E(G)$, there exists a magnifier glass $M_{m,2n+1} \cong P_m + C_{2n+1}$ in G with $m+n \leq N$ and $e \in E(M_{m,2n+1})$. (The definition of $M_{m,n}$ was given in Section 2.2.) Also for any essentially non-bipartite graph G, it holds that $\lambda_0(G) + \lambda_\infty(G) < 2$. Details are omitted here.

4. Line graphs, subdivisions and para-line graphs

Here we focus on the various aspects of para-line graphs, which may be unfamiliar. In Section 4.1, we give the definitions of line graph, subdivision and para-line graph, and then discuss the relationship among them. We also show the homological structure of para-line graphs. In Section 4.2, we show the spectral property of para-line graphs. In Section 4.3, we show the injectiveness of the above graph-operations as maps among graphs. Finally, as an application, we determine the spectrum of infinite Sierpiński lattice through the structure of para-line graphs.

In this section, we always assume that a transition probability p is associated with the simple random walk on G, that is, $p(e) = (\deg_G o(e))^{-1}$.

4.1. Covering structure for line graphs and para-line graphs. Line graph is a well-known notion in graph theory while para-line graph, which was first introduced in [**9**], is not. Now we give the definitions of line graphs and para-line graphs. The definition of line graphs is usually given for simple graphs. Here we give the definitions for general graphs.

The line graph $L(G)$ of G is defined by

$$\begin{aligned} V(L(G)) &= E(G) = \{[e] \mid e \in A(G)\}, \\ A(L(G)) &= \{(e_1, e_2) \in A(G) \times A(G) \mid e_1 \neq e_2, o(e_1) = o(e_2)\}, \end{aligned}$$

where the incidence of an edge is given as $o((e_1, e_2)) = [e_1]$, $t((e_1, e_2)) = [e_2]$, and $\overline{(e_1, e_2)} = (e_2, e_1)$.

The para-line graph $P(G)$ of G is defined by

$$\begin{aligned} V(P(G)) &= A(G), \\ A(P(G)) &= \{(e_1, e_2) \in A(G) \times A(G) \mid o(e_1) = o(e_2), e_1 \neq e_2\} \\ &\quad \sqcup \{(e, \overline{e})_\circ \mid e \in A(G)\}, \end{aligned}$$

where $o((e_1, e_2)) = e_1$, $t((e_1, e_2)) = e_2$ and $\overline{(e_1, e_2)} = (e_2, e_1)$, and $o((e, \overline{e})_\circ) = e$, $t((e, \overline{e})_\circ) = \overline{e}$ and $\overline{(e, \overline{e})_\circ} = (\overline{e}, e)_\circ$. The symbol $\sqcup$ stands for the disjoint union. If there exists a self-loop e in G, the oriented edge $(e, \overline{e})$ and $(e, \overline{e})_\circ$ appears in the definition of $A(P(G))$. In such a case we regard them as double edges.

In order to mention the relationship between line graphs and para-line graphs, we introduce subdivision graphs that are also well-known in graph theory. The subdivision graph $S(G)$ of G is defined by

$$\begin{aligned} V(S(G)) &= V(G) \sqcup E(G), \\ A(S(G)) &= \{(o(e), e), (e, o(e)) \mid e \in A(G)\}, \end{aligned}$$

where $o((o(e), e)) = o(e)$, $t((o(e), e)) = [e]$ and $\overline{(o(e), e)} = (e, o(e))$.

REMARK 4.1. Let G be a d-regular graph. Then $L(G)$ is $(2d-2)$-regular, $P(G)$ is d-regular and $S(G)$ is $(d, 2)$-semiregular. Here G is said to be (d_1, d_2)-semiregular if G is bipartite with bipartition $V_1 \cup V_2$ and $\deg_G x \equiv d_i$ on V_i $(i = 1, 2)$.

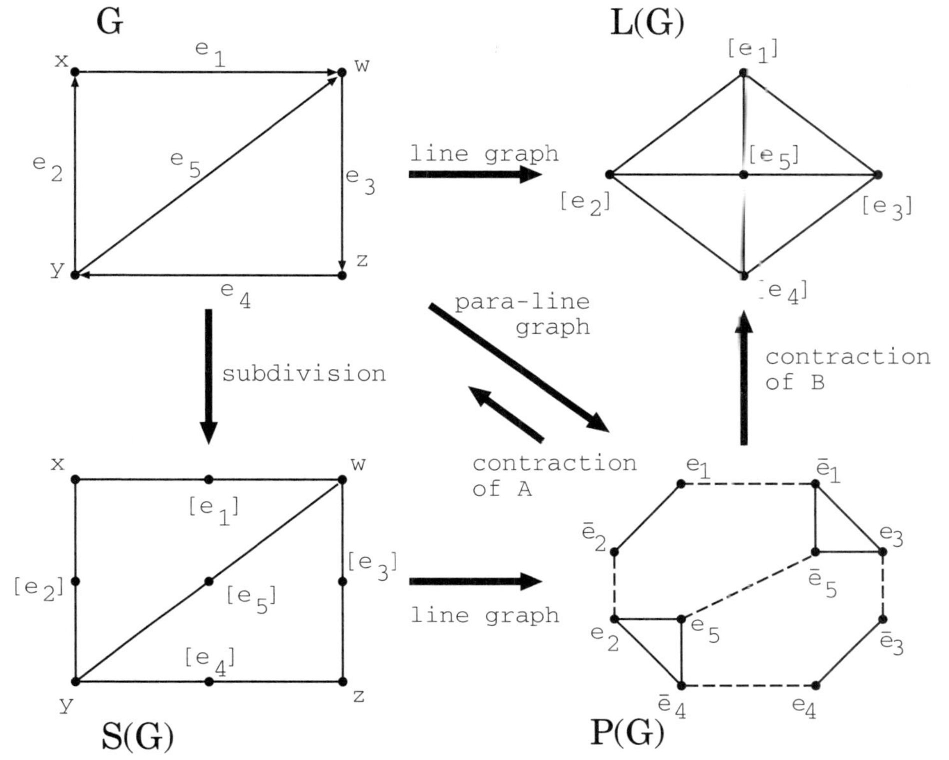

FIGURE 2. Line, para-line and subdivision graphs.

The relationship between G, $L(G)$, $S(G)$ and $P(G)$ is as follows. See also Figure 2 and Remark 4.3.

PROPOSITION 4.2. *Let G be a graph. Then the para-line graph $P(G)$ of G is isomorphic to the line graph of the subdivision graph of G, that is, $P(G) \cong L(S(G))$.*

PROOF. By the above definitions, it is easy to see that

$$V(L(S(G))) = \{[(o(e), e)] \mid e \in A(G)\}$$

and

$$\begin{aligned} A(L(S(G))) &= \{((o(e_1), e_1), (o(e_2), e_2)) \mid e_1, e_2 \in A(G), o(e_1) = o(e_2), e_1 \neq e_2\} \\ &\quad \cup \{((e, o(e)), (\bar{e}, o(\bar{e}))) \mid e \in A(G)\}, \end{aligned}$$

where the origin vertices, the terminal ones and the inverse edges are naturally defined. If we identify a vertex $[(o(e), e)]$ with e, an edge $((o(e_1), e_1), (o(e_2), e_2))$ with (e_1, e_2) and an edge $((e, o(e)), (\bar{e}, o(\bar{e})))$ with $(e, \bar{e})_\circ$, it is obvious that $L(S(G))$ is isomorphic to $P(G)$. □

REMARK 4.3. In Figure 2, A and B are the sets of the unoriented edges which are obtained by forgetting the orientation in $\{(e_1, e_2) \in A(G) \times A(G) \mid o(e_1) = o(e_2), e_1 \neq e_2\}$ (solid edges) and $\{(e, \bar{e})_\circ \mid e \in A(G)\}$ (dashed edges) in $P(G)$, respectively. By the definitions, there exist bijections from $E(L(G))$ to A and from $E(G)$ to B. It is easy to see that two graphs G and $L(G)$ are obtained from $P(G)$

by contracting A and B, respectively. Here it is called a contraction of an edge $[e]$ in G to remove $[e] \in E(G)$ and identify $o(e)$ and $t(e)$. We simply say a contraction of A when we contract all edges in a edge set A.

When G is a covering graph, $L(G)$ is also a covering graph with the same transformation group. We recall that a covering map $\pi : G \to M$ with covering transformation group Γ is normal if Γ acts transitively on each fiber $\pi^{-1}(x)$ for $x \in V(M)$.

LEMMA 4.4 ([**13**]). *Let G be a normal covering graph of a finite graph M with covering transformation group Γ, then $L(G)$ is a normal covering graph of $L(M)$ with the same covering transformation group Γ.*

The same assertion holds for para-line graphs as a corollary of the above lemma.

LEMMA 4.5. *Let G be a normal covering graph of a finite graph M with covering transformation group Γ, then $P(G)$ is a normal covering graph of $P(M)$ with the same covering transformation group Γ.*

PROOF. The action of Γ on G induces actions on $S(G)$ and $P(G)$ in a natural manner. The assertion of the lemma for subdivision graphs are almost trivial, and hence Lemma 4.5 follows immediately from Proposition 4.2 and Lemma 4.4. □

Now let $\phi_i : C_i(P(M), \mathbf{Z}) \to C_i(M, \mathbf{Z})$ for $i = 0, 1$ be the homomorphism defined as

$$\begin{cases} \phi_0(e) = o(e), \\ \phi_1((e_1, e_2)) = 0, \quad \phi_1((e, \bar{e})_\circ) = e. \end{cases}$$

Then we have the following:

LEMMA 4.6. *(ϕ_0, ϕ_1) is a chain map, that is, $\phi_0 \circ \partial = \partial \circ \phi_1$.*

PROOF. It follows from

$$\phi_0(\partial(e_1, e_2)) = \phi_0(e_2 - e_1) = o(e_2) - o(e_1) = 0 = \partial(\phi_1((e_1, e_2)))$$

and

$$\phi_0(\partial(e, \bar{e})_\circ) = \phi_0(\bar{e} - e) = o(\bar{e}) - o(e) = t(e) - o(e) = \partial(\phi_1((e, \bar{e})_\circ)).$$

□

LEMMA 4.7. *Let $\phi_1 : H_1(P(M), \mathbf{Z}) \to H_1(M, \mathbf{Z})$ be the induced homomorphism. Then it is surjective.*

PROOF. An element $\sigma \in H_1(M, \mathbf{Z})$ is represented by a closed path $c = (e_1, \ldots, e_n)$ in M such that $\bar{e}_i \neq e_{i+1}$ for $i = 1, 2, \ldots, n$. Define a closed path c' in $P(M)$ by

$$c' = ((e_1, \bar{e}_1)_\circ, (\bar{e}_1, e_2), (e_2, \bar{e}_2)_\circ, \ldots, (\bar{e}_{n-1}, e_n), (e_n, \bar{e}_n)_\circ, (\bar{e}_n, e_1))$$

It is obvious that $\phi_1(c') = c$. □

PROPOSITION 4.8. *Let G be the abelian covering graph of a finite graph M with transformation group Γ. Let η and η^P be the surjective homomorphisms associated with the covering maps $\pi : G \to M$ and $\pi^P : P(G) \to P(M)$, respectively. Then the homomorphism $\eta^P : H_1(P(M), \mathbf{Z}) \to \Gamma$ coincides with the composition $\eta \circ \phi_1$:*

$$H_1(P(M), \mathbf{Z}) \xrightarrow{\phi_1} H_1(M, \mathbf{Z}) \xrightarrow{\eta} \Gamma.$$

PROOF. Let c' be a closed path in $P(M)$ and its lift to $P(G)$ be $\widetilde{c}'$, and write

$$\widetilde{c}' = (\widetilde{\mathbf{e}}_1, \widetilde{\mathbf{e}}_2, \dots, \widetilde{\mathbf{e}}_n),$$

where $\widetilde{\mathbf{e}}_i = (\widetilde{e}_i, \widetilde{e}_{i+1})$ or $(\widetilde{e}_i, \widetilde{e}_{i+1})_\circ \in A(P(M))$. Replacing $\widetilde{\mathbf{e}}_i$ with two edges $\widetilde{e}_i, \overline{\widetilde{e}_i}$ if $\widetilde{\mathbf{e}}_i = (\widetilde{e}_i, \widetilde{e}_{i+1})$ and with one edge $\widetilde{e}_i$ if $\widetilde{\mathbf{e}}_i = (\widetilde{e}_i, \widetilde{e}_{i+1})_\circ$ in the expression of $\widetilde{c}'$ for every $i = 1, 2, \dots, n$, we obtain a new sequence of edges of G and denote it by $\widetilde{c}$. It is easy to see that $\widetilde{c}$ is a path in G such that the projection $c = \pi(\widetilde{c})$ is a closed path in M with $\phi_1(c') = c$. Now we recall that $\eta : H_1(M, \mathbf{Z}) \to \Gamma$ and $\eta^P : H_1(P(M), \mathbf{Z}) \to \Gamma$ are given by $\eta(c)o(\widetilde{c}) = t(\widetilde{c})$ for a closed path c in M with a lift $\widetilde{c}$ in G and $\eta^P(c')o(\widetilde{c}') = t(\widetilde{c}')$ for a closed path c' in $P(M)$ with a lift $\widetilde{c}'$ in $P(G)$, respectively. It is easy to check that $\eta^P(c')o(\widetilde{c}) = t(\widetilde{c})$. Since the Γ-action is free, we obtain $\eta^P = \eta \circ \phi_1$. $\square$

REMARK 4.9. The surjective homomorphism ϕ_1 induces the injective homomorphism $\phi_1^* : H^1(M, \mathbf{R}) \to H^1(P(M), \mathbf{R})$ through

$$\phi_1^*\theta((e_1, e_2)) = \theta(\phi_1((e_1, e_2))) = 0, \quad \phi_1^*\theta((e, \overline{e})_\circ) = \theta(\phi_1((e, \overline{e})_\circ)) = \theta(e).$$

We can identify $\widehat{\Gamma}$ with a subset of $J(M)$ and $J(P(M))$ by the maps $\widehat{\eta} : \widehat{\Gamma} \to J(M)$ and $\widehat{\eta}^P : \widehat{\Gamma} \to J(P(M))$. By the above lemma, the relationship between $\widehat{\eta}$ and $\widehat{\eta}^P$ is given by $\widehat{\eta}^P = \phi_1^* \circ \widehat{\eta}$. Hence the cohomology class of $\phi_1^*\theta \in C^1(P(M), \mathbf{R})$ is naturally identified with a character $\chi = \chi_\theta \in \widehat{\Gamma}$.

4.2. Spectrum for line graphs and para-line graphs. Once we know the spectrum of G, we get the spectra of $L(G)$, $S(G)$ and $P(G)$. The proofs of the following three theorems can be found in [**18**]. Theorems 4.10 and 4.11 are also discussed from the supersymmetric point of view in [**17**]. Recall a transition probability $p(e)$ is assumed to be given by $(\deg_G o(e))^{-1}$.

THEOREM 4.10. *Let G be an infinite d-regular graph with $d \geq 3$ and $L(G)$ its line graph of G. Then,*

$$Spec(-\Delta_{L(G)}) = \frac{d}{2d-2} Spec(-\Delta_G) \cup \{\frac{d}{d-1}\},$$

where $\frac{d}{d-1}$ is an eigenvalue with infinite multiplicity.

THEOREM 4.11. *Let G be an infinite d-regular graph with $d \geq 3$ and $S(G)$ its subdivision graph. Let $\varphi_S(x) = 2x(2-x)$. Then*

$$Spec(-\Delta_{S(G)}) = \varphi_S^{-1}(Spec(-\Delta_G)) \cup \{1\},$$

where 1 is an eigenvalue with infinite multiplicity.

THEOREM 4.12. *Let G be an infinite d-regular graph with $d \geq 3$ and $P(G)$ its para-line graph, and $\varphi_P(x) = -d \cdot x^2 + (d+2)x$. Then*

$$Spec(-\Delta_{P(G)}) = \varphi_P^{-1}(Spec(-\Delta_G)) \cup \{1\} \cup \{\frac{d+2}{d}\},$$

where 1 and $\frac{d+2}{d}$ are eigenvalues with infinite multiplicity.

The corresponding results for laplacians on line graphs and subdivisions of finite graphs are found in e.g. [**3**]. Here we give the corresponding result for the relationship between twisted laplacians $L_{\theta,M}$ on a finite graph M and $L_{\theta,P(M)}$ on its para-line graph $P(M)$.

We recall the definition of $L_{\theta,M}$ (see also (2.2)). Since we treat only regular graphs, $L_{\theta,M} = A_{\theta,M}/d - I$ and $L_{\theta_P,P(M)} = A_{\theta_P,P(M)}/d - I$, where $A_{\theta,M}$ and $A_{\theta_P,P(M)}$ are the twisted adjacency operators defined by

$$A_{\theta,M}f(x) = \sum_{e\in A_x(M)} \exp(2\pi\sqrt{-1}\theta(e))f(t(e))$$

and

$$A_{\theta_P,P(M)}F(e) = \sum_{\substack{e'\in A(M)\\ o(e')=o(e),e'\neq e}} \exp(2\pi\sqrt{-1}\theta_P(e,e'))F(e') + \exp(2\pi\sqrt{-1}\theta_P((e,\overline{e})_\circ))F(\overline{e}).$$

For our theorem, we consider $\phi_1^*\theta \in C^1(P(M),\mathbf{R})$ as was defined in Remark 4.9 through $\theta \in C^1(M,\mathbf{R})$ by

$$\phi_1^*\theta((e_1,e_2)) = 0, \quad \phi_1^*\theta((e,\overline{e})_\circ) = \theta(e).$$

Then we have the following theorem.

THEOREM 4.13. *Let M be a finite connected d-regular graph with $d \geq 2$ and $P(M)$ its para-line graph, and $\varphi_P(x) = -d\cdot x^2 + (d+2)x$. Let $N = |V(M)|$ and $0 \leq \mu_1 \leq \mu_2 \leq \cdots \leq \mu_N \leq 2$ be the eigenvalues of $-L_{\theta,M}$. Then the characteristic polynomial of $-L_{\phi_1^*\theta,P(M)}$ is given by*

$$(\lambda-1)^{K-1}(\lambda - \frac{d+2}{d})^{K-1}\prod_{i=1}^{N}\{\varphi_P(\lambda) - \mu_i\}$$

where $K = \mathrm{rank}H_1(M,\mathbf{Z}) = \frac{d-2}{2}N + 1$.

REMARK 4.14. In the case where G is an abelian covering graph of a finite graph M, Theorem 4.13 together with Remark 4.9 recovers Theorem 4.12 since the spectrum of Δ_G is the union of the eigenvalues of $L_{\theta,M}$ (see (2.3)).

The multiplicities of the eigenvalues 1 and $\frac{d+2}{d}$ may be K since $\varphi_P^{-1}\{0\} = \{\frac{d+2}{d}, 0\}$ and $\varphi_P^{-1}\{2\} = \{1, \frac{2}{d}\}$. They depend on the bipartiteness of M and the integrals of θ over cycles:

i) M is bipartite and $\int_c \theta \in \mathbf{Z}$ for any closed path c.
ii) M is non-bipartite and $\int_c \theta \in \mathbf{Z}$ for any closed path c.
iii) M is non-bipartite and $\int_c \theta \in \mathbf{Z}$ for any closed path c of even length and $\int_c \theta \in \mathbf{Z} + 1/2$ for any closed path c of odd length.
iv) Otherwise.

The exact multiplicities of 1 and $\frac{d+2}{d}$ are determined according to the fact that $-L_{\theta,M}$ has the simple eigenvalue 0 in the cases i) and ii); the simple eigenvalue 2 in the cases i) and iii).

Here we give a sketch of the proof in the rest of this subsection. Let $\Phi_i : \ell^2(M) \to \ell^2(P(M))$ for $i = 0, 1$ be defined by

$$\Phi_0 f(e) = f(o(e)), \quad \Phi_1 f(e) = \exp(2\pi\sqrt{-1}\theta(e))f(t(e)).$$

It is easy to see that both Φ_0 and Φ_1 are injective and their dual operators are given by

$$\Phi_0^* F(x) = \sum_{e\in A_x} F(e), \quad \Phi_1^* F(x) = \sum_{e\in A_x} \exp(2\pi\sqrt{-1}\theta(e))F(\overline{e}).$$

Define two subspaces E and E_0 of $\ell^2(P(M))$ by

$$E = \Phi_0(\ell^2(M)) + \Phi_1(\ell^2(M)), \quad E_0 = \Phi_0(\ell^2(M)) \cap \Phi_1(\ell^2(M)).$$

If $F \in E_0$, there exist $f, g \in \ell^2(M)$ such that $F = \Phi_0 f = \Phi_1 g$, and then

$$f(t(e_2)) = \exp(-2\pi\sqrt{-1}(\theta(e_1) + \theta(e_2)))f(o(e_1)) \tag{4.1}$$

for any path (e_1, e_2) of length 2. From this equality, it is easy to see that $E_0 = \{0\}$ in the case iv). We fix a vertex x_0 and define the function s_{x_0} by

$$s_{x_0}(x) = \exp(-2\pi\sqrt{-1}\int_p \theta)s_{x_0}(x_0) \tag{4.2}$$

if there exists a path p of *even* length joining x_0 to x; otherwise $s_{x_0}(x) = 0$. In the cases i), ii) and iii), it is well-defined and satisfies $\Phi_0 s_{x_0} = \Phi_1 s_{x_0}$.

In the case i), let V_0 and V_1 be a bipartition, and take $x_0 \in V_0$ and $x_1 \in V_1$. Then we can define two linearly independent such functions s_{x_0} and s_{x_1} by (4.2) and so $E_0 = Span\{\Phi_0 s_{x_0}, \Phi_0 s_{x_1}\}$. In the cases ii) and iii), for any $x \in V(M)$ there exists a path p of even length joining to x_0 to x. Then $E_0 = Span\{\Phi_0 s_{x_0}\}$. We summarize as follows:

LEMMA 4.15. *Let s_{x_0} and s_{x_1} be defined as above. Then*

$$E_0 = \begin{cases} Span\{\Phi_0 s_{x_0}, \Phi_0 s_{x_1}\}, & \textit{case } \text{i}), \\ Span\{\Phi_0 s_{x_0}\}, & \textit{cases } \text{ii}), \text{iii}). \\ \{0\}, & \textit{case } \text{iv}). \end{cases} \tag{4.3}$$

In particular, $\dim E = 2|V(M)| - 2$ *in the case* i); $\dim E = 2|V(M)| - 1$ *in the cases* ii), iii); $\dim E = 2|V(M)|$ *in the case* iv).

Theorem 4.13 follows from Lemmas 4.16 and 4.17 together with Lemma 4.15.

LEMMA 4.16. *Let $A_{\theta,M}$ and $A_{\phi_1^*\theta,P(M)}$ be adjacency operators on $\ell^2(M)$ and $\ell^2(P(M))$, respectively. Then*

$$\begin{cases} A_{\phi_1^*\theta,P(M)}\Phi_0 = (d-1)\Phi_0 + \Phi_1, \\ A_{\phi_1^*\theta,P(M)}\Phi_1 = \Phi_0(A_{\theta,M} + I) - \Phi_1. \end{cases} \tag{4.4}$$

In particular, $A_{\phi_1^*\theta,P(M)}$ leaves the subspace E invariant.

PROOF. It follows from a direct computation. □

Now we define a self-adjoint operator $J_\theta : \ell^2(P(M)) \to \ell^2(P(M))$ by

$$J_\theta F(e) = \exp\left(2\pi\sqrt{-1}\theta(e)\right) F(\bar{e}).$$

It is easy to see that $J_\theta^2 = I$ and $J_\theta \Phi_0 = \Phi_1$. Hence J_θ can be restricted on $E^\perp$, the orthogonal complement of E in $\ell^2(P(M))$, and its eigenvalues on $E^\perp$ are either 1 or -1. We consider the two subspaces of $E^\perp$:

$$\begin{aligned} E_S^\perp &= \{F \in E^\perp \mid J_\theta F = F\}, \\ E_A^\perp &= \{F \in E^\perp \mid J_\theta F = -F\}. \end{aligned}$$

It is easy to see that $E^\perp = E_S^\perp \oplus E_A^\perp$.

LEMMA 4.17. *Let $E_S^\perp$ and $E_A^\perp$ be as above. Then $E_S^\perp$ and $E_A^\perp$ are the eigenspaces of the eigenvalues* 0 *and* -2 *of* $A_{\phi_1^*\theta,P(M)}$*, respectively, and*

$$\dim E_S^\perp = \begin{cases} K, & \textit{cases } \text{i), iii)}, \\ K-1, & \textit{cases } \text{ii), iv)}, \end{cases} \quad \textit{and} \quad \dim E_A^\perp = \begin{cases} K, & \textit{cases } \text{i), ii)}, \\ K-1, & \textit{cases } \text{iii), iv)}, \end{cases}$$

where $K = \mathrm{rank} H_1(M, \mathbf{Z}) = \frac{d-2}{2}N + 1$.

PROOF. The first assertion immediately follows from the formula

$$A_{\phi_1^*\theta,P(M)}F(e) = \Phi_0^* F(o(e)) + (J_\theta - I)F(e).$$

Let $\{\sigma_i\}_{i=1}^K$ be a $\mathbf{Z}$-basis of 1-homology group $H_1(M,\mathbf{Z})$. Every σ_i can be represented by a closed path in M, say $c_i = (e_1^{(i)}, \dots, e_{n_i}^{(i)})$, such that the edges $\{e_j^{(i)}\}_{j=1}^{n_i}$ in c_i are all distinct. Now we define functions $F_i, G_i \in \ell^2(P(M))(i = 1, 2, \dots, K)$ by

$$\begin{aligned} F_i(e) &= (-1)^{k-1}\exp(-2\pi\sqrt{-1}\int_{c_{i,k}} \theta), \\ G_i(e) &= \exp(-2\pi\sqrt{-1}\int_{c_{i,k}} \theta), \end{aligned}$$

if $e = e_k^{(i)}$ for some k $(1 \le k \le n_i)$, where $c_{i,k} = (e_1^{(i)}, \dots, e_{k-1}^{(i)})(k \ge 2)$ and $c_{i,1} = 0$; otherwise $F_i(e) = G_i(e) = 0$. Note that the definitions of F_i and G_i depend on the choice of an edge as $e_1^{(i)}$ in the closed path c_i. The functions $\{F_i\}_{i=1}^K$ are linearly independent and so are $\{G_i\}_{i=1}^K$. We also define Q_S and Q_A by $Q_S = I + J_\theta$ and $Q_A = I - J_\theta$. It is easy to see the following:

(1) For a closed path c_i of even length, $Q_S F_i$ is an eigenfunction with respect to 0 if and only if $\int_{c_i}\theta \in \mathbf{Z}$; $Q_A G_i$ is an eigenfunction with respect to -2 if and only if $\int_{c_i}\theta \in \mathbf{Z}$.
(2) For a closed path c_i of odd length, $Q_A G_i$ is an eigenfunction with respect to -2 if and only if $\int_{c_i}\theta \in \mathbf{Z}$.
(3) For a closed path c_i of odd length, $Q_S F_i$ is an eigenfunction with respect to 0 if and only if $\int_{c_i}\theta \in \mathbf{Z} + \frac{1}{2}$.

Suppose that there exists a closed path, say c_1, such that $\Phi_0^* Q_A F_1(x) \ne 0$ for some $x \in V(c_1)$, and if it exists, then it must be $o(e_1^{(1)})$ by the definition of F_1. We may assume that $x \in V(c_i)$ and $x = o(e_1^{(i)})$ for any $i = 1, 2, \dots, K$. Indeed, if necessary, we can take a path $P = (e_1, e_2, \dots, e_\ell)$ such that $x = o(e_1)$ and $t(e_\ell) \in V(c_i)$ in M so that a new closed path $\tilde{c}_i = Pc_i\bar{P} = (e_1, e_2, \dots, e_\ell, e_1^{(i)}, \dots, e_{n_i}^{(i)}, \bar{e}_\ell, \bar{e}_{\ell-1}, \dots, \bar{e}_1)$ obviously satisfies that $x \in V(\tilde{c}_i)$, $x = o(e_1^{(1)})$ and $\int_{\tilde{c}_i}\theta = \int_{c_i}\theta$. We consider linear combinations of F_i's instead of themselves:

$$\widetilde{F}_i = (\Phi_0^* Q_A F_1(x))F_i - (\Phi_0^* Q_A F_i(x))F_1.$$

It is easy to see that $Q_A\widetilde{F}_i$ is an eigenfunction with respect to -2 for $i = 2, 3, \dots, K$. In the same way as above, suppose that there exists at least one closed path, say c_1, such that $\Phi_0^* Q_S G_1(y) \ne 0$ for some $y \in V(c_1)$. If we consider linear combinations of G_i's

$$\widetilde{G}_i = (\Phi_0^* Q_S G_1(y))G_i - (\Phi_0^* Q_S G_i(y))G_1,$$

then $Q_S\widetilde{G}_i$ is an eigenfunction with respect to 0 for $i = 2, 3, \ldots, K$.

Through all the discussion above, we get the dimensions of $E_S^{\perp}$ and $E_A^{\perp}$ in each case. □

4.3. Injectiveness of taking line graphs and para-line graphs. In this subsection we assume $G = (V(G), E(G))$ is a connected, simple graph, which may be infinite. Here a graph is said to be simple if it has no self-loops and no multiple edges. Also we often use the notation like $xy \in E(G)$ as an unoriented edge whose endvertices are x and y.

Here we show the following result, which follows from Proposition 4.21 and Corollaries 4.22 and 4.23 below.

THEOREM 4.18. *Let $\mathcal{G}$ be the set of all infinite simple connected graphs. Then the maps $L, S, P : \mathcal{G} \to \mathcal{G}$ are injective up to isomorphisms.*

Let us recall graph-isomorphisms. Two graphs G and G' are *isomorphic* ($G \cong G'$) if there exists a bijection φ from $V(G)$ to $V(G')$ satisfying the following: two vertices x_1 and x_2 are adjacent in G if and only if the vertices $\varphi(x_1)$ and $\varphi(x_2)$ are adjacent in G'. Such a map ϕ is called an isomorphism. In particular, if $G = G'$, it is called an automorphism.

Now we introduce the notion of edge-isomorphism. Further topics about edge-isomorphisms can be found in, e.g., [**2**]. We call G and G' are *edge-isomorphic* if there exists a bijection ψ from $E(G)$ to $E(G')$ satisfying the following: two edges $[e_1]$ and $[e_2]$ in G are adjacent (that is, $[e_1]$ and $[e_2]$ have a common vertex in their endvertices) if and only if $\psi([e_1])$ and $\psi([e_2])$ in G' are adjacent.

REMARK 4.19. G and G' are edge-isomorphic if and only if $L(G)$ and $L(G')$ are isomorphic.

If G and G' are isomorphic, then they are also edge-isomorphic, equivalently, $L(G)$ and $L(G')$ are isomorphic. In fact, if φ is an isomorphism from G to G', setting $\psi([e]) = \varphi(x)\varphi(y)$ for any edge $[e] = xy \in E(G)$, one can see that ψ is an edge-isomorphism. In this case, this ψ is said to be the *induced* edge-isomorphism (by the isomorphism φ). However, the converse direction is not true. Indeed, two graphs $X_1 = K_3$ and $X_2 = K_{1,3}$ are edge-isomorphic but not isomorphic. In addition, the following $X_i (i = 3, 4, 5)$ has an edge-automorphism (that is, an edge-isomorphism from X_i to itself) which is not induced by any automorphism: $X_3 = K_4$, $X_4 = K_4 -$ "one edge" and $X_5 = K_4 -$ "P_2". Fortunately, for simple and finite graphs, it is known that "bad" graphs are only five graphs X_i $(i = 1, 2, 3, 4, 5)$ mentioned above, which is proved by H.Whitney[**22**]. It can be generalized for infinite graphs.

PROPOSITION 4.20. *Let G_1 and G_2 be connected and simple graphs which may be infinite, and we assume that G_1 is not isomorphic to any X_i $(i = 1, 2, 3, 4, 5)$. If ψ is an edge-isomorphism from G_1 to G_2, then ψ is induced by an isomorphism from G_1 to G_2.*

PROOF. If graphs G_1 and G_2 are finite, Proposition 4.20 is obtained by Whitney ([**22**]). Thus we may assume G_1 and G_2 are infinite. From the assumption, there exists an edge-isomorphism ψ from G_1 to G_2. Let S_0 be a connected finite subgraph of G_1, which is induced by $V(S_0)$, with $|V(S_0)| \geq 5$ and set $T_0 = \langle \psi(E(S_0)) \rangle_{G_2}$, which is the subgraph of G_2 induced by the edge set $\psi(E(S_0))$. Denote the restriction of ψ to $E(S_0)$ by $\psi_0 = \psi|_{E(S_0)}$. Then it is obvious that ψ_0 is an edge-isomorphism from S_0 to T_0, and hence ψ_0 is induced by an isomorphism φ_0 from S_0 to T_0 since S_0 is finite and not isomorphic to any X_i $(i = 1, 2, 3, 4, 5)$.

For any subgraph K in G, we define a neighborhood $N_G(K)$ of K by the set $N_G(K) = \{x \in V(G) \setminus V(K) \mid xy \in E(G) \text{ for some } y \in V(K)\}$. Now G_1 is an infinite, connected graph, so there exists a vertex $w \in N_{G_1}(S_0)$ and one can choose two edges $uv \in E(S_0)$ and $vw \notin E(S_0)$. Since $vw \notin E(S_0)$ and ψ is an edge-isomorphism, $\psi(vw) \notin E(T_0)$ and is adjacent to $\psi(uv) = \varphi_0(u)\varphi_0(v)$. Then the edge $\psi(vw)$ in G_2 must be $\varphi_0(u)\tilde{w}$ or $\varphi_0(v)\tilde{w}$ for some $\tilde{w} \in V(G_2) \setminus V(T_0)$. Since S_0 is connected and $|V(S_0)| \geq 5$, there exists a vertex $a(\neq u, v)$ in S_0 such that $au \in E(S_0)$ or $av \in E(S_0)$. If $au \in E(S_0)$, since au and vw is not adjacent, $\psi(au) = \varphi_0(a)\varphi_0(u)$ and $\psi(vw)$ do not have a common vertex, and hence $\psi(vw)$ is $\varphi_0(v)\tilde{w}$; if $av \in E(S_0)$, since av and vw is adjacent, $\psi(av) = \varphi_0(a)\varphi_0(v)$ and $\psi(vw)$ have a common vertex, and hence $\psi(vw)$ is $\varphi_0(v)\tilde{w}$. In both cases, $\psi(vw)$ must be $\varphi_0(v)\tilde{w}$. Similarly, for any edge joining $b \in V(S_0)$ and w, it holds that $\psi(bw) = \varphi_0(b)\tilde{w}'$ for some $\tilde{w}' \notin V(S_0)$. Since $\psi(aw)$ and $\psi(bw)$ are adjacent in G_2, $\tilde{w}$ and $\tilde{w}'$ must coincide. Hence $\tilde{w}$ is determined uniquely by w. Therefore, setting

$$\varphi_1(x) = \begin{cases} \varphi_0(x), & \text{if } x \in V(S_0), \\ \tilde{w}, & \text{if } x = w, \end{cases}$$

one can easily check that φ_1 is an isomorphism from $S_1 = \langle V(S_0) \cup \{w\}\rangle_{G_1}$ to $T_1 = \langle V(T_0) \cup \{\tilde{w}\}\rangle_{G_2}$; it induces the edge-isomorphism $\psi|_{E(S_1)}$, the restriction of ψ to $E(S_1)$.

Continuing this procedure, we can define inductively an isomorphism φ_k from S_k to T_k so that $\{S_k\}_{k=0}^\infty$ and $\{T_k\}_{k=0}^\infty$ are increasing sequences of finite induced subgraphs of G_1 and G_2 with $V(G_1) = \bigcup_{i=0}^\infty V(S_i)$ and $V(G_2) = \bigcup_{i=0}^\infty V(T_i)$, respectively, and $\varphi_k|_{S_\ell} = \varphi_\ell$ if $k \geq \ell \geq 0$. We define a map $\varphi : V(G_1) \to V(G_2)$ by

$$\varphi(x) = \varphi_k(x) \quad \text{if } x \in S_k.$$

It is well-defined by the consistency of φ_k's. Then it is easy to see that φ is an isomorphism from G_1 to G_2 and induces the edge-isomorphism ψ, that is, $\psi(xy) = \varphi(x)\varphi(y) \in E(G_2)$ for any edge $xy \in E(G_1)$. This completes the proof. □

Remark 4.19 and Proposition 4.20 shows that the following proposition.

PROPOSITION 4.21. *Let G_1 and G_2 be two graphs such that G_1 is not isomorphic to $X_1 = K_3$ or $X_2 = K_{1,3}$, and G_2 is not isomorphic to $X_2 = K_{1,3}$ or $X_1 = K_3$, respectively. Then $L(G_1)$ is isomorphic to $L(G_2)$ if and only if G_1 is isomorphic to G_2.*

PROOF. By the definition of line graphs, there exists a bijection from $E(G)$ to $V(L(G))$ such that any two vertices in $L(G)$ are adjacent if and only if two of the corresponding edges in G are adjacent. As mentioned in Remark 4.19, $L(G_1)$ is isomorphic to $L(G_2)$ if G_1 is isomorphic to G_2.

Assume that $L(G_1)$ is isomorphic to $L(G_2)$ and that both of G_1 and G_2 are not isomorphic to K_3 and $K_{1,3}$. By Remark 4.19, there exists an edge-isomorphism ψ from G_1 to G_2. If G_1 is not isomorphic to any X_i $(i = 3, 4, 5)$, it follows from Proposition 4.20 that ψ is induced by an isomorphism from G_1 to G_2. Thus

$G_1 \cong G_2$. It is easy to see that G is the one and only one graph which has $L(G)$ as its line graph if G is isomorphic to X_i $(i = 3, 4, 5)$. □

As corollaries, we can obtain the following.

COROLLARY 4.22. *Let G_1 be a subdivision graph of a graph. If there exists an edge-isomorphism ψ from G_1 to a graph G_2, then ψ is induced by an isomorphism from G_1 to G_2.*

PROOF. Since a subdivision graph G_1 is bipartite and each vertex in one of the partite sets has degree 2, G_1 is not isomorphic to any X_i $(i = 1, 2, 3, 4, 5)$. Thus the conclusion holds. □

COROLLARY 4.23. *Let G_1 and G_2 be two graphs. Then $P(G_1)$ is isomorphic to $P(G_2)$ if and only if G_1 is isomorphic to G_2.*

PROOF. Recall that $P(G) \cong L(S(G))$. By the previous results, we only have to see the relationship between G and $S(G)$. It is obvious by definition that $S(G_1) \cong S(G_2)$ if $G_1 \cong G_2$. Thus it is sufficient to show that $G_1 \cong G_2$ if $S(G_1) \cong S(G_2)$. So we suppose $S(G_1) \cong S(G_2)$.

If every vertex in $S(G_1)$ has degree 2, then $S(G_1)$ is isomorphic to C_{2n} $(n = 1, 2, \dots)$ or an infinite path. In such cases, it is obvious that $G_1 \cong C_n$ or $G_1 \cong S(G_1)$. It also holds for $S(G_2)$ since every vertex in $S(G_2)$ has degree 2 by an isomorphism. Hence $G_1 \cong G_2$.

Assume that $S(G)$ has at least one vertex whose degree is not 2. Since $S(G)$ is bipartite, we may take its bipartition such that one partite set has a vertex whose degree is not 2 and the degree of every vertex in the other partite set is 2. So we can determine which partite set is $V(G)$ (or $E(G)$). By the assumption, there exists a bijection φ mapping $V(G_1)$ to $V(G_2)$ and $E(G_1)$ to $E(G_2)$ in such a way that x and $[e]$ are adjacent in $S(G_1)$ if and only if $\varphi(x)$ and $\varphi([e])$ are adjacent in $S(G_2)$. On the other hand, two vertices x and y are adjacent and joined by an edge e in a graph G if and only if x and y are adjacent to $[e]$ in $S(G)$. Hence it is shown that $G_1 \cong G_2$ if $S(G_1) \cong S(G_2)$. □

4.4. Spectrum for infinite Sierpiński lattice. Spectral analysis for laplacians on fractals has been studied by many authors. Among many works, we focus on the spectrum of the laplacian on an infinite Sierpiński lattice. M. Fukushima and T. Shima computed the eigenvalues and the density of states of the laplacian on an infinite Sierpiński lattice by using Dirichlet form theory and showed that the oscillatory behaviors of the integrated density of states [**5**]. A. Teplyaev [**21**] showed that the set of eigenfunctions with compact support for the laplacian on the infinite Sierpiński lattice with Neumann boundary condition is complete and this is also the case where the Sierpiński lattice has no boundary, which implies the absence of continuous part of the spectrum.

We define an infinite Sierpiński lattice (see Figure 3). Let $\{e_i\}_{i=1}^d$ be the standard basis of $\mathbf{R}^d$ and e_0 be the 0 vector. Define $f_i : \mathbf{R}^d \to \mathbf{R}^d$ by $f_i(x) = \frac{1}{2}(x + e_i)$ $(0 \le i \le d)$. Furthermore we define V_n inductively as follows:

$$V_0 = \bigcup_{0 \le i < j \le d} \{(1-t)e_i + te_j \in \mathbf{R}^d \mid 0 \le t \le 1\}$$

FIGURE 3. Infinite Sierpiński lattice ($d = 2$).

and

$$V_n = f_0^{-1}\left(\bigcup_{i=0}^{d} f_i(V_{n-1})\right), \quad n \geq 1.$$

We regard $\widetilde{S}_d = \cup_{n\geq 0} V_n$ as an infinite graph which is $2d$-regular except at the origin and the degree of the origin is d. Here the set of vertices of V_0 is identified with $\{e_i\}_{i=0}^d$ and $V(\widetilde{S}_d)$ with the set of all vertices defined inductively. We prepare two copies of an infinite graph $\widetilde{S}_d$ and identify the vertices (the origins) of degree d. We call the infinite $2d$-regular graph constructed here the d-dimensional Sierpiński lattice and denote it by S_d.

The theorem is the following.

THEOREM 4.24 ([**5**], [**21**]). *Let S_d be the d-dimensional infinite Sierpiński lattice with $d \geq 2$ and Δ_{S_d} the laplacian associated with the simple random walk on S_d. Then*

$$Spec(-\Delta_{S_d}) = \overline{\bigcup_{k=0}^{\infty}\left\{\{\rho^{-k}(\frac{d+1}{2d})\} \cup \{\rho^{-k}(\frac{d+3}{2d})\}\right\}} \cup \{\frac{d+1}{d}\},$$

where $\rho(x) = -2dx^2 + (d+3)x$.

Here we give another proof of Theorem 4.24. Our method is based on the relationship between the spectrum of the laplacian for an infinite graph G and that for a para-line (line) graph of G as was discussed in Section 4.2 and on the self-similar structure of the Sierpiński lattice.

Now, in the same way as in the above, we define an auxiliary graph, an infinite Sierpiński pre-lattice (see Figure 4). Define $g_i : \mathbf{R}^d \to \mathbf{R}^d$ by $g_i(x) = \frac{1}{3}(x + 2e_i)$ $(0 \leq i \leq d)$, where $\{e_i\}_{i=0}^d$ are the same as before. Furthermore we define W_n inductively as follows:

$$W_0 = \bigcup_{0\leq i<j\leq d} \{(1-t)e_i + te_j \in \mathbf{R}^d \mid 0 \leq t \leq 1\}$$

and

$$W_n = g_0^{-1}\left(\bigcup_{i=0}^{d} g_i(W_{n-1})\right), \quad n \geq 1.$$

We regard $\widetilde{G}_d = \cup_{n\geq 0} W_n$ as an infinite graph which is $(d+1)$-regular except at the origin and the degree of the origin is d. We prepare two copies of an infinite graph $\widetilde{G}_d$ and join the vertices (the origins) of degree d by an edge. We call the infinite $(d+1)$-regular graph constructed here the d-dimensional Sierpiński pre-lattice and denote it by G_d^*. The next observation is crucial.

LEMMA 4.25. (1) Let $\mathcal{G}_d$ be the set of all infinite, simple connected, d-regular graphs. Then the d-dimensional Sierpiński pre-lattice G_d^* is a fixed point of the map $P : \mathcal{G}_d \to \mathcal{G}_d$, that is, $P(G_d^*) = G_d^*$.
(2) The d-dimensional Sierpiński lattice S_d is the line graph of the graph G_d^*, that is, $S_d = L(G_d^*)$.

PROOF. (1) If we regard W_0 as a graph, the corresponding graph is isomorphic to a complete graph K_d. Let $K_d^{(1)}$ be the graph which corresponds to W_1, and let L_d be the para-line graph of a star graph $K_{1,d}$. The graph L_d consists of a K_d and d edges each of which attaches to each vertex of K_d. It is easy to see that the para-line graph of L_d is isomorphic to the graph obtained from L_d by replacing K_d with $K_d^{(1)}$. Thus $P(G_d^*)$ is nothing but the graph obtained from G_d^* by replacing every K_d with $K_d^{(1)}$, and hence $P(G_d^*)$ is isomorphic to G_d^*.

(2) By Remark 4.3, $L(G_d^*)$ is obtained from $P(G_d^*) \cong G_d^*$ by contracting all the edges joining K_d's. Thus the resultant graph is obviously isomorphic to S_d. $\square$

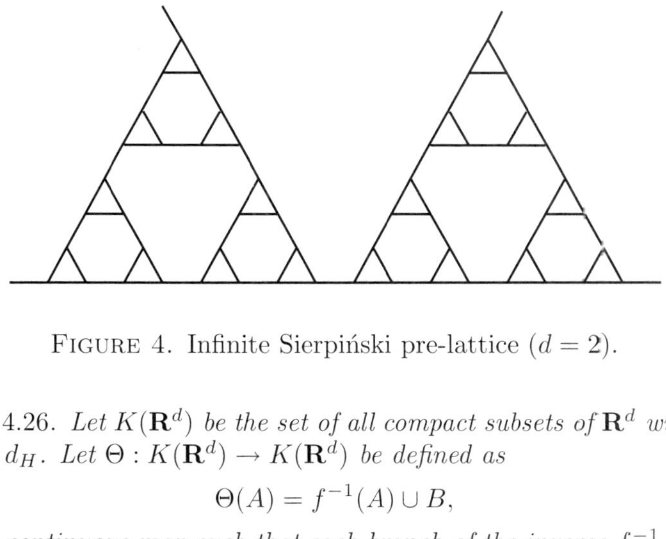

FIGURE 4. Infinite Sierpiński pre-lattice ($d = 2$).

LEMMA 4.26. *Let $K(\mathbf{R}^d)$ be the set of all compact subsets of $\mathbf{R}^d$ with the Hausdorff metric d_H. Let $\Theta : K(\mathbf{R}^d) \to K(\mathbf{R}^d)$ be defined as*

$$\Theta(A) = f^{-1}(A) \cup B,$$

where f is a continuous map such that each branch of the inverse f^{-1} is a contraction map on $\mathbf{R}^d$ and B is a compact subset of $\mathbf{R}^d$. Then Θ is a contraction map on $(K(\mathbf{R}^d), d_H)$ and, in particular, Θ has the unique fixed point $A^* \in K(\mathbf{R}^d)$.

PROOF. It is well-known that $(K(\mathbf{R}^d), d_H)$ is a complete metric space. Hence in order to show the existence and uniqueness of the fixed point of the map Θ, it is sufficient to prove that Θ is a contraction. In fact, it follows from the following: for any contraction map q on $\mathbf{R}^d$ satisfying that $|q(x) - q(y)| \leq C|x - y|$ with $0 < C < 1$, it holds that

$$d_H(q(A), q(B)) \leq C d_H(A, B),$$
$$d_H(A_1 \cup A_2, B_1 \cup B_2) \leq \max(d_H(A_1, B_1), d_H(A_2, B_2)).$$

□

PROOF OF THEOREM 4.24. Since G_d^* is $(d+1)$-regular, by Theorem 4.12 and by Lemma 4.25, we obtain

$$\begin{aligned} Spec(-\Delta_{G_d^*}) &= Spec(-\Delta_{P(G_d^*)}) \\ &= \varphi_P^{-1}(Spec(-\Delta_{G_d^*})) \cup \{1\} \cup \{\frac{d+3}{d+1}\}, \end{aligned} \tag{4.5}$$

where $\varphi_P(x) = -(d+1)x^2 + (d+3)x$. Lemma 4.26 shows that the above equation (4.5) has a unique solution, and it is given by

$$Spec(-\Delta_{G_d^*}) = \overline{\bigcup_{k=0}^{\infty} \left\{ \{\varphi_P^{-k}(1)\} \cup \{\varphi_P^{-k}(\frac{d+3}{d+1})\} \right\}}.$$

Since S_d is the line graph of G_d^* (Lemma 4.25), we obtain

$$\begin{aligned} Spec(-\Delta_{S_d}) &= \frac{d+1}{2d} \left(\overline{\bigcup_{k=0}^{\infty} \left\{ \{\varphi_P^{-k}(1)\} \cup \{\varphi_P^{-k}(\frac{d+3}{d+1})\} \right\}} \right) \cup \{\frac{d+1}{d}\} \\ &= \overline{\bigcup_{k=0}^{\infty} \left\{ \{\rho^{-k}(\frac{d+1}{2d})\} \cup \{\rho^{-k}(\frac{d+3}{2d})\} \right\}} \cup \{\frac{d+1}{d}\}, \end{aligned}$$

where $\rho(x) = (\tau \circ \varphi_P \circ \tau^{-1})(x) = -2dx^2 + (d+3)x$ with $\tau(x) = \frac{(d+1)x}{2d}$. □

REMARK 4.27. The spectrum set of any fixed point of the map $P : \mathcal{G}_d \to \mathcal{G}_d$ should be the same as that of the infinite Sierpiński pre-lattice.

5. Problems

In this section, we collect some problems around the topics in this paper. First three problems are concerned with FSP.

PROBLEM 5.1. Does M^{ab} always have FSP when M is regular? If M is even-regular, M^{ab} has FSP by Proposition 2.5. If M is odd-regular and bipartite, then, by the classical fact in graph theory (cf. [**1**],[**2**]), M has a 1-factor. As a result, M has an even factor satisfying the condition (2) in Proposition 2.5, therefore M^{ab} has FSP. It is remained for regular graphs to show that M^{ab} has FSP even if M is a non-bipartite odd-regular graph.

PROBLEM 5.2. Full spectrum conjecture has been inspired by several topics in (cf. [**16, 20**]) for a non-compact manifold X such that X has an isometry group Γ acting discontinuously on X with a compact quotient space $\Gamma\backslash X$. One might say that an maximal abelian covering manifold of a compact manifold has FSP if the spectrum of the Laplace-Beltrami operator on it is equal to $[0, \infty)$. The method used in our proof of Proposition 2.5 depends on graph structure. Can it be transplanted to the case of manifolds?

PROBLEM 5.3. In Definition 2.1 we defined FSP as a graph property by fixing the simple random walk. When we replace the simple random walk with another reversible random walk, what is a suitable formulation of FSP? For example, we consider the asymmetric reversible random walk on $\mathbf{Z}^1$ defined by $p(e) = p$ and $p(\overline{e}) = q$ for every $e \in A(G)$ with $o(e) = x$ and $t(e) = x+1$. Then the spectrum set is $[1 - 2\sqrt{pq}, 1 + 2\sqrt{pq}]$. Taking account of Theorem 3.10, it might be said to have

FSP in a sense, even if p is not associated with the simple random walk. Likewise, when we consider a non-amenable covering graph, what is a suitable formulation of FSP for such a covering graph? In this case, the spectrum is also away from 0 (cf. [**20**]).

When graphs are hyperbolic, there are few examples whose spectra are completely determined. Once one finds a graph for which one can compute the spectrum, by using Theorems 4.10, 4.11 and 4.12, one can make many examples.

PROBLEM 5.4. We have obtained the isoperimetric constant for a (d, f)-regular planar graph in Theorem 3.1. Determine the spectrum for a (d, f)-regular planar graph. At least determine $\lambda_0(G)$ for (d, f)-regular planar graphs.

The next problem, which is originally raised in [**8**] as a conjecture, is also related to the result in Section 3:

PROBLEM 5.5. Let G be a planar graph. When $\kappa_G(x) > 0$ for every vertex x in G, is G always a finite graph? It is a discrete version of the well-known Myers theorem in Riemannian geometry which says that a complete Riemannian manifold with Ricci curvature bounded below by a positive number is compact.

The next two problems are concerned with the results in Section 3.

PROBLEM 5.6. Let $\mathcal{F}$ be the set of all the square summable functions on $V(G)$ taking values only in $\{-1, 0, 1\}$. We denote $\alpha_0(G, \mathcal{F})$ and $\alpha_\infty(G, \mathcal{F})$ for the $\mathcal{F}$ by $h(G)$ and $h_\infty(G)$, respectively. As mentioned in Example 3.7, $h(G)$ is nothing but the isoperimetric constant. For a finite graph G, both $\lambda_0(G)$ and $h(G)$ are equal to 0 and $\lambda_\infty(G)$, of course, can be directly computed as the maximal eigenvalue. When G is bipartite, $h_\infty(G) = 2$ by Proposition 3.8. For general non-bipartite graph, what is $h_\infty(G)$? Are there any algorithms for computing $h_\infty(G)$?

PROBLEM 5.7. By Theorem 3.10 and Remark 3.12 it holds that $\lambda_0(G) + \lambda_\infty(G) < 2$ when G is essentially non-bipartite. Does it also hold that $h(G) + h_\infty(G) < 2$ when G is essentially non-bipartite?

The following is concerned with results in Sections 4.3 and 4.4.

PROBLEM 5.8. Do there exist any fixed points of the map $P : \mathcal{G}_d \to \mathcal{G}_d$ other than the infinite Sierpiński pre-lattice (see Remark 4.27)? In the proof of Theorem 4.24 which says the spectrum set is Cantor-like, we gave the set equation for the spectrum by using the fact that the infinite Sierpiński pre-lattice is a fixed point of the map P. By emphasizing this fact can one show the absence of continuous singular spectrum to give another proof of some results in [**21**]?

References

[1] B. Bollobás, Modern Graph Theory, Springer-Verlag, 1998.

[2] M. Behzad, G. Chartround and L. Lesniak-Foster, Graphs and digraphs, Prindle, Weber & Schmidt, 1979.

[3] D. M. Cvetković, M. Doob and H. Sachs. Spectra of graphs—theory and application, Deutscher Verlag der Wissenschaften, Berlin, and Academic Press, New York, 1979.

[4] J. Dodziuk and L. Karp, *Spectral and function theory for combinatorial Laplacians*, Contemp. Math. **73** (1988), 25–40.

[5] M. Fukushima and T. Shima, *On the spectral analysis for the Sierpinski gasket*, Potential Analysis **1** (1992), 1–35.

[6] M. Gromov *Hyperbolic groups*, Essays in group theory, S. M. Gersten, (Editor), M.S.R.I. Publ. **8**, Springer, 1987, pp.75–263.

[7] O. Häggström, J. Jonasson and R. Lyons *Explicit isoperimetric constants and phase transitions in the random-cluster model*, Ann. Probab. **30** (2002), 443–473.

[8] Yu. Higuchi, *Combinatorial curvature for planar graphs*, J. Graph Theory **38** (2001), 220–229.

[9] Yu. Higuchi, *Isoperimetric inequality and random walks on an infinite graph and its line graph*, preprint.

[10] Yu. Higuchi and T. Shirai, *A remark on the spectrum of magnetic Laplacian on a graph*, the proceedings of TGT10, Yokohama Math. J. Vol. **47** (1999), Special issue, 129–142.

[11] Yu. Higuchi and T. Shirai, *Isoperimetric constants of (d,f)-regular planar graphs*, Interdisp. Inform. Sci. **9** (2003), 221–228.

[12] Yu. Higuchi and T. Shirai, *Non-bipartiteness of graphs and the upper bounds of Dirichlet forms*, preprint.

[13] M. Kotani and T. Sunada, *Jacobian tori associated with a finite graph and its abelian covering graphs*, Adv. in Appl. Math. **24** (2000), 89–110.

[14] M. Kotani, T. Shirai and T. Sunada, *Asymptotic behavior of the transition probability of a random walk on an infinite graph*, J. Funct. Anal. **159** (1998), 664–689.

[15] S. B. Myers, *Riemannian manifolds with positive mean curvature*, Duke Math. J. **8** (1941), 401–404.

[16] T. Kobayashi, K. Ono and T. Sunada, *Periodic Schrödinger operators on a manifold*, Forum Math. **1** (1989), 69–79.

[17] O. Ogurisu, *Supersymmetric analysis of the spectral theory on infinite graphs*, available from mp-arc 02-242.

[18] T. Shirai, *The spectrum of infinite regular line graphs*, Trans. Amer. Math. Soc. **352** (2000), 115–132

[19] T. Shirai, *Asymptotic behavior of the transition probability of a simple random walk on a line graph*, J. Math. Soc. Japan **52** (2000), 99–108.

[20] T. Sunada *Fundamental groups and Laplacians*, In: Proc. of Taniguchi Symposium on Geometry and Analysis on Manifolds 1987, Lect. Notes Math. **1339**, 248–277, Springer, Berlin 1988.

[21] A. Teplyaev, *Spectral analysis on infinite Sierpinski gaskets*, J. Funct. Anal. **159** (1998), 537-567.

[22] H. Whitney, *Congruent graphs and the connectivity of graphs*, Amer. J. Math. **54** (1932), 150–168.

MATHEMATICS LABORATORIES, COLLEGE OF ARTS AND SCIENCES, SHOWA UNIVERSITY
E-mail address: higuchi@cas.showa-u.ac.jp

DEPARTMENT OF COMPUTATIONAL SCIENCE, KANAZAWA UNIVERSITY
E-mail address: shirai@kenroku.kanazawa-u.ac.jp

Contemporary Mathematics
Volume **347**, 2004

Asymptotic behavior of a transition probability for a random walk on a nilpotent covering graph

Satoshi Ishiwata

ABSTRACT. First, we introduce some results for the behavior of random walk on a nilpotent covering graph. By using these results, we prove that the Riesz transform on a nilpotent covering graph is bounded on L^p, $1 < p < \infty$.

1. Introduction

Our interest is the geometric properties in the behavior of random walks on nilpotent covering graphs. These graphs are considered as a generalization of finitely generated nilpotent groups and also crystal lattices, namely abelian coverings of finite graphs. There are many results for the behavior of random walks on groups (cf. Alexopoulos [**1**], Hebisch and Saloff-Coste [**7**], Varopoulos, Saloff-Coste and Coulhon [**16**] and Woess [**17**]). However, it is difficult to extend these results to infinite graphs in general because the symmetry of the Cayley graph plays an important role. To consider covering graphs is the first step for this generalization.

In a classical case, as a central limit theorem, the transition probability on the $\mathbb{Z}$-lattice graph converges to the Gaussian kernel on the Euclidean space $\mathbb{R}$ in a sense. Similarly, on a nilpotent covering graph X, by using a realization of X to a nilpotent Lie group G_Γ, the transition probability converges to a heat kernel on G_Γ as the time goes to infinity (see [**8**] and [**9**]). In this article, first, we will consider the speed of this convergence. It is called a Berry-Esseen type theorem. Next, we introduce Gaussian bounds for the kernel function of the transition operator. In section 3, by using these results, we will prove the L^p boundedness of the Riesz transform.

Acknowledgment. The author would like to express his gratitude to Professor T. Sunada for his constant encouragement and valuable advices. He would like to thank Professor M. Kotani for her valuable advices and comments. He would also like to thank Professor G. Alexopoulos and Professor T. Coulhon for their helpful suggestions during this symposium.

2000 *Mathematics Subject Classification.* Primary 22E25, 60J10; Secondary 22E30.
Key words and phrases. Covering graph, Berry-Esseen type theorem, Gaussian estimate, Riesz transform.

2. Some results

Let $X = (V, E)$ be a locally finite connected graph, V being the set of vertices and E being the set of oriented edges. For $e \in E$, the origin and the terminus of e are denoted by $o(e)$ and $t(e)$, respectively, and the inverse edge is denoted by $\overline{e}$. We suppose that X is a *nilpotent covering graph*, namely a covering of a finite graph X_0 whose covering transformation group Γ is a finitely generated nilpotent group. Furthermore, we assume that Γ is torsion free.

A *symmetric random walk* on X with a weight $m : V \to \mathbb{R}_{>0}$ is given by a positive valued function p on E satisfying $\sum_{e \in E_x} p(e) = 1$ and $p(e)m(o(e)) = p(\overline{e})m(t(e))$, where $E_x = \{e \in E \,|\, o(e) = x\}$. We assume that m and p are Γ-invariant. We consider $p(e)$ the probability that a particle placed at $o(e)$ moves to the terminus $t(e)$ along the edge e in one unit time. The transition probability that a particle starting at x reaches y at time n is given by

$$p_n(x, y) = \sum_{c=(e_1, e_2, \ldots, e_n)} p(e_1)p(e_2)\cdots p(e_n),$$

where the sum is taken over all paths $c = (e_1, e_2, \ldots, e_n)$ of length n whose origin $o(c) = x$ and terminus $t(c) = y$. The transition operator L associated with the random walk is the operator acting on functions on V defined by

$$Lf(x) = \sum_{e \in E_x} f(t(e))p(e).$$

Then the discrete Laplacian Δ can be written by $I - L$. It is easy to check that the function $k_n(x, y) = p_n(x, y)m(y)^{-1}$ is the kernel function of L^n, namely $L^n f(x) = \sum_{y \in V} k_n(x, y)f(y)m(y)$. The hypothesis of m and p implies $k_n(x, y) = k_n(y, x)$.

Let G_Γ be a nilpotent Lie group in which Γ can be embedded as a cocompact lattice. Given our assumptions on Γ, such a nilpotent Lie group exists by a theorem of A. I. Mal'cev [**13**]. Let $\mathfrak{g}$ be the Lie algebra of G_Γ and $n_1, \ldots, n_r$ be the descending series, namely $n_1 = \mathfrak{g}$ and $n_{k+1} = [\mathfrak{g}, n_k]$. Since $\mathfrak{g}$ is nilpotent, we have a filtration:

$$n_1 \supset n_2 \supset \cdots \supset n_r \neq \{0\} \supset n_{r+1} = \{0\}.$$

We denote by $\mathfrak{g}^{(1)}, \ldots, \mathfrak{g}^{(r)}$ subspaces of $\mathfrak{g}$ such that $n_k = \mathfrak{g}^{(k)} \oplus n_{k+1}$. Let $\{X_1^{(k)}, \ldots, X_{d_k}^{(k)}\}$ be a basis of $\mathfrak{g}^{(k)}$. Then we have an identification of G_Γ with $\mathbb{R}^n$ as differential manifold given by

$$(x_{d_r}^{(r)}, x_{d_r-1}^{(r)}, \ldots, x_1^{(1)}) \mapsto \exp x_{d_r}^{(r)} X_{d_r}^{(r)} \cdot \exp x_{d_r-1}^{(r)} X_{d_r-1}^{(r)} \cdot \cdots \cdot \exp x_1^{(1)} X_1^{(1)},$$

which is called the *canonical coordinates of the second kind.* For $x \in G_\Gamma$, we denote $P_i^{(k)}(x) = x_i^{(k)}$.

Let $\Phi : X \to G_\Gamma$ be a Γ-equivariant map such that

$$\sum_{e \in E_x} p(e) \exp^{-1} \left(\Phi(o(e))^{-1}\Phi(t(e))\right)\Big|_{\mathfrak{g}^{(1)}} = 0$$

for all $x \in V$. We note that this property of Φ is equivalent to $\exp^{-1} \Phi\big|_{\mathfrak{g}^{(1)}} : X \to \mathfrak{g}^{(1)}$ is a harmonic realization (see [**8**]). Hence, by using the theory of harmonic map from a graph to a Riemannian manifold by Kotani and Sunada [**11**], there exists such a map and it is uniquely defined on $\mathfrak{g}^{(1)}$ up to parallel transform (see [**8**]).

The sub-Laplacian Ω for the *Albanese metric* on $\mathfrak{g}^{(1)}$ is defined by

$$\Omega = -\frac{1}{2m(X_0)} \sum_{e \in E_0} m(e) X_e^2,$$

where $m(e) = p(e)m(o(e))$ and X_e is the left invariant vector field on G_Γ associated to $\exp^{-1}\left(\Phi(o(e))^{-1}\Phi(t(e))\right)\big|_{\mathfrak{g}^{(1)}}$. Here, the Albanese metric on $\mathfrak{g}^{(1)}$ is defined by the following diagram (see [**8**]):

$$\begin{array}{ccccccc} \mathfrak{g}^{(1)} & \simeq & \Gamma/\left([G_\Gamma, G_\Gamma] \cap \Gamma\right) \otimes \mathbb{R} & \longleftarrow & \Gamma/[\Gamma,\Gamma] \otimes \mathbb{R} & \longleftarrow & \mathrm{H}_1(X_0, \mathbb{R}) \\ \updownarrow \text{dual} & & \updownarrow \text{dual} & & \updownarrow \text{dual} & & \updownarrow \text{dual} \\ \mathrm{Hom}(\mathfrak{g}^{(1)}, \mathbb{R}) & \simeq & \mathrm{Hom}(\Gamma/\left([G_\Gamma, G_\Gamma] \cap \Gamma\right), \mathbb{R}) & \hookrightarrow & \mathrm{Hom}(\Gamma/[\Gamma,\Gamma], \mathbb{R}) & \hookrightarrow & \mathrm{H}^1(X_0, \mathbb{R}), \end{array}$$

where $\mathrm{H}^1(X_0, \mathbb{R})$ is the first cohomology space of the quotient graph X_0. By using Hodge-Kodaira's theorem, $\mathrm{H}^1(X_0, \mathbb{R})$ is identified with the space of harmonic 1-forms on X_0. Hence we have a metric on $\mathrm{H}^1(X_0, \mathbb{R})$:

$$\langle\langle \omega, \eta \rangle\rangle = \frac{1}{2} \sum_{e \in E} \omega(e)\eta(e)m(e) \qquad \omega, \eta \in \mathrm{H}^1(X_0, \mathbb{R}).$$

By inducing this metric, we can define a metric on $\mathfrak{g}^{(1)}$. This induced metric is called Albanese metric. Let h_t be the heat kernel of Ω. Then we have the following Berry-Esseen type theorem.

THEOREM 1 ([**9**]). *For simplicity, we assume that a nilpotent covering graph X is a non-bipartite graph. Let $D \geq 0$ be an integer which satisfies for some $c > 0$ and for all $n \in \mathbb{N}$ and $x \in V$,*

$$\frac{1}{c} n^D \leq m(B(x,n)) \leq c n^D,$$

where $B(x,n) = \{y \in V \,:\, d_X(x,y) \leq n\}$. Then for any $0 < \epsilon < 1/2$, there exists a constant C_ϵ such that

$$\sup_{x,y \in V} \left| p_n(x,y)m(y)^{-1} - \frac{|G_\Gamma/\Gamma|}{m(X_0)} h_n(\Phi(x), \Phi(y)) \right| \leq C_\epsilon n^{-\frac{D+1/2-\epsilon}{2}}. \tag{2.1}$$

As a related result of this estimate, we remark that when X is the Cayley graph of Γ, Alexopoulos proved a Berry-Esseen type theorem:

THEOREM (Alexopoulos [**1**]).

$$\sup_{x,y \in \Gamma} |p_n(x,y) - |G_\Gamma/\Gamma| h_n(x,y)| \leq C n^{-\frac{D+1}{2}}. \tag{2.2}$$

On the other hand, when X is a crystal lattice, that is, an abelian covering of a finite graph, Kotani and Sunada obtained a local central limit theorem (see [**12**]). We remark that the estimate in (2.1) is slower than (2.2). In our proof, we have not been able to improve this estimate, in general. However, if

1:

$$\sum_{e \in E_x} p(e) \exp^{-1}\left(\Phi(o(e))^{-1}\Phi(t(e))\right)\big|_{\mathfrak{g}^{(2)}} = 0 \quad (x \in V)$$

and

2: the second order differential operator on G_Γ

$$\sum_{e \in E_x} p(e) X_e^2$$

is independent of the choice of $x \in V$,

are satisfied, then the speed of convergence is estimated by $Cn^{-\frac{D+1}{2}}$ which is the same as (2.2) (see [**9**]). In view of these results, we have two problems.

PROBLEM 1. *Is the estimate in* (2.1) *best possible?*

If the estimate in (2.1) is best or at least different from the estimate for the Cayley graph, the following problem makes sense.

PROBLEM 2. *What are conditions* 1 *and* 2*? In other words, we want to know the geometric meaning of these conditions.*

For Problem 2, we can check these conditions for many examples. For example, the simple random walks on Cayley graphs of finitely generated nilpotent groups satisfy these conditions. The simple random walks on the triangular lattice and the hexagonal lattice (see [**12**]) also satisfy them. However, there exist examples which do not satisfy these conditions. For example, the simple random walk on the Kagome lattice (see [**12**]) does not satisfy them.

In the proof of Theorem 1, we use Gaussian upper bounds for the kernel function k_n of L^n and its gradient on a nilpotent covering graph. The definition of a gradient of a function f on V is given as follows:

$$\nabla f(x) = \left(\frac{1}{2} \sum_{e \in E_x} p(e) |f(t(e)) - f(o(e))| \right)^{1/2}.$$

We remark that Hebisch and Saloff-Coste gave Gaussian bounds for k_n and ∇k_n on a Cayley graph of Γ in [**7**]. Furthermore, Pittet and Saloff-Coste showed that the long run behavior of the probability of return after $2n$-steps to the beginning is left invariant by *quasi-isometry* in [**14**]. Since a nilpotent covering graph X and its transformation group Γ are quasi-isometric, the Gaussian upper bound for k_n on X (Theorem 2, (2.3)) is deduced from their results.

In [**7**], the symmetry $\mu^{*n}(x) = \mu^{*n}(x^{-1})$ for a probability measure μ on Γ plays a crucial role. In our case, instead of this symmetry, we use an invariance for the action of Γ and a symmetry of k_n, namely $k_n(\gamma x, \gamma y) = k_n(x, y)$ and $k_n(x, y) = k_n(y, x)$, respectively. For $x, y \in V$, let $d_X(x, y)$ be the length of the shortest path from x to y. Then we have

THEOREM 2 ([**9**], [**14**] cf. [**7**]). *We assume that* X *is a non-bipartite graph. There exist two constants* C *and* $C' > 0$ *such that*

$$k_n(x, y) \leq Cn^{-\frac{D}{2}} e^{-d_X(x,y)^2/C'n} \tag{2.3}$$

$$\nabla^y k_n(x, y) \leq Cn^{-\frac{D+1}{2}} e^{-d_X(x,y)^2/C'n} \tag{2.4}$$

for all $x, y \in V$, *and all* $n = 1, 2, \ldots$.

We note that these estimates are closely related to Harmonic analysis on graphs. By a theorem of Coulhon and Grigor'yan [**4**], the Gaussian upper estimate (2.3) is equivalent to the relative Faber-Krahn inequality, namely there exist constants $a > 0$, $\nu > 0$ such that, for any $x \in V$, any $r > 0$ and any finite and non-empty set K in $B_V(x,r)$, the first eigenvalue $\lambda_1(K)$ of the Dirichlet Laplacian on K is estimated by

$$\lambda_1(K) \geq \frac{a}{r^2}\left[\frac{m(B_V(x,r))}{m(K)}\right]^\nu.$$

On the other hand, the Berry-Esseen type estimate (Theorem 1) is related to the L^p boundedness of the Riesz transform on X which is defined by $\nabla\Delta^{-1/2}$. When X is a Cayley graph of Γ, Alexopoulos proved the L^p boundedness for $1 < p < \infty$ and weak-$(1,1)$ in [**1**]. When X is an oriented graph with the doubling volume property and the Gaussian upper bound (2.3) holds, Russ proved that the Riesz transform is bounded on L^p for $1 < p \leq 2$ and weak-$(1,1)$ in [**15**]. We remark that nilpotent covering graphs satisfy the assumptions of Russ's theorem.

In this article, we prove the following result:

THEOREM 3. *Let X be a nilpotent covering graph and assume that X is non-bipartite. Then the Riesz transform is bounded on L^p for $1 < p < \infty$ and weak-$(1,1)$, which means that there exists a constant $C_p > 0$ such that for all finitely supported functions f on V,*

$$\|\nabla f\|_p \leq C_p\|\Delta^{1/2} f\|_p, \quad 1 < p < \infty$$

and

$$\sup_{\lambda>0} \lambda m\left(\{x \in V : |\nabla f(x)| > \lambda\}\right) \leq C_1\|\Delta^{1/2}f\|_1.$$

REMARK. There are some developments for the Riesz transform on manifolds. In [**5**], Dungey proves that the Riesz transform is L^p bounded for $1 < p < \infty$ on *nilpotent covering manifolds*. This can be adapted for nilpotent covering graphs. Moreover, Coulhon and Duong prove that if a complete Riemannian manifold satisfies doubling volume property, the Gaussian upper estimate for the heat kernel and the Gaussian upper estimate for the heat kernel on 1-forms then the Riesz transform is L^p bounded for $1 < p < \infty$ in [**3**].

3. Riesz transform

For a nilpotent covering graph $X = (V, E)$ and $1 \leq p < \infty$, let

$$L^p(V) := \{f : V \to \mathbb{R} \mid \|f\|_p < \infty\}$$

and $L^p(E)$ the space of L^p 1-forms on E defined by

$$L^p(E) := \{\omega : E \to \mathbb{R} \mid \omega(\overline{e}) = -\omega(e),\ \|\omega\|_p < \infty\},$$

where

$$\|f\|_p = \left(\sum_{x\in V} |f(x)|^p m(x)\right)^{1/p}$$

and

$$\|\omega\|_p = \left(\frac{1}{2}\sum_{e\in E} |\omega(e)|^p m(e)\right)^{1/p}.$$

In particular, inner products on $L^2(V)$ and $L^2(E)$ are defined by

$$\langle f, g\rangle = \sum_{x\in V} f(x)g(x)m(x), \quad f, g \in L^2(V)$$

and

$$\langle\langle \omega, \eta \rangle\rangle = \frac{1}{2} \sum_{e \in E} \omega(e)\eta(e)m(e), \quad \omega, \eta \in L^2(E),$$

respectively. Let d be the *coboundary operator* from functions on V to 1-forms on E defined by $df(e) = f(t(e)) - f(o(e))$. Since $\|\nabla f\|_p \sim \|df\|_p$, $1 < p < \infty$ and $|\nabla f(x)| \leq \sup_{e \in E_x} |df(e)|$ for all finitely supported functions f on V, we will consider the L^p boundedness of $R = d\Delta^{-1/2}$. By the spectral theory, the kernel of R can be written by $r(e, x) = \sum_{n \geq 0} a_n dk_n(e, x)$ in L^2, where a_n is given by

$$(1 - x)^{-1/2} = \sum_{n=0}^{\infty} a_n x^n.$$

Let $R^* : L^2(E) \to L^2(V)$ be the adjoint operator of R and $r^*(x, e)$ its kernel. We note that $r^*(x, e) = r(e, x)$. According to the Calderon-Zygmund theory (cf. [**6**]), the L^p boundedness of R follows from an estimate of r.

First we consider an estimate for the derivative of the kernel $k_n(x, y)$. We use the Berry-Esseen type estimate (Theorem 1), the Gaussian estimate for k_n (Theorem 2), and the following lemma.

LEMMA 3.1 ([**9**] cf. [**1**], [**8**] and [**10**]). *For $u \in C^\infty(\mathbb{R}_{\geq 0} \times G_\Gamma)$, let $\partial_N u(t, g) = u(t + N, g) - u(t, g)$ and $\Phi^* u(t, x) = u(t, \Phi(x))$. Then for any $J \geq 4$, there exists a constant $C > 0$ such that*

$$\begin{aligned} &\left|\left(\partial_N + (I - L^N)\right) \Phi^* u(t, x) - N\left(\partial_t + \Omega\right) u(t, \Phi(x))\right| \\ &\leq C \sup_{\theta \in [0,1],\, g \in U_N} \Bigg(N^2 \left| \frac{\partial^2}{\partial t^2} u(t + \theta N, \Phi(x)) \right| + X^2 u(t, \Phi(x)) \\ &\qquad + \sum_{j=3}^{J-1} N^{j-1} X^j u(t, \Phi(x)) + \sum_{k=J}^{Jr} N^k X^k u(t, \Phi(x) g) \Bigg), \end{aligned}$$

where

$$X^k u(t, \Phi(x)) = \sum_{l=1}^{k} \sum_{k_1 + k_2 + \cdots + k_l = k} \left| X_{i_1}^{(k_1)} X_{i_2}^{(k_2)} \cdots X_{i_l}^{(k_l)} u(t, \Phi(x)) \right|$$

and U_N is a set of all $g \in G_\Gamma$ satisfying that there exists $c \in C_{x,N}$ such that

$$\left|P_i^{(k)}(g)\right| \leq \left|P_i^{(k)}(\Phi(x)^{-1}\Phi(t(c)))\right|$$

for all (i, k).

By the same argument as Alexopoulos [**1**, Theorem 13], we have

LEMMA 3.2. *Let $S_n(x, y) = \frac{|G_\Gamma/\Gamma|}{m(X_0)} p_n(\Phi(x), \Phi(y))$, $x, y \in V$. For any $0 < \epsilon < 1/2$, there exists a constant $C_\epsilon > 0$ such that*

$$\sup_{x \in V, e \in E} |dk_n(x, e) - dS_n(x, e)| \leq C_\epsilon n^{-\frac{D + 3/2 - \epsilon}{2}}.$$

PROOF. Let $V_n(x, y) = k_n(x, y) - S_n(x, y)$ and denote

$$V_n \cdot k_T(x, y) = \sum_{z \in V} V_n(x, z) k_T(z, y) m(z).$$

For $N < n$, let I be the quotient of n by N. Then we have

$$\begin{aligned}
V_n(x,y) &= \sum_{0\le i<[\frac{I}{2}]} \Big(V_{n-iN}\cdot k_{iN}(x,y) - V_{n-(i+1)N}\cdot k_{(i+1)N}(x,y)\Big) \\
&\quad + V_{n-[\frac{I}{2}]N}\cdot k_{[\frac{I}{2}]N}(x,y) \\
&= \sum_{0\le i<[\frac{I}{2}]} \Big(V_{n-iN} - V_{n-(i+1)N}\cdot k_N\Big)\cdot k_{iN}(x,y) \\
&\quad + V_{n-[\frac{I}{2}]N}\cdot k_{[\frac{I}{2}]}(x,y),
\end{aligned}$$

where $[I/2]$ is the greatest integer not greater than $I/2$. Since

$$\begin{aligned}
dV_n(x,e) &= V_n(x,t(e)) - V_n(x,o(e)) \\
&= \sum_{0\le i<[\frac{I}{2}]} \Big(V_{n-iN} - V_{n-(i+1)N}\cdot k_N\Big)\cdot dk_{iN}(x,e) \\
&\quad + V_{n-[\frac{I}{2}]N}\cdot dk_{[\frac{I}{2}]N}(x,e),
\end{aligned}$$

we have

$$\begin{aligned}
|dV_n(x,e)| &\le \sum_{0\le i<[\frac{I}{2}]} \Big\|\Big(V_{n-iN} - V_{n-(i+1)N}\cdot k_N\Big)(x,\cdot)\Big\|_\infty \|dk_{iN}(\cdot,e)\|_1 \\
&\quad + \|V_{n-[\frac{I}{2}]N}(x,\cdot)\|_\infty \|dk_{[\frac{I}{2}]N}(\cdot,e)\|_1 \\
&= C\Big\|\Big\{\partial_N + (I - L^N)\Big\}S_{n-N}(x,\cdot)\Big\|_\infty \\
&\quad + \sum_{0<i<[\frac{I}{2}]} \Big\|\Big\{\partial_N + (I-L^N)\Big\}S_{n-(i+1)N}(x,\cdot)\Big\|_\infty \|dk_{iN}(\cdot,e)\|_1 \\
&\quad + \|V_{n-[\frac{I}{2}]N}(x,\cdot)\|_\infty \|dk_{[\frac{I}{2}]N}(\cdot,e)\|_1.
\end{aligned}$$

By Lemma 3.1,

$$\begin{aligned}
|dV_n(x,e)| &\le C\Big(N^2(n-N)^{-\frac{D+3}{2}} + \sum_{j=4}^{4r} N^j(n-N)^{-\frac{D-j}{2}}\exp(\frac{N^2}{c(n-N)})\Big) \\
&\quad + C\sum_{0<i<[\frac{I}{2}]} \Big(N^2(n-(i+1)N)^{-\frac{D+4}{2}} + (n-(i+1)N)^{-\frac{D+2}{2}} \\
&\quad + N^2(n-(i+1)N)^{-\frac{D+3}{2}} \\
&\quad + \sum_{j=4}^{4r} N^j(n-(i+1)N)^{-\frac{D+j}{2}}\exp(\frac{N^2}{c(n-(i+1)N)})\Big)(iN)^{-\frac{1}{2}} \\
&\quad + C_\epsilon\,(n-[I/2]\,N)^{-\frac{D+1/2-\epsilon}{2}}\,([I/2]\,N)^{-\frac{1}{2}}.
\end{aligned}$$

By choosing $N \sim n^{1/4}$, we conclude

$$
\begin{aligned}
|dV_n(x,e)| &\le C_\epsilon n^{-\frac{D+3/2-\epsilon}{2}} \\
&\quad + C \sum_{1\le i\le[\frac{I}{2}]} \Big(n^{-\frac{D+2}{2}} + \sum_{j=4}^{4r} n^{-\frac{D+j/2}{2}}\Big) N^{-1/2} i^{-1/2} \\
&\le C_\epsilon n^{-\frac{D+3/2-\epsilon}{2}} \\
&\quad + C n^{-\frac{D+2}{2}} N^{-1/2} \big(n/N\big)^{1/2} \\
&\le C'_\epsilon n^{-\frac{D+3/2-\epsilon}{2}}.
\end{aligned}
$$

□

By using this lemma, we show that the kernel $r^*(x,e)$ is a *standard kernel* (see [**6**]).

LEMMA 3.3. *Let assume that* $d_X(x_1,x_2) \le \frac{1}{2} d_X(x_1,o(e))$ *and* $d_X(x_1,o(e)) > 2$. *For any* $0 < \epsilon < 1/2$ *and* $0 < \delta < 1$, *there exists* $C > 0$ *such that*

$$
|r^*(x_1,e) - r^*(x_2,e)| \le C \left(\frac{1}{d_X(x_1,o(e))^{D+\delta(1/2-\epsilon)}} + \sum_{l=1}^{r} \frac{d_X(x_1,x_2)^{l\delta}}{d_X(x_1,o(e))^{D+l\delta}} \right).
$$

PROOF. First, we prove that

$$
\begin{aligned}
(3.1)\quad & |dk_n(x_1,e) - dk_n(x_2,e)| \\
&\le C \left(n^{\frac{1/2+\epsilon}{2}} + \sum_{l=1}^{r} d_X(x_1,x_2)^l n^{-\frac{l-1}{2}} \right)^{\delta} n^{-\frac{D+1+\delta}{2}} \exp(-\frac{d_X(x_1,o(e))^2}{cn}).
\end{aligned}
$$

By using the Gaussian estimate of the gradient of k_n (Theorem 2) and by the assumptions on x_1, x_2 and e, it is easy to obtain that

$$
(3.2)\qquad |dk_n(x_1,e) - dk_n(x_2,e)| \le C n^{-\frac{D+1}{2}} \exp\big(-\frac{d_X(x_1,o(e))^2}{cn}\big).
$$

On the other hand, by the previous lemma, we have

$$
\begin{aligned}
|dk_n(x_1,e) - dk_n(x_2,e)| &\le |dk_n(x_1,e) - dS_n(x_1,e)| + |dS_n(x_1,e) - dS_n(x_2,e)| \\
&\quad + |dS_n(x_2,e) - dk_n(x_2,e)| \\
&\le C n^{-\frac{D+3/2-\epsilon}{2}} + |dS_n(x_1,e) - dS_n(x_2,e)|.
\end{aligned}
$$

By Taylor's formula, there exist $g_1, g_2, g_3 \in G_\Gamma$ such that

$$\begin{aligned} dS_n(x_1,e) - dS_n(x_2,e) &= \frac{|G_\Gamma/\Gamma|}{m(X_0)}\Big(\sum_{\substack{k,l\\ i,j}} X_j^{(l)} X_i^{(k)} p_n(\Phi(x_1)g_3, \Phi(o(e))) \\ &\times P_j^{(l)}(\Phi(x_1)^{-1}\Phi(x_2)) P_i^{(k)}(\Phi(o(e))^{-1}\Phi(t(e))) \\ &+\frac{1}{2}\sum_{\substack{k_1,k_2\\ i_1,i_2}} X_{i_1}^{(k_1)} X_{i_2}^{(k_2)} p_n(\Phi(x_1), \Phi(o(e))g_1) \\ &\times P_{i_1}^{(k_1)}(\Phi(o(e))^{-1}\Phi(t(e))) P_{i_2}^{(k_2)}(\Phi(o(e))^{-1}\Phi(t(e))) \\ &-\frac{1}{2}\sum_{\substack{k_1,k_2\\ i_1,i_2}} X_{i_1}^{(k_1)} X_{i_2}^{(k_2)} p_n(\Phi(x_2), \Phi(o(e))g_2) \\ &\times P_{i_1}^{(k_1)}(\Phi(o(e))^{-1}\Phi(t(e))) P_{i_2}^{(k_2)}(\Phi(o(e))^{-1}\Phi(t(e)))\Big). \end{aligned}$$

The Gaussian estimates for p_n by Varopoulos [**16**] imply

$$|dk_n(x_1,e) - dk_n(x_2,e)| \le C\left(n^{\frac{1/2+\epsilon}{2}} + \sum_{l=1}^{r} d_X(x_1,x_2)^l n^{-\frac{l-1}{2}}\right) n^{-\frac{D+2}{2}}. \tag{3.3}$$

By interpolating (3.2) and (3.3), we obtain (3.1). Finally we have

$$\begin{aligned} &|r^*(x_1,e) - r^*(x_2,e)| \\ &\le C\Big(\frac{1}{d_X(x_1,o(e))^{D+\delta(1/2-\epsilon)}} \\ &\quad\times \sum_{n=1}^{\infty} a_n n^{-1/2}\left(\frac{d_X(x_1,o(e))^2}{n}\right)^{\frac{D+\delta(1/2-\epsilon)}{2}} \exp\Big(-\frac{d_X(x_1,o(e))^2}{cn}\Big) \\ &\quad+\sum_{l=1}^{r}\frac{d_X(x_1,x_2)^{l\delta}}{d_X(x_1,o(e))^{D+l\delta}} \\ &\quad\times \sum_{n=1}^{\infty} a_n n^{-1/2}\left(\frac{d_X(x_1,o(e))^2}{n}\right)^{\frac{D+l\delta}{2}} \exp\Big(-\frac{d_X(x_1,o(e))^2}{cn}\Big)\Big) \\ &\le C\left(\frac{1}{d_X(x_1,o(e))^{D+\delta(1/2-\epsilon)}} + \sum_{l=1}^{r}\frac{d_X(x_1,x_2)^{l\delta}}{d_X(x_1,o(e))^{D+l\delta}}\right). \end{aligned}$$

□

Proof of Theorem 3. We will show the L^p boundedness of the adjoint operator R^* for $1 < p \le 2$. We can use the same argument to obtain the L^p boundedness of R for $1 < p \le 2$. It is easy to see that R^* is bounded on L^2 and $L^p \subset L^2$ for $1 \le p \le 2$. By Marcinkiewicz interpolation (cf. [**6**]), it suffices to show that the adjoint operator R^* is weak-$(1,1)$:

$$m\big(\{x \in V \,:\, |R^*\omega(x)| > \lambda\}\big) \le \frac{C}{\lambda}\|\omega\|_1.$$

Let $E = E_1 \coprod E_2$ be a decomposition such that $\overline{E_1} = E_2$. Then the fundamental domain $F \subset E$ for the action of Γ is decomposed to $F = F_1 \coprod F_2$. Then, each ω in

$L^1(E)$ can be written by $\omega = \sum_{e_1 \in F_1} \omega^{e_1}$, where

$$\omega^{e_1}(e) = \begin{cases} \omega(e) & \text{if } e \in \Gamma e_1 \cap \Gamma \overline{e_1} \\ 0 & \text{otherwise.} \end{cases}$$

We remark that ω^{e_1} can be identified with an element in $L^1(\Gamma)$ by $\omega^{e_1}(\gamma) = \omega^{e_1}(\gamma e_1)$. Let $S = \{s_1, \ldots, s_n\}$ be a symmetric finite generator of Γ. A distance d_Γ on Γ is defined by

$$d_\Gamma(\gamma_1, \gamma_2) := \min\{k \in \mathbb{N} \, : \, \gamma_1 = s_{i_1} s_{i_2} \cdots s_{i_k} \gamma_2, \; s_{i_j} \in S, \, 1 \le i_j \le n\}.$$

We denote

$$B_\Gamma(\gamma, d) := \{\eta \in \Gamma \, : \, d_\Gamma(\gamma, \eta) \le d\}.$$

Here we apply the following theorem:

THEOREM 4 (Coifman and Weiss [**2**]). *There exists a constant $C > 0$ such that, for any $\omega^{e_1} \in L^1(\Gamma)$ and $\lambda > 0$, ω^{e_1} is decomposed by $g^{e_1} + b^{e_1}$ with $b^{e_1} = \sum_{i \in I} b_i^{e_1}$ so that*

- **(a):** $|g^{e_1}(\gamma)| \le C\lambda$, $\gamma \in \Gamma$.
- **(b):** *For any $i \in I$, there exists $B_\Gamma(\gamma_i, d_i)$ so that the support of b_i is contained in $B_\Gamma(\gamma_i, d_i)$, $\sum_{\gamma \in \Gamma} |b_i^{e_1}(\gamma)| \le C\lambda |B_\Gamma(\gamma_i, d_i)|$ and $\sum_{\gamma \in \Gamma} b_i^{e_1}(\gamma) = 0$.*
- **(c):** $\sum_{i \in I} |B_\Gamma(\gamma_i, d_i)| \le \frac{C}{\lambda} \|\omega^{e_1}\|_1$.

We denote

$$M = \sup_{x \in V, \gamma \in \Gamma} \frac{d_X(x, \gamma x)}{d_\Gamma(id, \gamma)}$$

and $A^{e_1} = \cup_{i \in I} B_V(\gamma_i o(e_1), 2Md_i) = \cup_{i \in I} B_i$. Hence we have

$$\begin{aligned} m(\{x \in V \, : \, |R^*\omega(x)| > \lambda\}) &\le \sum_{e_1 \in F_1} m\left(\left\{x \in V \, : \, |R^*\omega^{e_1}(x)| > \frac{\lambda}{\#F_1}\right\}\right) \\ &\le \sum_{e_1 \in F_1} \left[m\left(\left\{x \in V \, : \, |R^* g^{e_1}(x)| > \frac{\lambda}{2\#F_1}\right\}\right)\right. \\ &\qquad \left. + m\left(\left\{x \in V \, : \, |R^* b^{e_1}(x)| > \frac{\lambda}{2\#F_1}\right\}\right)\right]. \end{aligned}$$

Then

$$m\left(\left\{x \in V \, : \, |R^* g^{e_1}(x)| > \frac{\lambda}{2\#F_1}\right\}\right) \le \left(\frac{2\#F_1}{\lambda}\right)^2 \sum_{x \in V} |R^* g^{e_1}(x)|^2 m(x).$$

Since R^* is bounded on L^2, we obtain

$$m\left(\left\{x \in V \, : \, |R^* g^{e_1}(x)| > \frac{\lambda}{2\#F_1}\right\}\right) \le \frac{C}{\lambda} \|\omega^{e_1}\|_1.$$

Next we consider $m\left(\left\{x\in V\,:\,|R^*b^{e_1}(x)|>\frac{\lambda}{2\#F_1}\right\}\right)$. By the assumption of b^{e_1}, we have

$$\begin{aligned}
&m\left(\left\{x\in V\,:\,|R^*b^{e_1}(x)|>\frac{\lambda}{2\#F_1}\right\}\right)\\
\leq\;& |A^{e_1}|+m\left(\left\{x\in V\backslash A^{e_1}\,:\,|R^*b^{e_1}(x)|>\frac{\lambda}{2\#F_1}\right\}\right)\\
\leq\;& |A^{e_1}|+\frac{2\#F_1}{\lambda}\sum_{x\in V\backslash A^{e_1}}|R^*b^{e_1}(x)|m(x)\\
\leq\;& \frac{C}{\lambda}\|\omega^{e_1}\|_1+\frac{2\#F_1}{\lambda}\sum_{i\in I}\sum_{x\in V\backslash B_i}\left|\sum_{e\in E_1}r^*(x,e)b_i^{e_1}(e)m(e)\right|m(x)\\
=\;& \frac{C}{\lambda}\|\omega^{e_1}\|_1+\frac{2\#F_1}{\lambda}\sum_{i\in I}\sum_{x\in V\backslash B_i}\left|\sum_{\gamma\in\Gamma}r^*(x,\gamma e_1)b_i^{e_1}(\gamma e_1)m(e_1)\right|m(x).
\end{aligned}$$

Since $b_i^{e_1}$ has a zero integral, we have

$$\begin{aligned}
m\left(\left\{x\in V\,:\,|R^*b^{e_1}(x)|>\frac{\lambda}{2\#F_1}\right\}\right)\leq\;& \frac{C}{\lambda}\|\omega^{e_1}\|_1\\
&+\frac{2\#F_1}{\lambda}\sum_{i\in I}\sum_{\gamma\in B_\Gamma(\gamma_i,d_i)}|b_i^{e_1}(\gamma e_1)|m(e_1)\\
&\times\sum_{x\in V\backslash B_i}|r^*(x,\gamma e_1)-r^*(x,\gamma_i e_1)|m(x).
\end{aligned}$$

By Lemma 3.3,

$$\begin{aligned}
&|r^*(x,\gamma e_1)-r^*(x,\gamma_i e_1)|\\
&\leq\; C\left(\frac{1}{d_X(x,\gamma_i o(e_1))^{D+\delta(1/2-\epsilon)}}+\sum_{l=1}^{r}\frac{M^{l\delta}d_i^{l\delta}}{d_X(x,\gamma_i o(e_1))^{D+l\delta}}\right).
\end{aligned}$$

Consequently we have

$$\begin{aligned}
&\frac{2\#F_1}{\lambda}\sum_{i\in I}\sum_{\gamma\in B_\Gamma(\gamma_i,d_i)}|b_i^{e_1}(\gamma e_1)|m(e_1)\sum_{x\in V\backslash B_i}|r^*(x,\gamma e_1)-r^*(x,\gamma_i e_1)|m(x)\\
\leq\;& \frac{C'}{\lambda}\sum_{i\in I}\sum_{\gamma\in B_\Gamma(\gamma_i,d_i)}|b_i^{e_1}(\gamma e_1)|m(e_1)\\
&\times\sum_{x\in V\backslash B_i}\left(\frac{1}{d_X(x,\gamma_i o(e_1))^{D+\delta(1/2-\epsilon)}}+\sum_{l=1}^{r}\frac{M^{l\delta}d_i^{l\delta}}{d_X(x,\gamma_i o(e_1))^{D+l\delta}}\right)\\
\leq\;& \frac{C}{\lambda}\|\omega^{e_1}\|_1.
\end{aligned}$$

□

References

[1] G. Alexopoulos, *Convolution powers on discrete groups of polynomial volume growth*, Canad. Math. Soc. Conf. Proc. **21** (1997), 31-57.

[2] R. Coifman and G. Weiss, *Analyse harmonique non-commutative sur certains espaces homog'enes*, Lecture Notes in Math. **242** (1971).

[3] T. Coulhon and X. T. Duong, *Riesz transform and related inequalities on non-compact Riemannian manifolds*, preprint.
[4] T. Coulhon and A. Grigor'yan, *Random walks on graphs with regular volume growth*, GAFA, **8** (1998), 656-701.
[5] N. Dungey, *Heat kernel estimates and Riesz transforms on some Riemannian covering manifolds*, preprint.
[6] J. Duoandikoetxea, *Fourier Analysis*, Graduate Studies in Mathematics **29**, American Mathematical Society (2001).
[7] W. Hebisch and L. Saloff-Coste, *Gaussian estimates for Markov chains and random walks on groups*, Ann. Probab. **21** (1993), no. 2, 673-709.
[8] S. Ishiwata, *A central limit theorem on a covering graph with a transformation group of polynomial growth*, to appear in J. Math. Soc. Japan.
[9] S. Ishiwata, *A Berry-Esseen type theorem on nilpotent covering graphs*, preprint.
[10] M. Kotani, *A central limit theorem for magnetic transition operators on a crystal lattice*, J. London Math. Soc. **65** (2002), 464–482.
[11] M. Kotani and T. Sunada, *Standard realizations of crystal lattices via harmonic maps*, Trans. Amer. Math. Soc. **353** (2001), no. 1, 1–20.
[12] M. Kotani and T. Sunada, *Albanese maps and off diagonal long time asymptotics for the heat kernel*, Comm. Math. Phys. **209** (2000), 633-670.
[13] A. I. Mal'cev, *On a class of homogeneous spaces*, Amer. Math. Soc. Transl. **39** (1951).
[14] C. Pittet, L. Saloff-Coste, *On the stability of the behavior of random walks on groups*, J. Geom. Anal. **10** (2000), no. 4, 713–737.
[15] E. Russ, *Riesz Transforms on Graphs for* $1 \leq p \leq 2$, Math. Scand. **87** (2000), 133-160.
[16] N. Th. Varopoulos, L. Saloff-Coste and T. Coulhon, *Analysis and geometry on groups*, Cambridge Tracts in Mathematics, 100. Cambridge University Press, Cambridge, 1992.
[17] W. Woess, *Random walks on infinite graphs and groups*, Cambridge Tracts in Mathematics, 138. Cambridge University Press, Cambridge, 2000.

Mathematical Institute, Tohoku University, Sendai 980-8578, Japan
E-mail address: `s98m03@math.tohoku.ac.jp`

Contemporary Mathematics
Volume **347**, 2004

Non-Commutative Poisson Boundaries

Masaki Izumi

ABSTRACT. We present a brief survey of the notion of non-commutative Poisson boundaries. We discuss amenability of the Poisson boundary for a random walk on a discrete group from the viewpoint of an injective operator system.

1. Introduction

The notion of non-commutative Poisson boundaries was introduced as a tool to investigate problems in operator algebras, more precisely, those in compact quantum group actions [**11**] and in subfactors [**12**]. Given a Markov operator (synonymously, a unital normal completely positive map) acting on a von Neumann algebra, one can equip the fixed point set with an abstract von Neumann algebra structure, which we call the non-commutative Poisson boundary. In this note we present a brief survey of non-commutative Poisson boundaries for non-specialists giving the simplest possible, but still significant, non-commutative example. In purely operator algebraic problems, the non-commutative Poisson boundaries are particularly useful for capturing structure appearing only after taking closure in the weak operator topology. But we refrain from discussing these applications here and we just give an incomplete list of references for them [**19, 3, 4, 11, 12**].

To justify the term "non-commutative Poisson boundary", we begin our exposition with the most classical situation. Let $H^\infty(D)$ be the set of all bounded harmonic functions on the unit disc $D \subset \mathbb{C}$, which is a Banach space as a closed subspace of $L^\infty(D)$, the set of essentially bounded measurable functions on D. Then the classical Poisson integral formula

$$f(re^{i\theta}) = \frac{1}{2\pi}\int_{\mathbb{T}} \tilde{f}(e^{i\varphi})\frac{1-r^2}{1-2r\cos(\theta-\varphi)+r^2}d\varphi$$

holds for $f \in H^\infty(D)$, where $\tilde{f}(e^{i\theta})$ denotes the boundary value of f at $e^{i\theta}$, which makes sense for almost all $e^{i\theta}$ in the unit circle $\mathbb{T}$.

Here, we emphasize the following functional analytic fact: the map

$$H^\infty(D) \ni f \mapsto \tilde{f} \in L^\infty(\mathbb{T})$$

1991 *Mathematics Subject Classification.* Primary 46L53, 60J50; Secondary 46L07.

Key words and phrases. Poisson boundary, von Neumann algebras, random walk on a group, amenable action.

Supported in part by JSPS.

gives an *isometry* between the two Banach spaces $H^\infty(D)$ and $L^\infty(\mathbb{T})$ and the inverse is given by the Poisson integral. Note that these two Banach spaces are equipped with extra structure other than Banach space structure. For example, the pointwise product of two functions in $L^\infty(\mathbb{T})$ belongs to $L^\infty(\mathbb{T})$ again: in other words, $L^\infty(\mathbb{T})$ has an algebra structure (or more precisely, a von Neumann algebra structure, see Section 2). It is this algebra structure that recovers geometry of the problem, and in principle, all information of the algebra structure should be encoded in the Banach space $H^\infty(D)$ with an extra structure.

The above mysterious correspondence of the two Banach spaces is only the tip of the iceberg. To generalize the situation, the space of bounded harmonic functions $H^\infty(D)$ should be regarded as the set of fixed points in $L^\infty(D)$ under the action of the Markov semigroup $\{P_t\}_{t\in\mathbb{R}_+}$ generated by the Laplacian (with respect to the Poincaré metric). An important observation here is that the structure of the von Neumann algebra $L^\infty(\mathbb{T})$ is completely determined by the pair $(L^\infty(D), \{P_t\}_{t\in\mathbb{R}_+})$, and no geometry is required for that. There are varieties of generalization of this classical situation in the literature, and we restrict ourselves to the discrete time case for simplicity.

For a given Markov operator P acting on $L^\infty(S,\mu)$, where (S,μ) is a measure space, we denote by $H^\infty(S,P)$ the set of fixed points in $L^\infty(S,\mu)$ under P, whose elements are called P-harmonic. Then one can construct, in a purely measure theoretic way [**13**], a probability space (Ω,ν) and a Banach space isometry

$$\theta : H^\infty(S,P) \to L^\infty(\Omega,\nu)$$

preserving constant functions and sending the set of positive harmonic functions in $H^\infty(S,P)$ onto the set of positive functions in $L^\infty(\Omega,\nu)$. The probability space (Ω,ν) is called the *Poisson boundary* for (S,μ,P), and in fact an analogue of the Poisson integral holds, which gives the inverse of the isometry θ. Note that although (Ω,ν) is regarded as the "boundary" of S, in general it is not obtained geometrically unlike the case of D having the Poisson boundary $\mathbb{T}$. Readers are referred to [**13**] for details of the theory of the Poisson boundaries of general Markov chains (in the commutative case). The central issue of this note is a further generalization of the situation to the non-commutative case, which actually gives a functional analytic substitute of the measure theoretic construction mentioned above (which was already hinted in [**15**] in the commutative case).

While the Poisson boundary is built to describe the set of bounded harmonic functions, there is a topological version of it, the Martin boundary, to describe the set of positive (not necessarily bounded) harmonic functions. Although we will not discuss this topic in this note, we just mention that a non-commutative version of the Martin boundary is also discussed in [**1, 2, 7, 17**].

2. Operator algebras

We summarize basic facts about C^*-algebras and von Neumann algebras in this section. Our basic reference is [**21**].

2.1. C^*-algebras. For a (complex) Hilbert space H, we denote by $B(H)$ the set of bounded linear operators of H, which has a natural algebra structure over $\mathbb{C}$. $B(H)$ is a Banach space with the operator norm

$$||x|| = \sup\{||x\xi||;\ \xi \in H,\ ||\xi|| \le 1\}, \quad x \in B(H).$$

For $x \in B(H)$, x^* denotes the adjoint operator of x determined by the relation

$$\langle x^*\xi, \eta\rangle = \langle \xi, x\eta\rangle, \quad \forall \xi, \eta \in H.$$

A subalgebra $A \subset B(H)$ is said to be a $*$-subalgebra of $B(H)$ if $x \in A$ implies $x^* \in A$.

DEFINITION 2.1. A C^*-algebra is a $*$-subalgebra of $B(H)$ closed in the norm topology.

Here are a few examples of C^*-algebras: The set of complex numbers $\mathbb{C}$ is a (trivial) C^*-algebra regarded as the scalar operators. $B(H)$ itself is a C^*-algebra, which is not separable unless H is finite dimensional. The set of compact operators $K(H)$ is a typical example of a C^*-algebra without unit (unless H is finite dimensional).

Strictly speaking, the C^*-algebra defined above should be called a concrete C^*-algebra, and there is a notion of an abstract C^*-algebra [**21**, Chapter I, Definition 1.2]: an abstract C^*-algebra A is a Banach $*$-algebra satisfying the C^*-condition:

$$||x^*x|| = ||x||^2, \quad \forall x \in A.$$

However, it is known that every abstract C^*-algebra has a faithful (and necessarily isometric) representation in $B(H)$ for some Hilbert space H [**21**, Chapter I, Theorem 9.18]. For example, the set of continuous functions $C(\Omega)$ on a compact Hausdorff space Ω is an abstract C^*-algebra with the uniform norm

$$||f|| = \sup\{|f(\omega)|;\ \omega \in \Omega\}, \quad f \in C(\Omega),$$

and the $*$-operation $f^*(\omega) = \overline{f(\omega)}$. In fact, the Gelfand-Naimark theorem [**21**, Chapter I, Theorem 4.4] says that every commutative C^*-algebra with unit is isomorphic to $C(\Omega)$ for some Ω, which is unique up to homeomorphism.

2.2. von Neumann algebras. Recall that the weak operator topology of $B(H)$ is defined in terms of convergence of nets: a net $\{x_i\}_{i\in I}$ in $B(H)$ converges to $x \in B(H)$ in the weak operator topology if

$$\lim_{i\in I}\langle x_i\xi, \eta\rangle = \langle x\xi, \eta\rangle, \quad \forall \xi, \eta \in B(H).$$

DEFINITION 2.2. A von Neumann algebra is a C^*-subalgebra of $B(H)$ closed in the weak operator topology.

In the above definition, the weak operator topology can be replaced with any natural topology of $B(H)$ except for the norm topology. Among the examples of C^*-algebras given above, $B(H)$ and $\mathbb{C}$ are von Neumann algebras while $K(H)$ is not.

Let (Ω, ν) be a measure space. Then $L^\infty(\Omega, \nu)$ has a representation on $L^2(\Omega, \nu)$ as the multiplication operators. This gives $L^\infty(\Omega, \nu)$ a von Neumann algebra structure. In fact, it is well-known that every commutative von Neumann algebra is isomorphic to $L^\infty(\Omega, \nu)$ for some measure space (Ω, ν) [**21**, Chapter III, Theorem 1.18]. The Gelfand-Naimark theorem and this fact indicates an essential difference between C^*-algebras and von Neumann algebras: it is similar to the difference between topological spaces and measure spaces.

Other than the norm topology, the most natural topology of a von Neumann algebra is the σ-weak topology. Let $C_1(H)$ be the set of trace class operators in

$B(H)$, which is a Banach space with the trace norm

$$||y||_1 = \mathrm{Tr}(|y|), \quad y \in C_1(H),$$

where Tr stands for the usual trace. It is well-known that $B(H) = C_1(H)^*$ holds as a Banach space [**21**, Chapter II, Theorem 1.8], where the pairing of $x \in B(H)$ and $y \in C_1(H)$ is given by $\mathrm{Tr}(xy)$. The σ-weak topology of $B(H)$ is defined to be the weak* topology as the dual Banach space of $C_1(H)$. The σ-weak topology coincides with the weak operator topology on bounded subsets of $B(H)$ [**21**, Chapter II, Lemma 2.5] though the former is strictly stronger than the latter in general. It is known that every $*$-isomorphism between two von Neumann algebras is automatically a homeomorphism in the σ-weak topology (and also in the norm topology) [**21**, Chapter III, Corollary 3.10], which means that the σ-weak topology of a von Neumann algebra is determined by its $*$-algebra structure. This observation makes it possible to introduce the notion of an abstract von Neumann algebra free from a particular representation in $B(H)$, which is usually called a W^*-algebra [**21**, Chapter III, Definition 3.1].

For a von Neumann algebra M, we denote by M_* the set of σ-weakly continuous linear functionals on M, which is called the predual of M. The bipolar theorem implies that $(M_*)^* = M$ and that the σ-weak topology of M coincides with the weak $*$-topology as the dual space of M_*. For example, we have $B(H)_* = C_1(H)$ and $L^\infty(\Omega, \mu)_* = L^1(\Omega, \mu)$. In fact, Sakai's theorem characterizes von Neumann algebras as C^*-algebras with preduals [**21**, Chapter III, Theorem 3.5]. In other word, if a C^*-algebra A happens to be the dual Banach space of some Banach space X, A is $*$-isomorphic to some von Neumann algebra M and the isomorphism is a homeomorphism in the weak* topology of $X^* = A$ and the σ-weak topology of M. This is a key observation for introducing the non-commutative Poisson boundaries using the Choi-Effros product in the next section.

3. Non-commutative Poisson boundaries

We now introduce the notion of a Markov operator of a von Neumann algebra and its Poisson boundary, which coincides with the classical notion described in Introduction when the von Neumann algebra is commutative.

A self-adjoint operator $x \in B(H)$ is said to be positive and is denoted by $x \geq 0$ if the spectrum of x is contained in the non-negative real line $\mathbb{R}_{\geq 0}$. A linear map between C^*-algebras is said to be positive if it sends every positive operator to a positive operator. A positive linear functional of norm one is said to be a state. A positive map is said to be faithful if $x \geq 0$ and $\phi(x) = 0$ imply $x = 0$.

Let $A \subset B(H)$ be a C^*-algebra. We denote by $M_n(A)$ the set of n by n matrices whose entries are in A. As $M_n(A)$ acts on $H^{\oplus n}$, it has a natural C^*-algebra structure coming from $B(H^{\oplus n})$. Let ϕ be a linear map from a C^*-algebra A to a C^*-algebra B. Then ϕ induces the map $\phi^{(n)} : M_n(A) \to M_n(B)$ defined by

$$\phi^{(n)}(a)_{ij} = \phi(a_{ij}), \quad \forall a = (a_{ij}) \in M_n(A).$$

ϕ is said to be n-positive if $\phi^{(n)}$ is positive, and ϕ is said to be completely positive if ϕ is n-positive for every natural number n. If either A or B is commutative, complete positivity is equivalent to positivity [**21**, Chapter IV, Corollary 3.5, Proposition 3.9]. However, even in the case $A = B = M_2(\mathbb{C})$, the difference between these two conditions is so big.

A positive linear map between two von Neumann algebras is said to be normal if it is continuous in the σ-weak topology.

DEFINITION 3.1. A Markov operator P of a von Neumann algebra M is a unit preserving normal completely positive map from M to itself. An element $x \in M$ is said to be P-harmonic if $P(x) = x$ holds. We denote by $H^\infty(M, P)$ the set of P-harmonic elements in M.

Let S be a countable set and regard $\ell^\infty(S)$ as a von Neumann algebra acting on $\ell^2(S)$ as before. We denote by δ_s the Dirac function concentrating at $s \in S$. We show that the above definition of a Markov operator coincides with the usual one. For a Markov operator P in our sense, $P(\delta_s)$ is a non-negative function in $\ell^\infty(S)$ and there exist non-negative numbers $p(t, s)$ satisfying

$$P(\delta_s)(t) = p(t, s), \quad \forall s, t \in S.$$

As the equation

$$\sum_{s \in S} \delta_s = 1$$

holds in the σ-weak operator topology, we get

$$1 = P(1)(t) = P(\sum_{s \in S} \delta_s)(t) = \sum_{s \in S} P(\delta_s)(t) = \sum_{s \in S} p(t, s), \quad \forall t \in S.$$

Therefore $(p(t, s))_{t,s}$ is a transition matrix. Since every element $f \in \ell^\infty(S)$ can be expanded in the σ-weak topology as

$$f = \sum_{s \in S} f(s)\delta_s,$$

we get

$$P(f)(t) = P(\sum_{s \in S} f(s)\delta_s)(t) = \sum_{s \in S} f(s)P(\delta_s)(t) = \sum_{s \in S} p(t, s)f(s),$$

which shows that P is a Markov operator in a classical sense.

To introduce a von Neumann algebra structure into $H^\infty(M, P)$, we need to appeal to the theory of operator systems. A linear subspace $X \subset B(H)$ is said to be an operator system if X is closed under the $*$-operation and it contains the scalar operators. $H^\infty(M, P)$ is an example of a σ-weakly closed operator system.

Through their study of injective operator systems, Choi and Effros proved the following theorem [**5**, Theorem 3.1, Corollary 3.2]:

THEOREM 3.2 (Choi-Effros). *Let X be an operator system in a von Neumann algebra M. We assume that there exists a completely positive projection $E : M \to X$ onto X. Then X is an (abstract) C^*-algebra with a new product $x \circ y$ given by $E(xy)$, (which we will call the Choi-Effros product in the sequel). Moreover, if X is closed in the σ-weak topology, X is a von Neumann algebra.*

The first part follows from a very simple but ingenious manipulation of a 2 by 2 operator matrix. If X is closed in the σ-weak topology, it has a predual thanks to the bipolar theorem. Therefore Sakai's theorem mentioned in Section 2 implies that X has a von Neumann algebra structure.

Now we show that $H^\infty(M, P)$ has a von Neumann algebra structure.

THEOREM 3.3. *Let P be a Markov operator of a von Neumann algebra M. Then the set of harmonic element $H^\infty(M,P)$ has a unique von Neumann algebra structure.*

PROOF. Let $\omega \in \beta\mathbb{N} \setminus \mathbb{N}$ be a free ultrafilter [**23**, Chapter XIV, Section 4] and define $E_\omega : M \to H^\infty(M,P)$ by

$$E_\omega(x) = \lim_{n\to\omega} \frac{1}{n}\sum_{k=0}^{n-1} P^k(x), \quad x \in M,$$

where the limit is taken in the weak operator topology. (In fact, the limit exists without choosing ω in most of the concrete examples of our interest.) Then E_ω is a completely positive projection from M onto $H^\infty(M,P)$, and the Choi-Effros product gives $H^\infty(M,P)$ a von Neumann algebra structure. Uniqueness follows from the fact that every 2-positive isometry between von Neumann algebras is an isomorphism [**6**, Corollary 3.2]. □

When M is a commutative von Neumann algebra $L^\infty(S,\mu)$, the Choi-Effros product $x \circ y$ in $H^\infty(M,P)$ is commutative and $H^\infty(M,P)$ has a commutative von Neumann algebra structure thanks to the above theorem. Therefore the structure theorem of commutative von Neumann algebras mentioned in Section 2 implies that there exists a measure space (Ω,ν) such that $H^\infty(M,P)$ has a faithful representation onto $L^\infty(\Omega,\nu)$. This argument can substitute for the usual measure theoretic construction of the existence of the Poisson boundary.

In the definition below, a concrete realization of $H^\infty(M,P)$ means a faithful representation of the abstract von Neumann algebra $H^\infty(M,P)$ onto a concrete von Neumann algebra.

DEFINITION 3.4. Let P be a Markov operator on a von Neumann algebra M. A concrete realization of the von Neumann algebra structure of $H^\infty(M,P)$ is called the Poisson boundary of (M,P).

In general, it is not so easy to find a concrete realization of $H^\infty(M,P)$ for a given Markov operator even in the commutative case (Kaimanovich calls it an identification problem [**14**]). There are only two cases where concrete realizations are known in the non-commutative case [**11, 12**]. We will present the third example in the next section.

4. Markov operators associated with discrete groups

4.1. Random walks on discrete groups. Let Γ be a discrete group and let μ be a probability measure on Γ. We assume that the support of μ generates Γ as a semigroup. We denote by P_μ the right convolution operator acting on $\ell^\infty(\Gamma)$, which is a Markov operator. P_μ describes a random walk with transition probability from γ to σ given by $\mu\{\gamma^{-1}\sigma\}$. We use the notation $H^\infty(\Gamma,\mu)$ for $H^\infty(\ell^\infty(\Gamma),P_\mu)$. In this case, it is known that the Choi-Effros product is given by the pointwise limit [**15**] (the limit exists in the strong operator topology as well)

$$f \circ g = \lim_{n\to\infty} P_\mu^n(fg), \quad f,g \in H^\infty(\Gamma,\mu).$$

We will construct the Poisson boundary and an analogue of the Poisson integral only using Theorem 3.3.

If $f \in H^\infty(\Gamma,\mu)$ is a non-negative function satisfying $f(e)=0$, the assumption of μ implies that f is constantly 0. Therefore the point evaluation at e

$$H^\infty(\Gamma,\mu) \ni g \mapsto g(e)$$

gives a faithful normal state of $H^\infty(\Gamma,\mu)$ treated as a von Neumann algebra. Thus the structure theorem of commutative von Neumann algebras mentioned in Section 2 implies that there exists a measure space (Ω,ν) with a probability measure ν satisfying

(PB1) There exists an isometry $\theta : H^\infty(\Gamma,\mu) \to L^\infty(\Omega,\nu)$ that preserves constant functions and sends positive harmonic functions to positive bounded functions. θ satisfies

$$\theta(f \circ g) = \theta(f)\theta(g), \quad \forall f, g \in H^\infty(\Gamma,\mu).$$

(PB2) For every $f \in H^\infty(\Gamma,\mu)$, the following equation holds:

$$f(e) = \int_\Omega \theta(f)(\omega)d\nu(\omega).$$

As P_μ commutes with the left translation operator, Γ acts on $H^\infty(\Gamma,\mu)$ as the left translation, and so it acts on $H^\infty(\Gamma,\mu)$ treated as a von Neumann algebra. Since every $*$-automorphism of the von Neumann algebra $L^\infty(\Omega,\nu)$ is realized by a non-singular transformation of the measure space (Ω,ν) **[16]**, that is, an almost everywhere defined bimeasurable bijection preserving the measure class of ν, we get the following:

(PB3) There exists a Γ-action on Ω that preserves the measure class of ν such that

$$\theta(\alpha_\gamma(f)) = \tilde{\alpha}_\gamma(\theta(f)), \quad \forall f \in H^\infty(\Gamma,\mu), \ \forall \gamma \in \Gamma,$$

where $\alpha_\gamma(f)(\sigma) = f(\gamma^{-1}\sigma)$ for $\sigma \in \Gamma$ and $\tilde{\alpha}_\gamma(g)(\omega) = g(\gamma^{-1}\cdot\omega)$ for $g \in L^\infty(\Omega,\nu)$ and $\omega \in \Omega$.

In fact, the three properties (PB1), (PB2), and (PB3) completely characterize the Poisson boundary. An analogue of the Poisson integral is obtained easily from these properties. Indeed, for $f \in H^\infty(\Gamma,\nu)$ and $\gamma \in \Gamma$ we get

$$\begin{aligned} f(\gamma) &= \alpha_{\gamma^{-1}}(f)(e) = \int_\Omega \theta(\alpha_{\gamma^{-1}}(f))(\omega)d\nu(\omega) = \int_\Omega \tilde{\alpha}_{\gamma^{-1}}(\theta(f))(\omega)d\nu(\omega) \\ &= \int_\Omega \theta(f)(\gamma\cdot\omega)d\nu(\omega) = \int_\Omega \theta(f)(\omega)d\nu(\gamma^{-1}\cdot\omega) \\ &= \int_\Omega \theta(f)(\omega)\frac{d\nu(\gamma^{-1}\cdot)}{d\nu}(\omega)d\nu(\omega). \end{aligned}$$

Therefore the Radon-Nikodym derivative above is regarded as an analogue of the Poisson kernel.

4.2. Non-commutative extension. Let ρ (respectively λ) be the right (respectively the left) regular representation of Γ on $\ell^2(\Gamma)$. Then the Markov operator P_μ is expressed as

$$P_\mu(f) = \sum_{\gamma\in\Gamma} \mu(\gamma)\rho(\gamma)f\rho(\gamma^{-1}), \quad f \in \ell^\infty(\Gamma),$$

where we regard $\ell^\infty(\Gamma)$ as a von Neumann subalgebra of $B(\ell^2(\Gamma))$ as multiplication operators. Note that the above formula makes sense even if f is replaced with a general operator in $B(\ell^2(\Gamma))$. Therefore we can introduce a Markov operator Q_μ of $B(\ell^2(\Gamma))$ by the same formula

$$Q_\mu(x) = \sum_{\gamma\in\Gamma} \mu(\gamma)\rho(\gamma)x\rho(\gamma^{-1}), \quad x \in B(\ell^2(\Gamma)).$$

What is the Poisson boundary for $(B(\ell^2(\Gamma)), Q_\mu)$? Although this question sounds like a too simple-minded generalization to the non-commutative case, it still has an interesting connection to Zimmer's notion of amenable group actions.

Other than the "classical part" $H^\infty(\Gamma,\mu)$, $H^\infty(B(\ell^2(\Gamma)), Q_\mu)$ contains the image of the left regular representation $\{\lambda(\gamma)\}_{\gamma\in\Gamma}$ because the right regular representation commutes with the left regular representation. For the same reason, for every Q_μ-harmonic element x and every $\gamma \in \Gamma$, the (ordinary) products $\lambda(\gamma)x$ and $x\lambda(\gamma)$ are Q_μ-harmonic again and they coincide with the Choi-Effros products. Moreover, the following relation holds:

$$\lambda(\gamma)f = \alpha_\gamma(f)\lambda(\gamma), \quad \forall f \in H^\infty(\Gamma,\mu),\ \forall\gamma \in \Gamma.$$

From this observation, operator algebraists can easily guess the answer to the above question: the Poisson boundary must be the crossed product $L^\infty(\Omega,\nu)\rtimes_{\tilde{\alpha}}\Gamma$. Indeed, it is!

Forming a crossed product is a machinery to translate dynamical information into algebraic relations and it is one of the most fundamental constructions in operator algebras (see [**21**, Chapter V, Definition 7.4] for our restricted case and [**22**, Chapter X] for more general cases.) Roughly speaking, $L^\infty(\Omega,\nu) \rtimes_{\tilde{\alpha}} \Gamma$ is the von Neumann algebra generated by a copy of $L^\infty(\Omega,\nu)$ and a copy of the regular representation of Γ satisfying the semidirect product-like relation above. More precisely, we identify the Hilbert space tensor product $L^2(\Omega,\nu)\otimes\ell^2(\Gamma)$ with the set of square integrable functions on $\Omega\times\Gamma$ and introduce a representation π of $L^\infty(\Omega,\nu)$ on $L^2(\Omega,\nu)\otimes\ell^2(\Gamma)$ by

$$(\pi(g)\xi)(\omega,\gamma) = \tilde{\alpha}_{\gamma^{-1}}(g)(\omega)(\omega,\gamma), \quad g \in L^\infty(\Omega,\nu),\ \xi \in L^2(\Omega,\nu)\otimes\ell^2(\Gamma).$$

The crossed product $L^\infty(\Omega,\nu) \rtimes_{\tilde{\alpha}} \Gamma$ is the von Neumann algebra generated by $\pi(L^\infty(\Omega,\nu))$ and $\{1\otimes\lambda(\gamma)\}_{\gamma\in\Gamma}$.

THEOREM 4.1. *The crossed product $L^\infty(\Omega,\nu) \rtimes_{\tilde{\alpha}} \Gamma$ is a concrete realization of the von Neumann algebra structure of $H^\infty(B(\ell^2(\Gamma)), Q_\mu)$.*

Although the proof is very easy, to state it rigorously we need to use a purely operator algebraic language. Therefore, we make the only exception in this note here and we will use basic results in operator algebras in the proof freely.

PROOF. Let $\mathcal{E}$ be the unique normal conditional expectation from $B(\ell^2(\Gamma))$ onto $\ell^\infty(\Gamma)$ [**22**, Chapter IX, Section 4], which is the projection onto the diagonal part with respective to the orthonormal basis $\{\delta_\gamma\}_{\gamma\in\Gamma}$. Then $\mathcal{E}$ commutes with $\mathrm{Ad}(\rho(\gamma))$ and $\mathrm{Ad}(\lambda(\gamma))$, where $\mathrm{Ad}(u)(x) = uxu^*$ for a unitary $u \in B(\ell^2(\Gamma))$. Therefore it is easy to show

$$\mathcal{E}(H^\infty(B(\ell^2(\Gamma)), Q_\mu)) = H^\infty(\Gamma,\mu),$$

$$\mathcal{E}(f_1\circ x\circ f_2) = f_1\circ\mathcal{E}(x)\circ f_2, \quad \forall f_1, f_2 \in H^\infty(\Gamma,\mu),\ \forall x \in H^\infty(B(\ell^2(\Gamma)), Q_\mu).$$

For $x \in H^\infty(B(\ell^2(\Gamma)), Q_\mu)$ and $\gamma\in\Gamma$, we define

$$x_\gamma = \mathcal{E}(\lambda(\gamma)^*x) \in H^\infty(\Gamma,\mu).$$

Our basic idea is to use the formal expansion

$$x\text{``}=\text{''}\sum_{\gamma\in\Gamma}\lambda(\gamma)x_\gamma,$$

though it is well-known that the series does not converge in any reasonable operator algebra topology in general.

Let φ be a faithful normal state of $H^\infty(B(\ell^2(\Gamma)),Q_\mu)$ (treated as a von Neumann algebra) obtained by composing $\mathcal{E}$ and the point evaluation at $e\in\Gamma$. We denote by $(\pi_\varphi,H_\varphi,\xi_\varphi)$ the GNS triple for φ [**21**, Chapter I, Section 9]. Since φ is a faithful normal state, π_φ is a faithful normal representation, and so it suffices to show that $\pi_\varphi(H^\infty(B(\ell^2(\Gamma)))$ is isomorphic to the crossed product. Simple computation yields the following: for every $f_1,f_2\in H^\infty(\Gamma,\mu)$ and every $\gamma_1,\gamma_2\in\Gamma$

$$\begin{aligned}\langle\pi_\varphi(\lambda(\gamma_1)f_1)\xi_\varphi,\pi_\varphi(\lambda(\gamma_2)f_2)\xi_\varphi\rangle &= \varphi(f_2^*\circ\lambda(\gamma_2^{-1}\gamma_1)\circ f_1)\\ &= \varphi(f_2^*\circ\mathcal{E}(\lambda(\gamma_2^{-1}\gamma_1))\circ f_1)\\ &= \delta_{\gamma_1,\gamma_2}\varphi(f_2^*\circ f_1)\\ &= \delta_{\gamma_1,\gamma_2}\int_\Omega\theta(f_1)(\omega)\overline{\theta(f_2)(\omega)}d\nu(\omega).\end{aligned}$$

Therefore there exists a unitary U from H_φ onto $L^2(\Omega,\nu)\otimes\ell^2(\Gamma)$ satisfying

$$U\pi_\varphi(\lambda(\gamma)f)\xi_\varphi=\theta(f)\otimes\delta_\gamma,\quad\forall f\in H^\infty(\Gamma,\mu),\ \forall\gamma\in\Gamma.$$

It is straightforward to show

$$\mathrm{Ad}(U)\cdot\pi_\varphi(f)=\pi(\theta(f)),$$

$$\mathrm{Ad}(U)\cdot\pi_\varphi(\lambda(\gamma))=1\otimes\lambda(\gamma),$$

and so the image of $\mathrm{Ad}(U)\cdot\pi_\varphi$ contains the crossed product $L^\infty(\Omega,\nu)\rtimes_{\tilde{\alpha}}\Gamma$. Thus to prove the theorem it suffices to show that $\mathrm{Ad}(U)\cdot\pi_\varphi(x)$ is in the double commutant of the crossed product for every $x\in H^\infty(B(\ell^2(\Gamma)),Q_\mu)$ [**21**, Chapter II, Theorem 3.9].

Let v be the unitary representation of Γ on $L^2(\Omega,\nu)$ defined by

$$(v(\gamma)g)(\omega)=\sqrt{\frac{d\nu(\gamma^{-1}\cdot)}{d\nu}}(\omega)g(\gamma^{-1}\cdot\omega),\quad g\in L^2(\Omega,\nu).$$

Then it is known that the commutant of the crossed product $L^\infty(\Omega,\nu)\rtimes_{\tilde{\alpha}}\Gamma$ is generated by $L^\infty(\Omega,\nu)\otimes\mathbb{C}$ and $\{v(\gamma)\otimes\rho(\gamma)\}_{\gamma\in\Gamma}$ [**22**, Chapter X, Theorem 1.21]. We claim that for every $x\in H^\infty(B(\ell^2(\Gamma),Q_\mu)$, every $g_1,g_2\in L^\infty(\Omega,\nu)$, and every $\gamma_1,\gamma_2\in\Gamma$, the following holds (in fact, the only term with $\gamma=\gamma_2\gamma_1^{-1}$ survives in the right-hand side):

$$\begin{aligned}&\langle\mathrm{Ad}(U)\cdot\pi_\varphi(x)(g_1\otimes\delta_{\gamma_1}),g_2\otimes\delta_{\gamma_2}\rangle\\ &=\sum_{\gamma\in\Gamma}\langle\mathrm{Ad}(U)\cdot\pi_\varphi(\lambda(\gamma)x_\gamma)(g_1\otimes\delta_{\gamma_1}),g_2\otimes\delta_{\gamma_2}\rangle.\end{aligned}$$

Indeed, the right-hand side equals

$$\begin{aligned}
&\sum_{\gamma\in\Gamma}\langle(1\otimes\lambda(\gamma))\pi(\theta(x_\gamma))(g_1\otimes\delta_{\gamma_1}),g_2\otimes\delta_{\gamma_2}\rangle\\
&\quad=\sum_{\gamma\in\Gamma}\langle\pi(\theta(x_\gamma))(g_1\otimes\delta_{\gamma_1}),g_2\otimes\delta_{\gamma^{-1}\gamma_2}\rangle\\
&\quad=\sum_{\gamma\in\Gamma}\delta_{\gamma_1,\gamma^{-1}\gamma_2}\langle\tilde{\alpha}_{\gamma_1^{-1}}(\theta(x_\gamma))g_1),g_2\rangle\\
&\quad=\langle\tilde{\alpha}_{\gamma_1^{-1}}(\theta(x_{\gamma_2\gamma_1^{-1}}))g_1,g_2\rangle\\
&\quad=\varphi(\theta^{-1}(g_2)^*\circ\alpha_{\gamma_1^{-1}}(x_{\gamma_2\gamma_1^{-1}})\circ\theta^{-1}(g_1)).
\end{aligned}$$

On the other hand, the left-hand side equals

$$\begin{aligned}
&\langle\pi_\varphi(x)\pi_\varphi(\lambda(\gamma_1)\theta^{-1}(g_1))\xi_\varphi,\pi_\varphi(\lambda(\gamma_2)\theta^{-1}(g_2))\xi_\varphi\rangle\\
&\quad=\varphi(\theta^{-1}(g_2^*)\lambda(\gamma_2^{-1})\circ x\circ\lambda(\gamma_1)\theta^{-1}(g_1))\\
&\quad=\varphi(\theta^{-1}(g_2^*)\circ\lambda(\gamma_2^{-1})x\lambda(\gamma_1)\circ\theta^{-1}(g_1))\\
&\quad=\varphi(\theta^{-1}(g_2^*)\circ\mathcal{E}(\lambda(\gamma_2^{-1})x\lambda(\gamma_1))\circ\theta^{-1}(g_1))\\
&\quad=\varphi(\theta^{-1}(g_2^*)\circ\lambda(\gamma_1^{-1})\mathcal{E}(\lambda(\gamma_1\gamma_2^{-1})x)\lambda(\gamma_1)\circ\theta^{-1}(g_1))\\
&\quad=\varphi(\theta^{-1}(g_2^*)\circ\alpha_{\gamma_1^{-1}}(x_{\gamma_2\gamma_1^{-1}})\circ\theta^{-1}(g_1)),
\end{aligned}$$

which shows the claim. For every $g\in L^\infty(\Omega,\nu)$, the claim implies

$$\begin{aligned}
&\langle\mathrm{Ad}(U)\cdot\pi_\varphi(x)(g\otimes 1)(g_1\otimes\delta_{\gamma_1}),g_2\otimes\delta_{\gamma_2}\rangle\\
&\quad=\langle\mathrm{Ad}(U)\cdot\pi_\varphi(x)(gg_1\otimes\delta_{\gamma_1}),g_2\otimes\delta_{\gamma_2}\rangle\\
&\quad=\sum_{\gamma\in\Gamma}\langle\mathrm{Ad}(U)\cdot\pi_\varphi(\lambda(\gamma)x_\gamma)(gg_1\otimes\delta_{\gamma_1}),g_2\otimes\delta_{\gamma_2}\rangle\\
&\quad=\sum_{\gamma\in\Gamma}\langle\mathrm{Ad}(U)\cdot\pi_\varphi(\lambda(\gamma)x_\gamma)(g\otimes 1)(g_1\otimes\delta_{\gamma_1}),g_2\otimes\delta_{\gamma_2}\rangle\\
&\quad=\sum_{\gamma\in\Gamma}\langle(g\otimes 1)\mathrm{Ad}(U)\cdot\pi_\varphi(\lambda(\gamma)x_\gamma)(g_1\otimes\delta_{\gamma_1}),g_2\otimes\delta_{\gamma_2}\rangle\\
&\quad=\sum_{\gamma\in\Gamma}\langle\mathrm{Ad}(U)\cdot\pi_\varphi(\lambda(\gamma)x_\gamma)(g_1\otimes\delta_{\gamma_1}),g^*g_2\otimes\delta_{\gamma_2}\rangle\\
&\quad=\langle\mathrm{Ad}(U)\cdot\pi_\varphi(x)(g_1\otimes\delta_{\gamma_1}),g^*g_2\otimes\delta_{\gamma_2}\rangle\\
&\quad=\langle(g\otimes 1)\mathrm{Ad}(U)\cdot\pi_\varphi(x)(g_1\otimes\delta_{\gamma_1}),g_2\otimes\delta_{\gamma_2}\rangle,
\end{aligned}$$

which shows that $\mathrm{Ad}(U)\cdot\pi_\varphi(x)$ commutes with $(g\otimes 1)$.

For fixed $\gamma\in\Gamma$, we choose $g_1,g_2\in L^\infty(\Omega,\nu)$ satisfying $v(\gamma)g_1$ and $v(\gamma^{-1})g_2$ are in $L^\infty(\Omega,\nu)$. Note that the set all such g_1 (and also the set all such g_2) is dense in $L^2(\Omega,\nu)$. The same computation as above implies

$$\begin{aligned}
&\langle\mathrm{Ad}(U)\cdot\pi_\varphi(x)v(\gamma)(g_1\otimes\delta_{\gamma_1}),g_2\otimes\delta_{\gamma_2}\rangle\\
&\quad=\langle v(\gamma)\mathrm{Ad}(U)\cdot\pi_\varphi(x)(g_1\otimes\delta_{\gamma_1}),g_2\otimes\delta_{\gamma_2}\rangle.
\end{aligned}$$

Thus $\mathrm{Ad}(U)\cdot\pi_\varphi(x)$ commutes with $v(\gamma)$, and we are done. □

EXAMPLE 4.2. When Γ is the free group $\mathbb{F}_N$ with N generators $\{\gamma_i\}_{i=1}^N$ and μ is the uniform distribution on $\{\gamma_i,\gamma_i^{-1}\}_{i=1}^N$. Then the structure of the crossed product $L^\infty(\Omega,\nu)\rtimes_{\tilde{\alpha}}\Gamma$ is investigated in [**20**], and Theorem 4.1 implies that the

Poisson boundary for $(B(\ell^2(\Gamma)), Q_\mu)$ is the Powers factor of type $\mathrm{III}_{1/(2N-1)}$. For the case with more general groups and probability measures, see [**18**].

PROBLEM 4.3. Does the conclusion of Theorem 4.1 hold for a general second countable locally compact group Γ with a reasonably good probability measure μ on Γ ?

REMARK 4.4. In [**9**], Connes and Woods discussed the Poisson boundary for a time dependent random walk on a locally compact group in relation to an approximately transitive transformation group. For a discrete group Γ, Theorem 4.1 holds also in this case with an appropriate modification of the statement. Let $\{\mu_n\}_{n=1}^\infty$ be a sequence of probability measures on Γ. We set

$$A = \bigoplus_{n=1}^{\infty} \ell^\infty(\Gamma),$$

$$M = \bigoplus_{n=1}^{\infty} B(\ell^2(\Gamma)),$$

where the direct sums are understood in the von Neumann algebra sense. We introduce Markov operators P on A and Q on M by

$$P(f)_n = P_{\mu_n}(f_{n+1}), \quad f = (f_n) \in A,$$

$$Q(x)_n = Q_{\mu_n}(x_{n+1}), \quad x = (x_n) \in M.$$

Then the Poisson boundary for (A, P) is nothing but the one discussed in [**9**]. As in the stationary random walk case, the Poisson boundary for (M, Q) is given by the crossed product. The only difference in the proof from that of Theorem 4.1 is that one needs to use a state of the from $\psi \cdot \tilde{\mathcal{E}}$ instead of φ where

$$\tilde{\mathcal{E}} = \bigoplus_{n=1}^{\infty} \mathcal{E} : M \to A,$$

and ψ is any faithful normal state of A.

4.3. Amenable actions. In [**26**, Definition 1.4], Zimmer introduced the notion of amenability of a group action on a measure space pursuing Mackey's idea of virtual groups for ergodic transformation groups. He showed that the natural action of a group on the Poisson boundary for a random walk on it is always amenable [**26**, Theorem 5.2] (see also [**27**] for topics related to amenable group actions). Since Zimmer's definition of amenable actions has a functional analytic nature anyway, instead of giving it here we adopt his characterization of amenability in terms of the crossed products [**24**, Theorem 2.1], [**25**, Theorem], as our definition of amenable actions.

A von Neumann algebra $M \subset B(H)$ is said to be injective if there exists a projection of norm one from $B(H)$ onto M. Celebrated Connes' theorem [**8**], [**23**, Chapter XVI, Theorem 1.9] says that M is injective if and only if it is approximately finite dimensional (hyperfinite), that is, there exists an increasing sequence of finite dimensional $*$-subalgebras whose union is dense in M in the weak operator topology.

THEOREM 4.5 (Zimmer). *Let Γ be a countable discrete group acting on a measure space (S, m) as a non-singular transformation group. Then the crossed product $L^\infty(S, m) \rtimes \Gamma$ is injective if and only if (S, m) is an amenable Γ-space.*

Adopting this characterization as the definition of amenable actions, we give a purely functional analytic proof of amenability of the Poisson boundary in the discrete group case based on Theorem 4.1. The same proof works also for the time dependent random walk case (see Remark 4.4).

THEOREM 4.6. *Let the notations be as in Theorem 4.1. Then the crossed product* $L^\infty(\Omega,\nu)\rtimes_{\tilde{\alpha}}\Gamma$ *is injective.*

PROOF. We denote by Φ the map $\mathrm{Ad}(U)\cdot\pi_\varphi$ defined in the proof of Theorem 4.1, which is a unit preserving completely positive isometry from $H^\infty(B(\ell^2(\Gamma)),Q_\mu)$ onto $L^\infty(\Omega,\nu)\rtimes_{\tilde{\alpha}}\Gamma$. [**5**, Theorem 3.1] implies that Φ^{-1} is also a completely positive isometry. Thanks to Arveson's Hahn-Banach type theorem [**10**, Theorem 5.1.7], Φ^{-1} extends to a unit preserving completely positive map

$$\Psi : B(L^2(\Omega,\nu)\otimes\ell^2(\Gamma))\to B(\ell^2(\Gamma)).$$

As we saw in the proof of Theorem 3.3, there exists a completely positive projection

$$E : B(\ell^2(\Gamma))\to H^\infty(B(\ell^2(\Gamma)),Q_\mu).$$

Then $\Phi\cdot E\cdot\Psi$ is a completely positive projection from $B(L^2(\Omega,\nu)\otimes\ell^2(\Gamma))$ onto $L^\infty(\Omega,\nu)\rtimes_{\tilde{\alpha}}\Gamma$. □

References

[1] Ph. Biane, *Équation de Choquet-Deny sur le dual d'un groupe compact.* Probab. Theory Related Fields **94** (1992), 39–51.

[2] Ph. Biane, *Théorème de Ney-Spitzer sur le dual de* $SU(2)$. Trans. Amer. Math. Soc. **345** (1994), 179–194.

[3] D. Bisch, *Entropy of groups and subfactors.* J. Funct. Anal. **103** (1992), 190–208.

[4] D. Bisch; U. Haagerup, *Composition of subfactors: new examples of infinite depth subfactors.* Ann. Sci. École Norm. Sup. (4) **29** (1996), 329–383.

[5] M. D. Choi; E. Effros, *Injectivity and operator spaces.* J. Funct. Anal. **24** (1977), 156–209.

[6] M. D. Choi, *A Schwartz inequality for positive linear maps on* C^**-algebras.* Illinois J. Math. **18** (1974), 565-574.

[7] B. Collins, *Martin boundary theory of some quantum random walks.* preprint.

[8] A. Connes, *Classification of injective factors. Cases* II_1, II_∞, III_λ, $\lambda\neq 1$. Ann. of Math. (2) **104** (1976), 73–115.

[9] A. Connes; E. J. Woods, *Hyperfinite von Neumann algebras and Poisson boundaries of time dependent random walks.* Pacific J. Math. **137** (1989), 225–243.

[10] E. G. Effros; Zhong-Jin Ruan, *Operator spaces.* London Mathematical Society Monographs. New Series, 23. The Clarendon Press, Oxford University Press, New York, 2000.

[11] M. Izumi, *Non-commutative Poisson boundaries and compact quantum group actions.* Adv. Math. **169** (2002), 1–57.

[12] M. Izumi, *Non-commutative Markov Operators Arising from Subfactors.* To appear in the Proceedings of US-Japan Seminar, 1999.

[13] V. A. Kaimanovich, *Measure-theoretic boundaries of Markov chains, 0-2 laws and entropy* in "Harmonic analysis and discrete potential theory", M. A. Picardello (ed.), Plenum Press, New York, 1992, pp. 145–180.

[14] V. A. Kaimanovich, *Boundaries of invariant Markov operators: the identification problem.* in "Ergodic theory of Z^d actions" (Warwick, 1993–1994), 127–176, London Math. Soc. Lecture Note Ser., 228, Cambridge Univ. Press, Cambridge, 1996.

[15] V. A. Kaimanovich; A. M. Vershik, *Random walks on discrete groups: boundary and entropy.* Ann. Probab. **11** (1983), 457–490.

[16] Mackey, G. W. *Point realizations of transformation groups.* Illinois J. Math. **6** (1962), 327–335.

[17] S. Neshveyev; L. Tuset, *The martin boundary of a discrete quantum group.* preprint.

[18] R. Okayasu, *Type III Factors Arising from Cuntz-Krieger Algebras,* Proc. Amer. Math. Soc. **131** (2003), 2145–2153.

[19] S. Popa, *Sousfacteurs, actions des groupes et cohomologie.* C. R. Acad. Sci. Paris Sér. I Math. **309** (1989), 771–776.

[20] J. Ramagge; G. Robertson, *Factors from trees.* Proc. Amer. Math. Soc. **125** (1997), 2051–2055.

[21] M. Takesaki, *Theory of operator algebras. I.* Encyclopaedia of Mathematical Sciences, 124. Operator Algebras and Non-commutative Geometry, 5. Springer-Verlag, Berlin, 2002.

[22] M. Takesaki, *Theory of operator algebras. II.* Encyclopaedia of Mathematical Sciences, 125. Operator Algebras and Non-commutative Geometry, 6. Springer-Verlag, Berlin, 2003.

[23] M. Takesaki, *Theory of operator algebras. III.* Encyclopaedia of Mathematical Sciences, 127. Operator Algebras and Non-commutative Geometry, 8. Springer-Verlag, Berlin, 2003.

[24] R. J. Zimmer, *Hyperfinite factors and amenable ergodic actions.* Invent. Math. **41** (1977), 23–31.

[25] R. J. Zimmer, *On the von Neumann algebra of an ergodic group action.* Proc. Amer. Math. Soc. **66** (1977), 289–293.

[26] R. J. Zimmer, *Amenable ergodic group actions and an application to Poisson boundaries of random walks.* J. Funct. Anal. **27** (1978), 350–372.

[27] R. J. Zimmer, *Ergodic theory and semisimple groups.* Monographs in Mathematics, 81. Birkhauser Verlag, Basel, 1984.

DEPARTMENT OF MATHEMATICS, GRADUATE SCHOOL OF SCIENCE, KYOTO UNIVERSITY, SAKYO-KU, KYOTO 606-8502, JAPAN

E-mail address: izumi@kusm.kyoto-u.ac.jp

Contemporary Mathematics
Volume **347**, 2004

Boundary amenability of hyperbolic spaces

Vadim A. Kaimanovich

Abstract. It is well-known that a Kleinian group is amenable if and only if it is elementary. We establish an analogous property for equivalence relations and foliations with Gromov hyperbolic leaves: they are amenable if and only if they are elementary in the sense that one can assign (in a measurable way) to any leaf a finite subset of its hyperbolic boundary (as in the group case, such subsets cannot actually contain more than 2 points). The analogous result for actions of word hyperbolic groups with a quasi-invariant measure is that such an action is amenable if and only if it factorizes through the hyperbolic boundary or its symmetric square.

A byproduct of our approach is a proof of boundary amenability for isometry groups of Gromov hyperbolic spaces under the assumption that either the space is exponentially bounded or it is CAT(−1) and the group has a finite critical exponent. We also give examples showing that without these assumptions the boundary amenability may fail.

Introduction

The class of *amenable groups* is, from the analytical point of view, the most natural generalization of the class of compact groups. Amenable groups are those which admit an *invariant mean* (rather than an invariant probability measure, which is the case for compact groups). There are many different equivalent definitions of amenability. Probably, the most constructive definition (used for verifying amenability of a given group) is the one formulated in terms of existence of *approximatively invariant sequences of probability measures* on the group (*Reiter's condition*), whereas one of the main applications of amenability is the *fixed point property* for affine actions of amenable groups on compact spaces.

2000 *Mathematics Subject Classification.* Primary 37A20, 43A07, 53C99; Secondary 20F67, 28A20, 53C12.

Key words and phrases. Amenability, hyperbolic space, boundary, equivalence relation, measurability, foliation, transversal.

The paper was mostly written during author's stay at the Max-Planck-Institut (Bonn) in 2002. Its financial support and the excellent working conditions are gratefully acknowledged. I would also like to thank the organizers of the First JAMS Symposium "Discrete Geometric Analysis", where these results were presented, and, in particular, Professor Motoko Kotani, for creating an ideal atmosphere at this very fruitful meeting. I am grateful to Theo Bühler and to the anonymous referee for a number of useful comments and suggestions.

It turns out that non-amenable groups may still have actions which look like actions of amenable groups. This observation led Zimmer [**Zim77**], [**Zim78**] to introduce the notion of an *amenable action*. In the same way as with groups, there are several definitions of an amenable action. In particular, amenable actions can be characterized both in terms of a fixed point property (this was the original definition of Zimmer) and in terms of existence of a sequence of approximatively equivariant maps from the action space to the space of probability measures on the group (this is an analogue of Reiter's condition). Yet another generalization is the notion of amenability for *equivalence relations* and *foliations*. Actually, all these objects can be considered as *measured groupoids*, and the notion of amenability in each particular case is a specialization of a general notion of an *amenable groupoid*, see [**ADR00**], [**CHLI02**].

It is well-known that a Kleinian group is *amenable* if and only if it is *elementary*. We establish an analogous property for *equivalence relations and foliations with Gromov hyperbolic leaves*: under a suitable bounded geometry condition they are amenable if and only if they are elementary in the sense that one can assign (in a measurable way) to any leaf a finite subset of its hyperbolic boundary (actually, such subsets cannot consist of more than 2 points). As particular cases it contains and generalizes earlier results on *amenability of boundary actions* of isometry groups of hyperbolic spaces and on *amenability of the stable foliation* of the geodesic flow.

The main geometric ingredients of our approach — (i) *boundary convergence of geodesics* in hyperbolic spaces (in various guises), and (ii) the notion of the *barycenter* for hyperbolic spaces — are well-known and have been used before (e.g., see [**Ada96**], [**BM96**]). However, working with hyperbolic equivalence relations requires developing a new technique (based on the notion of a *measurable bundle of Banach spaces*) which allows one to deal with *measurable systems of* (possibly different) *leafwise hyperbolic boundaries*. Another technical novelty in this paper is the use of simple explicit constructions of approximatively invariant sequences of measures as *Cesaro averages for sandwiched sequences of sets with subexponential growth* (Theorem 1.33 and Lemma 1.35) and *Cesaro averages of pre-Patterson measures along geodesics* (Theorem 1.38).

The paper has the following structure. In Section 1 we collect the geometric properties of Gromov hyperbolic spaces used in the sequel. These are Proposition 1.3 on boundary convergence of geodesic rays, Proposition 1.18 on existence of (quasi)-barycenter sets and Theorems 1.33, 1.38 on existence of approximatively invariant boundary maps.

The main technical tool used to ensure measurability of leafwise application of these constructions in the setup of graphed equivalence relations with hyperbolic leaves is the measurable Banach bundle of leafwise functions with a continuous boundary extension (Section 2.1). By using this bundle we define measurable systems of boundary measures and show that a hyperbolic equivalence relation admits an invariant boundary system of probability measures if and only if it is elementary (Theorem 2.22). Further we establish an analogous result for measurable foliations (Theorem 2.31) by reducing them to equivalence relations.

We begin Section 3 with illustrating the key geometric ingredients of the relationship between amenability and hyperbolicity, namely, boundary convergence of geodesics and existence of quasi-barycenter sets, on the example of coincidence of the class of amenable groups of isometries of Gromov hyperbolic spaces with the class of elementary groups (under suitable bounded geometry type conditions). Further we proceed to establish analogues of this result for equivalence relations, foliations and actions (Theorem 3.4, Theorem 3.9 and Theorem 3.14, respectively) by taking stock of the results from the previous sections and following precisely the same scheme: amenability by the fixed point property implies existence of an invariant system of boundary measures, and therefore elementarity, whereas elementarity leads to existence of approximatively invariant sequences of probability measures, i.e., to amenability. As a byproduct we give a proof of the topological amenability of the boundary action of the group of isometries of a hyperbolic space under the assumption that either the space is exponentially bounded or the space is CAT(-1) and the group has a finite critical exponent (Theorem 3.15 and Theorem 3.17). Finally, we give examples showing that without these assumptions the boundary amenability (even in the weakest form) may well fail.

1. Asymptotic geometry of hyperbolic spaces

1.1. Hyperbolic spaces. A detailed discussion of the notion of δ-hyperbolicity and of the associated structures can be found in the seminal work of Gromov [**Gro87**] and in the notes [**GdlH90**]. Below are listed some of the definitions and properties used later on.

We shall choose the definition based on the *Rips condition*: a non-compact complete proper geodesic metric space $\mathcal{X}$ is *δ-hyperbolic* (with $\delta \geq 0$) if each of the sides of any geodesic triangle is contained in the δ-neighbourhood of the union of the other two sides (see, for instance, [**GdlH90**, Proposition 2.21] for a list of other equivalent definitions). The minimal number δ with this property is the *hyperbolicity constant* of $\mathcal{X}$. A graph is called δ-hyperbolic if the associated 1-complex with length 1 edges is δ-hyperbolic. Usually, we shall not be concerned with the precise value of the hyperbolicity constant, and call the above spaces just *hyperbolic*.

For the rest of this Section we shall fix a hyperbolic space $\mathcal{X}$ with metric d and the hyperbolicity constant δ. Denote by

$$(y|z)_x = \frac{1}{2}\left[d(x,y) + d(x,z) - d(y,z)\right] , \qquad x, y, z \in \mathcal{X}$$

the *Gromov product* on $\mathcal{X}$. Then by [**GdlH90**, Lemme 2.17 and Proposition 2.21] for any geodesic segment $[y,z]$ joining the points y and z

$$d(x,[y,z]) - 4\delta \leq (y|z)_x \leq d(x,[y,z]) \ . \tag{1.1}$$

The functions on $\mathcal{X} \times \mathcal{X}$

$$\varrho_x(y,z) = \begin{cases} 0 \, , & y = z \, , \\ e^{-(y|z)_x} \, , & \text{otherwise} \, , \end{cases} \tag{1.2}$$

determine a uniform structure "at infinity" on the space $\mathcal{X}$ (which does not depend on the choice of the point x). The completion $\overline{\mathcal{X}}$ of $\mathcal{X}$ with respect to this uniform structure is called the *hyperbolic compactification* of $\mathcal{X}$. Its boundary is denoted by $\partial\mathcal{X} = \overline{\mathcal{X}} \setminus \mathcal{X}$. The action of the group of isometries $\mathrm{Iso}(\mathcal{X})$ extends from $\mathcal{X}$ to a continuous action on $\partial\mathcal{X}$.

Any geodesic ray ξ converges in the hyperbolic compactification to a boundary point $\xi(\infty) \in \partial\mathcal{X}$. The definition of δ-hyperbolicity is easily seen to imply the following property (which is, for instance, essentially contained in [**GdlH90**, Proposition 7.2]):

PROPOSITION 1.3. *There exists a constant* $C_1 = C_1(\delta)$ *with the property that for any* δ*-hyperbolic space* $\mathcal{X}$ *and any two geodesic rays* ξ_1, ξ_2 *in* $\mathcal{X}$ *with* $\xi_1(\infty) = \xi_2(\infty)$ *there is a number* $T = T(\xi_1, \xi_2)$ *such that* $|T| \le d(\xi_1(0), \xi_2(0))$ *and*

$$d(\xi_1(t), \xi_2(t+T)) \le C_1 \qquad \forall t \ge d(\xi_1(0), \xi_2(0)) .$$

Although the functions ϱ_x (1.2) do not, in general, satisfy the triangle inequality, they are "almost" metrics. Namely, for $a = \frac{1}{15\delta}$ the inner metric

$$\rho_x(y,z) = \inf \left\{ \sum_{i=1}^{n} [\varrho_x(x_{i-1}, x_i)]^a : n \ge 1 \,,\; y = x_0, x_1, \dots, x_n = z \in \mathcal{X} \right\} \tag{1.4}$$

determined by the function ϱ_x^a satisfies the inequality

$$\frac{1}{2}(\varrho_x)^a \le \rho_x \le (\varrho_x)^a \tag{1.5}$$

(this is an adaptation of [**Woe00**, Proposition 22.8]). So, the hyperbolic compactification $\overline{\mathcal{X}}$ is the completion of $\mathcal{X}$ with respect to any of the metrics ρ_x.

NOTATION. Below we shall use the notation $B(o,r)$ (resp., $S(o,r)$) for the r-ball (resp., the r-sphere) in $\mathcal{X}$ of the hyperbolic metric d around a point $o \in \mathcal{X}$, and $B_x(o,r)$ (resp., $S_x(o,r)$) for the r-ball (resp., the r-sphere) in $\overline{\mathcal{X}}$ of the metric ρ_x around a point $o \in \overline{\mathcal{X}}$. We denote the space of probability measures on a topological space X by $\mathcal{P}(X)$, and the normalized restriction of a measure m to a positive measure set A by m_A.

1.2. The barycenter set of a boundary measure. Denote by

$$\beta_z(x,y) = d(y,z) - d(x,z) = d(x,y) - 2(y|z)_x \qquad x, y \in \mathcal{X} \tag{1.6}$$

the *distance cocycle* associated with a point $z \in \mathcal{X}$. The triangle inequality for the metric d and formula (1.5) imply

PROPOSITION 1.7. *For any* $x, y, z \in \mathcal{X}$ *the distance cocycle satisfies the inequalities*

$$|\beta_z(x,y)| \le d(x,y) ,$$

and (assuming $y \ne z$*)*

$$d(x,y) + \frac{2}{a}\log \rho_x(y,z) \le \beta_z(x,y) \le d(x,y) + \frac{2}{a}\log \rho_x(y,z) + C_2 ,$$

where

$$C_2 = C_2(\delta) = \frac{2\log 2}{a} = (30\log 2)\delta .$$

COROLLARY 1.8. *For any* $\gamma \in \partial\mathcal{X}$ *and* $x, y \in \mathcal{X}$

$$\limsup_{z\to\gamma} \beta_z(x,y) - \liminf_{z\to\gamma} \beta_z(x,y) \le C_2 .$$

Put

$$\overline{\beta}_\gamma(x,y) = \limsup_{z\to\gamma} \beta_z(x,y) \qquad x,y \in \mathcal{X},\ \gamma \in \partial\mathcal{X} .$$

As it follows from (1.6), the functions $\overline{\beta}_\gamma$ are Lipschitz with respect to each of the arguments. Although $\overline{\beta}_\gamma$ are not, generally speaking, cocycles, they still satisfy the cocycle identity with a uniformly bounded error (i.e., they are *quasi-cocycles*). More precisely, Proposition 1.7 and Corollary 1.8 imply

PROPOSITION 1.9. *For any $\gamma \in \partial\mathcal{X}$ the function $\overline{\beta}_\gamma$ has the following properties:*

(i) *$\overline{\beta}_\gamma$ is "jointly Lipschitz", i.e., $|\overline{\beta}_\gamma(x,y)| \le d(x,y)$ for all $x, y \in \mathcal{X}$, in particular, $\overline{\beta}_\gamma(x,x) \equiv 0$;*

(ii) *$0 \le \overline{\beta}_\gamma(x,y) + \overline{\beta}_\gamma(y,x) \le C_2$ for all $x, y \in \mathcal{X}$;*

(iii) *$0 \le \overline{\beta}_\gamma(x,y) + \overline{\beta}_\gamma(y,z) + \overline{\beta}_\gamma(z,x) \le 2C_2$ for all $x, y, z \in \mathcal{X}$.*

REMARK 1.10. Our definition of $\overline{\beta}_\gamma$ is different from the definition of the "Busemann function" in [**GdlH90**, Chapitre 8] (where the point z was allowed to converge to γ along geodesic rays only).

For any probability measure $\lambda \in \mathcal{P}(\partial\mathcal{X})$ put

$$\mathcal{B}_\lambda(x,y) = \int \overline{\beta}_\gamma(x,y)\, d\lambda(\gamma) \,.$$

The Lipschitz property of the functions β_γ implies that $\mathcal{B}_\lambda$ are also Lipschitz with respect to each of the arguments. Moreover, as it follows from Proposition 1.9, the functions $\mathcal{B}_\lambda$ are jointly Lipschitz, and they are quasi-cocycles with the same constants as in Proposition 1.9. In particular,

$$(1.11) \qquad \big|\mathcal{B}_\lambda(x,y) - \mathcal{B}_\lambda(x',y) - \mathcal{B}_\lambda(x,x')\big| \le 3C_2 \qquad \forall\, x, x', y \in \mathcal{X},\ \lambda \in \mathcal{P}(\partial\mathcal{X}) \,.$$

PROPOSITION 1.12. *If a sequence of probability measures $\lambda_n \in \mathcal{P}(\mathcal{X})$ is convergent in the weak* topology to a measure $\lambda \in \mathcal{P}(\mathcal{X})$, then*

$$\left|\mathcal{B}_\lambda(x,y) - \limsup_{n\to\infty} \int \beta_z(x,y)\, d\lambda_n(z)\right| \le C_2 \qquad \forall\, x, y \in \mathcal{X} \,.$$

PROOF. As it follows from Proposition 1.7, if $x \ne y$ then both terms in the left-hand side belong to the interval

$$\left[d(x,y) + \frac{2}{a}\int \log \rho_x(y,\gamma)\, d\lambda(\gamma),\ d(x,y) + \frac{2}{a}\int \log \rho_x(y,\gamma)\, d\lambda(\gamma) + C_2\right] \,.$$

□

PROPOSITION 1.13. *If a measure $\lambda \in \mathcal{P}(\partial\mathcal{X})$ does not have atoms of weight at least $\frac{1}{2}$, then for any $x \in X$*

$$\lim_y \mathcal{B}_\lambda(x,y) = \infty$$

if $y \to \infty$ in the space $\mathcal{X}$ (i.e., $d(x,y) \to \infty$).

PROOF. Without loss of generality we may assume that $y \to \gamma_0 \in \partial\mathcal{X}$. Choose $\varepsilon > 0$ in such a way that $\lambda B_x(\gamma_0, \varepsilon) < \frac{1}{2}$, and, by using the inequalities from Proposition 1.7 (which are satisfied for the functions $\overline{\beta}_\gamma$ as well), estimate $\overline{\beta}_\gamma(x,y)$ from below as

$$\overline{\beta}_\gamma(x,y) \ge \begin{cases} -d(x,y) \,, & \gamma \in B_x(\gamma_0, \varepsilon) \,, \\ d(x,y) + \frac{2}{a}\log \rho_x(y,\gamma) \,, & \gamma \in \partial\mathcal{X} \setminus B_x(\gamma_0,\varepsilon) \,. \end{cases}$$

Since $y \to \gamma_0$, we may assume that $\rho_x(y,\gamma) \ge \varepsilon/2$ for any $\gamma \in \partial\mathcal{X} \setminus B_x(\gamma_0, \varepsilon)$, whence integrating the above estimate yields the claim. □

REMARK 1.14. If $\mathcal{X}$ is a CAT(-1) space, then all functions $\overline{\beta}_\gamma$ and $\mathcal{B}_\lambda$ are actually cocycles, so that under conditions of Proposition 1.13 the function $\mathcal{B}_\lambda(x,\cdot)$ attains its minimum on a compact subset of $\mathcal{X}$ (the *barycenter set* of λ) which does not depend on the choice of the point $x \in \mathcal{X}$, see [**DE86**], [**BCG96**], [**BM96**]. Due to the usual technical complications connected with replacing the CAT(-1) property with δ-hyperbolicity, this is no longer true for a general δ-hyperbolic space, which is why we have to "quasify" the definition of the barycenter set.

DEFINITION 1.15. Given a number $r > 0$ the set

$$\mathcal{C}(\lambda, x, r) = \left\{ y \in \mathcal{X} : \mathcal{B}_\lambda(x,y) \le r + \inf_{z\in\mathcal{X}} \mathcal{B}_\lambda(x,z) \right\}$$

is called the *quasi-barycenter set* of the measure $\lambda \in \mathcal{P}(\partial\mathcal{X})$ with respect to the point $x \in \mathcal{X}$.

Inequality (1.11) and Proposition 1.13 imply

PROPOSITION 1.16. *If a measure $\lambda \in \mathcal{P}(\partial\mathcal{X})$ does not have atoms of weight at least $\frac{1}{2}$, then the infimum in Definition 1.15 is finite, the quasi-barycenter set $\mathcal{C}(\lambda, x, r)$ is compact for any $x \in \mathcal{X}$ and $r > 0$, and*

$$\mathcal{C}(\lambda, x, r) \subset \mathcal{C}(\lambda, x', r + 6C_2) \qquad \forall\, x, x' \in \mathcal{X} .$$

DEFINITION 1.17. We shall call the closure

$$\mathcal{C}(\lambda, r) = \overline{\bigcup_{x\in\mathcal{X}} \mathcal{C}(\lambda, x, r)} ,$$

the *quasi-barycenter set* of the measure $\lambda \in \mathcal{P}(\partial\mathcal{X})$.

Proposition 1.16 then implies (cf. [**Ada96**, Proposition 5.1]):

PROPOSITION 1.18. *If a measure $\lambda \in \mathcal{P}(\partial\mathcal{X})$ does not have atoms of weight at least $\frac{1}{2}$, then for any $r > 0$ the quasi-barycenter set $\mathcal{C}(\lambda, r)$ is compact, and the map $\lambda \mapsto \mathcal{C}(\lambda, r)$ is* Iso$(\mathcal{X})$*-equivariant.*

1.3. Bounded geometry conditions. Below we shall use several bounded geometry type conditions on a metric space $\mathcal{X}$.

DEFINITION 1.19. We say that a Radon measure m on a metric space $\mathcal{X}$ is (r_1, r_2)*-tempered* (with $0 < r_1 \le r_2 \le \infty$) if for any $r \in [r_1, r_2]$ the measures of closed r-balls in $\mathcal{X}$ are uniformly bounded from above and bounded away from 0, i.e., there exist constants $0 < \underline{C}(r) \le \overline{C}(r) < \infty$, $r \in [r_1, r_2]$ such that

$$\underline{C}(r) \le mB(x, r) \le \overline{C}(r) \qquad \forall\, x \in \mathcal{X} .$$

A measure is *tempered* if it is (r, ∞)-tempered for a certain r. A metric space $\mathcal{X}$ is *tempered* if it carries an Iso$(\mathcal{X})$-invariant tempered measure.

DEFINITION 1.20 (cf. [**Ada96**]). A metric space $\mathcal{X}$ is *exponentially ρ-bounded* (with $\rho > 0$) if there exists a constant $a = a(\rho)$ such that, for every $x \in \mathcal{X}$ and every $r > 0$, the ball $B(x, r)$ in $\mathcal{X}$ can contain at most a^r pairwise disjoint balls of radius ρ. We shall say that $\mathcal{X}$ is *exponentially bounded* if it is exponentially ρ-bounded for a certain ρ.

REMARK 1.21. Our terminology differs from that of [**Ada96**]. What we call "exponentially bounded" is "at most exponential" in [**Ada96**].

THEOREM 1.22. *A proper geodesic metric space $\mathcal{X}$ is exponentially bounded if and only if it is tempered.*

The proof consists in checking two following easy claims.

PROPOSITION 1.23. *If a geodesic metric space $\mathcal{X}$ carries a (r_1, r_2)-tempered measure with $r_2/r_1 > 3$, then it is exponentially r_1-bounded.*

PROOF. For simplicity put $\rho = r_1$ and $\varepsilon = r_2 - 3r_1$. For any $r \geq \rho$ let N_r be the supremum (over $x \in \mathcal{X}$) of cardinalities of disjoint systems of ρ-balls in the balls $B(x,r)$. We shall show that the growth of N_r is at most exponential.

Let $\{x_i\}_{i=1}^N$ be a maximal system of points in the ball $B(x,r)$ such that the balls $B(x_i, \rho)$ are pairwise disjoint. Since this system is maximal, for any point $x' \in B(x, r-\rho)$ there is a point x_i with $d(x', x_i) \leq 2\rho$, which means that 2ρ-balls centered at the points x_i cover the ball $B(x, r-\rho)$, and, therefore, r_2-balls centered at x_i cover the ball $B(x, r+\varepsilon)$ for any $\varepsilon > 0$. Thus,

$$mB(x, r+\varepsilon) \leq N\overline{C}(r_2) \leq N_r\overline{C}(r_2) .$$

On the other hand, the cardinality of any disjoint system of ρ-balls in $B(x, r+\varepsilon)$ obviously does not exceed the ratio $mB(x, r+\varepsilon)/\underline{C}(\rho)$, whence

$$\frac{N_{r+\varepsilon}}{N_r} \leq \frac{\overline{C}(r_2)}{\underline{C}(r_1)} ,$$

which implies the claim. □

PROPOSITION 1.24 (cf. [**Ada96**, Section 6.2]). *If a geodesic metric space $\mathcal{X}$ is exponentially ρ-bounded, then it carries a $(9\rho, \infty)$-tempered measure. Moreover, if $\mathcal{X}$ is proper, then such a measure can be chosen to be* $\mathrm{Iso}(\mathcal{X})$*-invariant.*

PROOF. Without the requirement of $\mathrm{Iso}(\mathcal{X})$-invariance the measure in question can be easily taken to be, for instance, the sum of δ-measures on a maximal 3ρ-separated system of points in $\mathcal{X}$. However, in order to obtain an $\mathrm{Iso}(\mathcal{X})$-invariant measure a little bit more work has to be done.

Since the space $\mathcal{X}$ is proper, the group $\mathrm{Iso}(\mathcal{X})$ is locally compact with respect to the pointwise convergence topology. For any orbit $\mathcal{O} = \mathrm{Iso}(\mathcal{X})o$, $o \in \mathcal{X}$, choose the left-invariant Haar measure $\mathfrak{m}_{\mathcal{O}}$ on $\mathrm{Iso}(\mathcal{X})$ normalized by the condition

$$\mathfrak{m}_{\mathcal{O}}\{g : d(o, go) \leq 6\rho\} = 1 , \tag{1.25}$$

and denote by $m_{\mathcal{O}}$ the image of $\mathfrak{m}_{\mathcal{O}}$ under the map $g \mapsto go$. Obviously, the measure $m_{\mathcal{O}}$ is $\mathrm{Iso}(\mathcal{X})$-invariant, it does not depend on the choice of the point o from the orbit $\mathcal{O}$, and

$$m_{\mathcal{O}}B(x, 6\rho) = 1 \qquad \forall x \in \mathcal{O} .$$

Now take a maximal system of $\mathrm{Iso}(\mathcal{X})$-orbits $\mathcal{O}_i$ in $\mathcal{X}$ with the property that the pairwise distances between different orbits from this system are at least 3ρ. The measure $m = \sum_i m_{\mathcal{O}_i}$ is clearly $\mathrm{Iso}(\mathcal{X})$-invariant. We claim that it is $(9\rho, \infty)$-tempered.

First of all, since the system $\{\mathcal{O}_i\}$ is maximal, any point $x \in \mathcal{X}$ lies within the distance 3ρ from at least one of the orbits $\mathcal{O}_i$. Therefore, the ball $B(x, 9\rho)$ contains a 6ρ-ball centered at a point $x' \in \mathcal{O}_i$, so that

$$mB(x, 9\rho) \geq m_{\mathcal{O}_i} B(x', 6\rho) \geq 1$$

by the definition of the measures $m_{\mathcal{O}_i}$.

In order to obtain an upper estimate for the measures $mB(x,r)$ first notice that the exponential ρ-boundedness of the space $\mathcal{X}$ implies that the ball $B(x,r)$ can be covered by at most e^{ar} balls $B(x_i, 3\rho)$ (see the beginning of the proof of Proposition 1.3). Thus, for any $\mathrm{Iso}(\mathcal{X})$-orbit $\mathcal{O}$ the intersection $B(x,r) \cap \mathcal{O}$ can be covered by at most e^{ar} balls of radius 6ρ centered at points from $\mathcal{O}$. Therefore,

$$m_{\mathcal{O}} B(x,r) \leq e^{ar} \ . \tag{1.26}$$

It remains to notice that the ball $B(x,r)$ can only intersect a uniformly bounded number of orbits $\mathcal{O}_i$. Indeed, choose a point x_i on each orbit $\mathcal{O}_i$ intersecting $B(x,r)$. Then the pairwise distances between the points x_i are at least 3ρ by the definition of the system $\{\mathcal{O}_i\}$. Therefore, the ρ-balls centered at the points x_i are all pairwise disjoint and contained in $B(x, r+\rho)$, so that by the exponential ρ-boundedness of the space $\mathcal{X}$ the number of these balls (and therefore of the orbits $\mathcal{O}_i$ intersecting $B(x,r)$) does not exceed $e^{a(r+\rho)}$. Combining this estimate with the inequality (1.26) gives

$$mB(x,r) \leq e^{a(2r+\rho)} \ .$$

□

Another bounded geometry type condition which can be imposed on a group of isometries of a metric space is that of *finiteness of the critical exponent.*

DEFINITION 1.27 ([**BM96**]). Given a positive Radon measure m on a metric space $\mathcal{X}$, the number (possibly infinite!)

$$\delta_{cr}(m) = \inf \left\{ \delta \geq 0 : \int_{\mathcal{X}} e^{-\delta d(o,x)} \, dm(x) < \infty \right\} \tag{1.28}$$

is independent of the choice of the reference point $o \in \mathcal{X}$, and is called the *critical exponent* of the measure m. The critical exponent $\delta_{cr}(G)$ of a closed subgroup $G \subset \mathrm{Iso}(\mathcal{X})$ is by definition the critical exponent of a positive G-invariant measure supported on a G-orbit in $\mathcal{X}$. Equivalently, for any two points $o, x \in \mathcal{X}$

$$\delta_{cr}(G) = \inf \left\{ \delta \geq 0 : \int_G e^{-\delta d(o,gx)} \, d\mathfrak{m}(g) < \infty \right\} ,$$

where $\mathfrak{m}$ is a left-invariant Haar measure on G.

Generally speaking, finiteness of the critical exponent of the group $\mathrm{Iso}(\mathcal{X})$ does not imply exponential boundedness of the space $\mathcal{X}$, because the orbits of $\mathrm{Iso}(\mathcal{X})$ may be "too small" to provide any information about the whole space $\mathcal{X}$. However, the converse implication is easily seen to be true:

PROPOSITION 1.29. *If the space $\mathcal{X}$ is exponentially bounded, then the critical exponent of any closed subgroup $G \subset \mathrm{Iso}(\mathcal{X})$ is finite.*

PROOF. Fix a G-invariant measure m on the G-orbit of a reference point $o \in \mathcal{X}$. By changing variables the integral (1.28) from the definition of the critical exponent can be presented as

$$\begin{aligned}\int_{\mathcal{X}} e^{-\delta d(o,x)}\, dm(x) &= \int_0^1 m\left\{x : e^{-\delta d(o,x)} \geq t\right\} dt \\ &= \delta \int_0^\infty mB(o,r)\, e^{-\delta r}\, dr\,.\end{aligned} \tag{1.30}$$

Since exponential boundedness of $\mathcal{X}$ implies that $mB(o,r)$ is dominated by an exponential function of r (cf. the proof of Proposition 1.24), the integral (1.30) is finite if δ is big enough. □

REMARK 1.31. In fact, it would be sufficient to look just at the full group $\mathrm{Iso}(\mathcal{X})$, because for any closed subgroup its critical exponent does not exceed that of the group $\mathrm{Iso}(\mathcal{X})$, see [**BM96**, Lemma 1.5].

REMARK 1.32. If the group $\mathrm{Iso}(\mathcal{X})$ acts on $\mathcal{X}$ cocompactly, then $\mathcal{X}$ is easily seen to be exponentially bounded (because the exponential rate of growth of the locally compact group $\mathrm{Iso}(\mathcal{X})$ is finite), which by Proposition 1.29 implies finiteness of the critical exponent of $\mathrm{Iso}(\mathcal{X})$ (cf. [**BM96**, Proposition 1.7]).

1.4. Boundary maps.

THEOREM 1.33. *For any exponentially bounded hyperbolic space $\mathcal{X}$ there exists a sequence of* $\mathrm{Iso}(\mathcal{X})$*-equivariant Borel maps $\lambda_n : \mathcal{X} \times \partial\mathcal{X} \to \mathcal{P}(\mathcal{X})$ such that*

$$\|\lambda_n(x,\gamma) - \lambda_n(x',\gamma)\| \to 0 \qquad \forall\, x, x' \in \mathcal{X},\ \gamma \in \partial\mathcal{X}\,. \tag{1.34}$$

For proving Theorem 1.33 we shall need the following simple inequality:

LEMMA 1.35. *Let $\{Z_k\}_{k=1}^\infty, \{Z'_k\}_{k=1}^\infty$ be two increasing sequences of measurable sets of finite positive measure in a measure space $(\mathcal{X}, m)$ such that for a certain integer $\tau > 0$*

$$Z_k \subset Z'_{k+\tau}\,, \qquad Z'_k \subset Z_{k+\tau} \qquad \forall\, k \geq 1\,, \tag{1.36}$$

and let

$$\lambda_n = \frac{1}{n}\sum_{k=1}^n m_{Z_k}\,, \qquad \lambda'_n = \frac{1}{n}\sum_{k=1}^n m_{Z'_k}$$

be the Cesaro averages of the sequences $\{m_{Z_k}\}, \{m_{Z'_k}\}$, respectively. Then

$$\|\lambda_n - \lambda'_n\| \leq \frac{2\tau}{n} + \frac{4(n-\tau)}{n}\left[1 - \left(\frac{mZ_1}{mZ_{n+\tau}}\right)^{\frac{2\tau}{n-\tau}}\right] \qquad \forall\, n > \tau\,.$$

PROOF. First notice that for any $k > \tau$

$$\|m_{Z_{k-\tau}} - m_{Z'_k}\| = 2\left(1 - \frac{mZ_{k-\tau}}{mZ'_k}\right) \leq 2\left(1 - \frac{mZ_{k-\tau}}{mZ_{k+\tau}}\right)\,,$$

and the same inequality holds for $\|m_{Z_{k-\tau}} - m_{Z_k}\|$, so that

$$\|m_{Z_k} - m_{Z'_k}\| \leq 4\left(1 - \frac{mZ_{k-\tau}}{mZ_{k+\tau}}\right) \qquad \forall\, k > \tau\,.$$

Then

$$\|\lambda_n - \lambda'_n\| \le \frac{1}{n}\sum_{k=1}^{n} \|m_{Z_k} - m_{Z'_k}\| \le \frac{2\tau}{n} + \frac{4}{n}\sum_{k=\tau+1}^{n}\left(1 - \frac{mZ_{k-\tau}}{mZ_{k+\tau}}\right)$$

$$\le \frac{2\tau}{n} + \frac{4(n-\tau)}{n}\left[1 - \left(\prod_{k=\tau+1}^{n}\frac{mZ_{k-\tau}}{mZ_{k+\tau}}\right)^{\frac{1}{n-\tau}}\right],$$

whence the claim. □

COROLLARY 1.37. *If in Lemma 1.35 the growth of the sets* Z_n *with respect to the measure* m *is subexponential, i.e.,* $(mZ_n)^{1/n} \to 1$, *then* $\|\lambda_n - \lambda'_n\| \to 0$.

PROOF OF THEOREM 1.33. Given points $x \in \mathcal{X}, \gamma \in \partial\mathcal{X}$ and positive integers $n \ge k \ge 1$ put

$$Y(x,\gamma,n,k) = \{\xi(n) : \xi \in \mathcal{R},\ d(x,\xi(0)) \le k,\ \xi(\infty) = \gamma\},$$

where $\mathcal{R}$ is the set of all geodesic rays in $\mathcal{X}$. We shall fix a (r,∞)-tempered Iso($\mathcal{X}$)-invariant measure m on $\mathcal{X}$ provided by Theorem 1.22, and denote by $Z(x,\gamma,n,k)$ the closed r-neighbourhood of $Y(x,\gamma,n,k)$.

As it follows from Proposition 1.3, for any ray η joining x and γ the set $Z(x,\gamma,n,k)$ is contained in the closed (C_1+r)-neighbourhood of the geodesic segment $\eta([n-k,n+k])$. Therefore, since the measure m is (r,∞)-tempered, the values $mZ(x,\gamma,n,k)$ are bounded away from 0 and uniformly bounded from above by an affine function of k.

For any $\tau \ge d(x,x')$ the finite sequences of sets

$$\{Z(x,\gamma,n,k)\}_{k=1}^{n} \qquad \text{and} \qquad \{Z(x',\gamma,n,k)\}_{k=1}^{n}$$

are τ-sandwiched in the sense of formula (1.36), whence, by putting

$$\lambda_n(x,\gamma) = \frac{1}{n}\sum_{k=1}^{n} m_{Z(x,\gamma,n,k)}$$

and applying Lemma 1.35 we obtain the claim. □

THEOREM 1.38. *If* G *is a closed group of isometries of a* CAT(-1) *space* $\mathcal{X}$ *with finite critical exponent, then there exists a sequence of* G*-equivariant Borel maps* $\lambda_n : \mathcal{X} \times \partial\mathcal{X} \to \mathcal{P}(\mathcal{X})$ *such that*

$$\|\lambda_n(x,\gamma) - \lambda_n(x',\gamma)\| \to 0 \qquad \forall\, x,x' \in \mathcal{X},\ \gamma \in \partial\mathcal{X}\,. \tag{1.39}$$

PROOF. Let us fix a number $\delta > \delta_{cr}$, a reference point $o \in \mathcal{X}$ and a G-invariant Radon measure m on the orbit Go. For any $x \in \mathcal{X}$ denote by $\nu_x = \nu_x^\delta$ the probability measure on the orbit Go with the density

$$\frac{e^{-\delta d(x,y)}}{\int_{\mathcal{X}} e^{-\delta d(x,y)}\,dm(y)}$$

with respect to the measure m. We shall call the measures ν_x *pre-Patterson* as their limits when δ tends to the critical exponent δ_{cr} provide the famous Patterson measures on the boundary $\partial\mathcal{X}$, see [**Pat76**], [**BM96**]. Obviously, the map $x \mapsto \nu_x$ is G-equivariant. This fact, together with the local Lipschitz continuity of the map $x \mapsto \nu_x$ and boundary convergence of geodesics (guaranteed by the CAT(-1) condition) yields the proof.

Since for any $x, x', y \in \mathcal{X}$

$$e^{-\delta d(x,x')} \le \frac{e^{-\delta d(x',y)}}{e^{-\delta d(x,y)}} \le e^{\delta d(x,x')} ,$$

the definition of the pre-Patterson measures implies that

$$\|\nu_x - \nu_{x'}\| \le 3\delta\, d(x,x') \tag{1.40}$$

provided $d(x,x')$ is small enough.

Now put

$$\lambda_n(x,\gamma) = \frac{1}{n}\int_0^n \nu_{\xi(t)}\, dt ,$$

where ξ is the geodesic ray joining x and γ. By the CAT(-1) property for any other point $x' \in \mathcal{X}$

$$d(\xi(t), \xi'(t+T)) \underset{t\to\infty}{\longrightarrow} 0 ,$$

where ξ' is the geodesic ray joining x' and γ, and T is the value of the Busemann cocycle $\beta_\gamma(x,x')$ (see Remark 1.14), which in combination with formula (1.40) implies the claim. □

REMARK 1.41. It can be easily seen from the proofs of Theorem 1.33 and Theorem 1.38 that for any compact subset $K \subset \mathcal{X}$ the convergence in formulas (1.34) and (1.39), respectively, is uniform on $K \times K \times \partial\mathcal{X}$.

2. Hyperbolic equivalence relations

2.1. Graphed equivalence relations. We begin with recalling the basic notions of the theory of discrete equivalence relations, see [**FM77**]. Let X be a standard Borel space. An equivalence relation R on X is called *standard* if it is a Borel subset of $X \times X$, and it is called *countable* if the equivalence class (the *leaf*) $[x] = R(x) = \{y : (x,y) \in R\}$ of any point $x \in X$ is at most countable. A standard countable equivalence relation is also called *discrete*.

PROPOSITION 2.1 (cf. [**FM77**, proof of Theorem 1]). *For any discrete equivalence relation R with infinite equivalence classes there exists a sequence $\{\alpha_n\}_{n=1}^\infty$ of measurable maps $\alpha_n : X \to X$ such that their graphs are pairwise disjoint and their union is R. In other words, for any $x \in X$ the sequence $\alpha_n(x)$ determines an ordering of the class $[x]$, and these orderings depend on x measurably.*

A standard equivalence relation R is called *non-singular* with respect to a Borel probability measure μ on X (or, equivalently, the measure μ is *quasi-invariant* with respect to R) if for any subset $A \subset X$ with $\mu(A) = 0$ its *saturation* $[A] = \bigcup_{x\in A}[x]$ also has measure 0. A non-singular equivalence relation (X,μ,R) is *ergodic* if there are no non-trivial saturated sets, it is *conservative* if all ergodic components are uncountable, and it is *dissipative* (or of type I) if any ergodic component consists of a single equivalence class, or, equivalently, there exists a leafwise constant measurable map $\varphi : X \to X$ with $(x,\varphi(x)) \in R$ for a.e. $x \in X$. In particular, dissipative ergodic equivalence relations are precisely those which consist of a single class. Any non-singular equivalence relation can be uniquely decomposed into a disjoint union of its conservative and dissipative parts.

Integrating the counting measures on the fibres of the left $(x,y) \mapsto x$ and the right $(x,y) \mapsto y$ projections from R onto X by the measure μ gives the *left* $d\mathrm{M}(x,y) = d\mu(x)$ and the *right* $d\check{\mathrm{M}}(x,y) = d\mathrm{M}(y,x) = d\mu(y)$ *counting measures* on R, respectively. The measures M and $\check{\mathrm{M}}$ are equivalent iff μ is quasi-invariant. If $\mathrm{M} = \check{\mathrm{M}}$, then the measure μ is called R-invariant.

A (non-oriented) *graph structure* on an equivalence relation (X, μ, R) is determined by a measurable subset $K \subset R$ which is symmetric and does not intersect the diagonal $\{(x, x)\} \subset R$. Two points $x, y \in X$ are then joined with an edge iff $(x, y) \in K$. We shall call (X, μ, R, K) a *graphed equivalence relation* [**Ada90**]. Denote by $[x]_K$ the equivalence class $[x]$ endowed with the graph structure K.

DEFINITION 2.2. A graphed equivalence relation (X, μ, R, K) is *hyperbolic* if a.e. leafwise graph $[x]_K$ is a hyperbolic graph (i.e., is Gromov hyperbolic with respect to the graph metric assigning length 1 to all edges). By $\overline{[x]}_K$ and $\partial[x]_K$ we denote the hyperbolic compactification of the graph $[x]_K$ and its boundary, respectively. As usual, $C(\overline{[x]}_K)$ denotes the space of continuous real valued functions on $\overline{[x]}_K$. Below we shall always deal with a fixed graph structure K and usually omit the corresponding subscript.

REMARK 2.3. The function assigning to any point $x \in X$ the hyperbolicity constant of the graph $[x]$ is easily seen to be measurable (because the hyperbolicity constant is defined in terms of geodesic triangles in $[x]$). Being leafwise constant, this function must therefore be constant on the ergodic components of R.

Examples of hyperbolic equivalence relations include, in particular,

(1) the leafwise Cayley graphs of a free action of a word hyperbolic group, see Example 2.16 for more details;
(2) equivalence relations on transversals to hyperbolic foliations, see below Section 2.6;
(3) *treed equivalence relations* (for which a.e. leafwise graph is a tree), see [**Ada90**];
(4) equivalence relations associated with certain fractal sets, see [**Kai02**].

In all these examples the hyperbolic boundaries of different equivalence classes can be naturally identified with (a subspace of) the same topological space (see the discussion of these examples at the end of Section 2.3). However, there is no reason for this to be the case in general (it would be interesting to have an explicit example of this kind), which is why we shall need a technique allowing one to deal with the general situation.

2.2. The bundle of leafwise continuous functions. Recall that a *measurable structure* on a bundle of Banach spaces $\{E_x\}$ over a Lebesgue ($\equiv$ standard measure) space (X, μ) is a family $\mathfrak{M}$ of sections (called *measurable*) $\sigma : x \mapsto \sigma_x \in E_x$ such that

(i) the function $x \mapsto \|\sigma_x\|$ is measurable for any section $\sigma \in \mathfrak{M}$;
(ii) the family $\mathfrak{M}$ is closed under addition, multiplication by measurable scalar functions and passing to pointwise limits in the norm topology.

The couple $(\{E_x\}, \mathfrak{M})$ is then called a *measurable bundle of Banach spaces* over (X, μ), see [**FD88**, Chapter II.4], [**ADR00**, Appendix A.3]. A measurable bundle $(\{E_x\}, \mathfrak{M})$ is *separable* if there exists a sequence of sections $\sigma^n \in \mathfrak{M}$ such that for a.e. $x \in X$ the sequence $\{\sigma^n_x\}$ is dense in E_x.

PROPOSITION 2.4. *Given a hyperbolic equivalence relation (X, μ, R, K), denote by $\mathfrak{M} = \mathfrak{M}(X, \mu, R, K)$ the space of all measurable functions $F : R \to \mathbb{R}$ such* *that for a.e. $x \in X$ the function $F(x,\cdot)$ extends to a continuous function on the completion $\overline{[x]}$. Then $\mathfrak{M}$ is a measurable structure on the bundle of Banach spaces $\{C(\overline{[x]})\}$, $x \in X$, and the measurable bundle $(\{C(\overline{[x]})\}, \mathfrak{M})$ is separable.*

PROOF. Verification of the conditions from the definition of a measurable structure being straightforward, we only have to check the separability. Since the leafwise hyperbolic compactifications $\overline{[x]}$ are completions of the equivalence classes $[x]$ with respect to the metrics ρ_x (1.4), and these metrics depend on x measurably, the functions

$$F_n(x,y) = \rho_x(y, \alpha_n(x)) ,$$

where α_n are the maps from Proposition 2.1, belong to $\mathfrak{M}$. For any $x \in X$ the family of functions $\{F_n(x,\cdot)\}$ separates points of $\overline{[x]}$. By adding to this family the constant function and applying the Weierstraß–Stone theorem we obtain the claim. □

The *dual bundle* $(\{E_x^*\}, \mathfrak{M}^*)$ of a separable measurable bundle $(\{E_x\}, \mathfrak{M})$ is the bundle of dual Banach spaces E_x^* endowed with the *weak* measurable structure* $\mathfrak{M}^*$ which consists of all sections $\sigma^* : x \mapsto \sigma_x^* \in E_x^*$ such that the map $x \mapsto \langle \sigma_x, \sigma_x^* \rangle$ is measurable for any section $\sigma \in \mathfrak{M}$, see [**ADR00**, Lemma A.3.7].

Returning to the measurable bundle $(\{C(\overline{[x]})\}, \mathfrak{M})$ of leafwise continuous functions, which is associated with the hyperbolic equivalence relation (X, μ, R, K), notice that for any space $C(\overline{[x]})$ its dual is the space $M(\overline{[x]})$ of signed Borel measures (equipped with the total variation norm) on the completion $\overline{[x]}$. Therefore, in view of Proposition 2.4 we have a measurable structure $\mathfrak{M}^*$ on the bundle $\{M(\overline{[x]})\}$. Denote by $\mathfrak{M}_\circ^*$ and $\mathfrak{M}_\partial^*$ the measurable structures induced by $\mathfrak{M}^*$ on the *interior subbundle* $\{M([x])\}$ and the *boundary subbundle* $\{M(\partial[x])\}$, respectively.

REMARK 2.5. One can easily check that the splitting of the bundle $\{M(\overline{[x]})\}$ into the direct sum of the interior and boundary subbundles is measurable in the sense that for any measurable section of $\{M(\overline{[x]})\}$ its interior and boundary parts are measurable too. Also, the *Hahn decomposition* of any measurable section of $\{M(\overline{[x]})\}$ into the difference of its positive and negative parts is measurable.

DEFINITION 2.6. Measurable sections of the bundle $(\{M([x])\}, \mathfrak{M}_\circ^*)$ (resp., of the bundle $(\{M(\partial[x])\}, \mathfrak{M}_\partial^*)$) are called *measurable systems of interior (resp., boundary) measures* on a hyperbolic equivalence relation (X, μ, R, K).

2.3. Measurable systems of boundary measures. Since the functions

$$\Delta_n(x,y) = \begin{cases} 1 , & y = \alpha_n(x) , \\ 0 , & \text{otherwise} , \end{cases}$$

where α_n are the maps from Proposition 2.1, obviously belong to $\mathfrak{M}$, the measurable structure $\mathfrak{M}_\circ^*$ of the interior bundle $\{M([x])\}$ consists precisely of those systems of measures $\{\lambda_x\}$ for which $\lambda_x(y)$ is measurable as a function on R. We shall now give an explicit description of the measurable structure $\mathfrak{M}_\partial^*$ of the boundary bundle $\{M(\partial[x])\}$.

THEOREM 2.7. *The measurable structure $\mathfrak{M}_\partial^*$ of the boundary bundle consists of pointwise weak* limits (in the spaces $M(\overline{[x]})$, $x \in X$) of sequences of measurable sections of the interior bundle $\{M([x])\}$.*

PROOF. The definition of the dual bundle implies that its measurable structure $\mathfrak{M}^*$ is closed with respect to passing to pointwise weak* limits, so that we only have to prove that any measurable system of boundary measures $\lambda \in \mathfrak{M}^*_\partial$ can be approximated by a sequence of measurable systems of interior measures $\lambda^n \in \mathfrak{M}^*_\circ$. In other words, for any boundary measure $\lambda_x \in M(\partial[x])$ we have to construct a sequence of interior measures $\lambda^n_x \in M([x])$ converging to λ_x, and to do it in a measurable way with respect to x.

Let us begin with the following observation. Any finite continuous partition of unity

$$\mathbf{1} = \sum_{i \in I} \varphi_i$$

on a compact space K determines a norm 1 linear map π from the space $M(K)$ of (signed) Borel measures on K to the space $M(I)$ of (signed) measures on the index set I by the formula

$$\pi\theta(i) = \langle \varphi_i, \theta \rangle \ . \tag{2.8}$$

Suppose now that for any $x \in X$ and any integer n we have a partition of unity on the completion $\overline{[x]}$ parameterized by the points from the n-sphere of the leafwise graph metric $S(x,n) \subset [x]$. These partitions then determine a sequence of maps π_n from $M\big(\overline{[x]}\big)$ to $M(S(x,n))$. We shall show that these partitions of unity can be constructed measurably with respect to x, and in such a way that $\pi_n\theta \to \theta$ for any boundary measure $\theta \in M(\partial[x])$.

For the moment we fix a point $x \in X$, an integer $n > 0$, and a real $\varepsilon > 0$. Recall that the hyperbolic compactification of the equivalence class $[x]$ is its completion with respect to the metric ρ_x (1.4). Now for any $s \in S = S(x,n)$ define the non-negative function $f_s \in C\big(\overline{[x]}\big)$ by putting for $z \in \overline{[x]}$

$$f_s(z) = \begin{cases} 1 - \dfrac{\rho_x(z,s) - \rho_x(z,S)}{\varepsilon} & , \quad \rho_x(z,s) \le \rho_x(z,S) + \varepsilon \ ; \\ 0 & , \quad \text{otherwise} \ . \end{cases}$$

Then the functions

$$\varphi_s = \frac{f_s}{\sum_{s \in S} f_s} \tag{2.9}$$

form a partition of unity on $\overline{[x]}$ parameterized by the points $s \in S$, and these functions have the property that

$$\varphi_s(z) = 0 \qquad \forall\, z \in \overline{[x]} : \rho_x(z,s) \ge \rho_x(z,S) + \varepsilon \ . \tag{2.10}$$

As it follows from the definition of the metric ρ_x (1.4),

$$\rho_x(\gamma, S) = e^{-an} \qquad \forall\, \gamma \in \partial[x] \ , \tag{2.11}$$

where $a = 1/15\delta$. Indeed, $\rho_x(y,s) \ge e^{-an}$ for any two points $y \in [x]$ and $s \in S$, so that $\rho_x(\gamma,s) \ge e^{-an}$. On the other hand, if ξ is a geodesic ray joining x and γ, then

$$\rho_x\big(\gamma, \xi(n)\big) = \lim_{m \to \infty} \rho_x\big(\xi(m), \xi(n)\big) = e^{-an} \ .$$

From now on we shall assume, for the sake of concreteness, that in the above construction

$$\varepsilon = \varepsilon(n) = e^{-an} \ .$$

Then formulas (2.10) and (2.11) imply that

$$\varphi_s(\gamma) = 0 \qquad \forall \gamma \in \partial[x] : \rho_x(\gamma, s) \geq 2e^{-an} . \tag{2.12}$$

Denote now by

$$\pi_n : M\big(\overline{[x]}\big) \to M(S)$$

the map (2.8) from the space of measures on $\overline{[x]}$ to the space of measures on $S = S(x,n)$ determined by the partition of unity $\{\varphi_s\}_{s\in S}$ (2.9). Then by (2.12)

$$\operatorname{supp} \pi_n \delta_\gamma \subset S \cap B_x(\gamma, 2e^{-an}) \qquad \forall \gamma \in \partial[x] , \tag{2.13}$$

and the linearity of the maps π_n implies the weak* convergence of the sequence $\pi_n\theta$ to θ for any boundary measure $\theta \in M(\partial[x])$. Obviously, the whole construction is measurable in x. □

PROPOSITION 2.14. *For any measurable system of boundary measures on a hyperbolic equivalence relation its continuous and atomic parts are also measurable.*

PROOF. In view of Remark 2.5 it is enough to establish the claim for systems of non-negative measures only. For any such system $\lambda = \{\lambda_x\}$ it is sufficient to show that the system $\widehat{\lambda} = \{\widehat{\lambda}_x\}$ which consists of maximal weight atoms of the measures λ_x is also measurable.

We shall use the approximation of the measures λ_x by the interior measures $\pi_n\lambda_x$ constructed in the proof of Theorem 2.7. Let us first notice that, as it follows from formulas (2.11) and (2.13), for any point $\gamma \in \partial[x]$ there exists $s \in S(x,n)$ such that

$$\operatorname{supp} \pi_n \delta_\gamma \subset S(x,n) \cap B_x(s, 3e^{-an}) . \tag{2.15}$$

Formulas (2.13) and (2.15) show that the maximal weight of the measure λ_x can be expressed in terms of the measures $\pi_n\lambda_x$ as

$$W(x) = \lim_{n\to\infty} \max_{s\in S(x,n)} (\pi_n\lambda_x)\big[B_x(s, 3e^{-an})\big] ,$$

and, in particular, the function $x \mapsto W(x)$ is measurable. Let us now denote by $\widehat{\lambda}_x^n$ the restriction of the measure $\pi_n\lambda_x$ to the $3e^{-an}$-neighbourhood (in the metric ρ_x) of the set

$$\Big\{ s \in S(x,n) : (\pi_n\lambda_x)\big[B_x(s, 3e^{-an})\big] \geq W(x) \Big\} .$$

Then for any $x \in X$ the sequence $\widehat{\lambda}_x^n$ converges (in the weak* topology) to the measure $\widehat{\lambda}_x$. Since for any n the system of measures $\{\widehat{\lambda}_x^n\}$ is measurable, by Theorem 2.7 the system $\widehat{\lambda}_x$ is also measurable. □

In the situations when the hyperbolic boundaries of different equivalence classes can be identified with (a subspace of) the same topological space, the notion of measurability introduced above coincides with the "usual" measurability as one can easily verify by using Theorem 2.7.

EXAMPLE 2.16. Recall that the classes of the *orbit equivalence relation* R_G of a measure class preserving action of a countable group G on a measure space (X,μ) are the orbits of the action. Any symmetric generating set $A \subset G$ determines the graph structure $K = \{(x, gx) : x \in X, g \in A\}$ on R_G. If the action of G is free (mod 0), then the map $g \mapsto g^{-1}x$ establishes an isomorphism of the (right) Cayley graph of G with a.e. graph $[x]$. If the group G is in addition *word hyperbolic* (i.e., the Cayley graph (G,A) is hyperbolic), then the orbit equivalence relation R_G is obviously hyperbolic, and the above isomorphism extends to a homeomorphism φ_x from the hyperbolic boundary ∂G of the group G to the hyperbolic boundary $\partial[x]$ of the equivalence class $[x]$ for a.e. point $x \in X$. Therefore, in this situation $M(\partial[x]) \cong M(\partial G)$ for a.e. $x \in X$, so that the boundary bundle is isomorphic to a constant Banach bundle, and Theorem 2.7 implies that the measurability of a system of boundary measures $\lambda_x \in M(\partial[x])$ in the sense of Definition 2.6 is equivalent to the measurability of the map $x \mapsto \varphi_x^{-1}(\lambda_x)$ from X to $M(\partial G)$.

EXAMPLE 2.17. Any simply connected Riemannian manifold M with pinched negative sectional curvatures is δ-hyperbolic, and the exponential map at any point $x \in M$ establishes a homeomorphism between the unit tangent sphere at the point x and the hyperbolic boundary of M (in this situation the hyperbolic compactification coincides with the visibility compactification). Therefore, given a hyperbolic Riemannian measurable foliation, the leafwise hyperbolic boundaries are homeomorphic to spheres, and the boundary bundle for the equivalence relation on any uniform system of transversals (see Section 2.6) is isomorphic to a constant Banach bundle. The measurability of a system of boundary measures on this equivalence relation is then equivalent to the measurability of the corresponding system of measures on the leafwise unit tangent spheres of the foliation.

EXAMPLE 2.18. Yet another example when the boundary bundle is isomorphic to a constant Banach bundle (more precisely, to a subbundle of a constant bundle) is provided by *treed equivalence relations* (see [**Ada90**]) for which a.e. graph $[x]$ is a tree. In this case any such tree can be embedded into the universal countable tree. Once again, Theorem 2.7 implies that the measurability of a system of boundary measures in the sense of Definition 2.6 is equivalent to the measurability of the corresponding map from X to the space of boundary measures of the universal tree.

2.4. Invariant systems of boundary measures.

DEFINITION 2.19. A measurable system $\lambda = \{\lambda_x\}$ of boundary measures on a hyperbolic equivalence relation (X, μ, R, K) is called *invariant* if it is a.e. constant, i.e., $\lambda_x = \lambda_y$ for M-a.e. $(x, y) \in R$.

THEOREM 2.20. *If $\lambda = \{\lambda_x\}$ is an invariant system of boundary measures on a conservative hyperbolic equivalence relation (X, μ, R, K), then a.e. measure λ_x is supported by at most 2 points.*

PROOF. As it follows from the proof of Proposition 2.14, the set X' of all points $x \in X$ with $\operatorname{card}\operatorname{supp}\lambda_x \geq 3$ is measurable, so that without loss of generality we may assume that $X' = X$. Further, again by the proof of Proposition 2.14, we may change the system λ so that the weights of atoms of a.e. measure λ_x be strictly less than $\frac{1}{2}$.

The claim is now a straightforward consequence of Proposition 1.16. The only thing we have to take care of is the measurability of the quasi-barycenter set. In order to check it we shall invoke Theorem 2.7 and approximate (in the leafwise weak* topology) the system λ by a sequence of measurable systems of interior measures $\lambda^n = \{\lambda_x^n\}$. Define the integer valued function φ on R as

$$\varphi(x, y) = \limsup_{n \to \infty} \sum_z \beta_z(x, y) \lambda_x^n(z) ,$$

where β_z is the distance cocycle (1.6). Let $Y(x) \subset [x]$ be the set of points where the function $\varphi(x, \cdot)$ attains its minimal value, and put

$$Z(x) = \bigcup_{x' \in [x]} Y(x') .$$

Clearly, the set

$$\{(x,y) : x \in X, y \in Z(x)\} \subset R$$

is measurable, and by Proposition 1.12 and Proposition 1.16 $Z(x)$ is contained in the finite set $\mathcal{C}(\lambda, x, 8C_2)$ for a.e. $x \in X$. Thus, we have a measurable map assigning to any point $x \in X$ a finite subset $Z(x)$ of the equivalence class $[x]$, which contradicts the conservativity of R. □

DEFINITION 2.21. A hyperbolic equivalence relation is *elementary* if there exists a measurable map assigning to a.e. class $[x]$ a finite subset $A[x] \subset \partial[x]$, i.e., if the system of uniform probability measures on the sets $A[x]$ is an invariant measurable system in the sense of Definition 2.6.

In the setup of this definition the map $x \mapsto \operatorname{card} A[x]$ is measurable (see the proof of Proposition 2.14). Therefore, if an elementary hyperbolic equivalence relation is ergodic, then almost all sets $A[x]$ must have the same cardinality. If $q \geq 1$ is its value, then we shall call the associated system of boundary measures and the equivalence relation *q-elementary*. Note that even if a hyperbolic equivalence relation is elementary, a.e. leafwise boundary may well be infinite (see the examples below). Theorem 2.20 implies

THEOREM 2.22. *Let (X, μ, R, K) be an ergodic hyperbolic equivalence relation which has more than one equivalence class. Then it admits an invariant measurable system of boundary probability measures if and only if it belongs to one of the following 3 classes:*

(i) *(X, μ, R, K) admits a unique invariant measurable system of boundary measures which is 1-elementary;*
(ii) *(X, μ, R, K) admits a unique invariant measurable system of boundary measures which is 2-elementary;*
(iii) *(X, μ, R, K) admits two disjoint 1-elementary systems of boundary measures, and any invariant system of boundary measures is their convex combination.*

2.5. Examples of elementary hyperbolic equivalence relations. We shall now give examples of hyperbolic equivalence relations from each of the types described in Theorem 2.22. Take a *non-elementary* word hyperbolic group G, i.e., such that its hyperbolic boundary ∂G consists of at least 3 points (which implies that ∂G is uncountable, see [**Gro87**]). For instance, one can take for G a non-abelian free group. Put

$$\partial^2 G = \partial G \times \partial G \setminus \operatorname{diag} = \{(\gamma_1, \gamma_2) : \gamma_i \in \partial G, \gamma_1 \neq \gamma_2\} ,$$
$$\partial^3 G = \{(\gamma_1, \gamma_2, \gamma_3) : \gamma_i \in \partial G, \gamma_i \neq \gamma_j\}$$

and, finally, let $\widetilde{\partial^2 G}$ be the *symmetrization* of $\partial^2 G$, i.e., the set of unordered pairs of elements $\gamma_1 \neq \gamma_2 \in \partial G$.

PROPOSITION 2.23. *The orbit equivalence relation R_G of a free measure class preserving action of a word hyperbolic group G on a measure space (X,μ) is 1-elementary (resp., 2-elementary) if and only if there exists a measurable G-equivariant map $\pi : X \to \partial G$ (resp., $\pi : X \to \widetilde{\partial^2 G}$).*

PROOF. First notice that the homeomorphisms $\varphi_x : \partial G \mapsto \partial[x]$ (see Example 2.16) have the property that $\varphi_{gx}(g\gamma) = \varphi_x(\gamma)$ for any $g \in G, \gamma \in \partial G$ and a.e. $x \in X$. Then equivariance of π implies that the map $x \mapsto \varphi_x \circ \pi(x) \in \partial[x]$ is leafwise constant, i.e., R_G is 1-elementary. Conversely, if R_G is 1-elementary, and $x \mapsto \Phi(x) \in \partial[x]$ is the corresponding leafwise constant map, then the map $\pi(x) = \varphi_x^{-1} \circ \Phi(x)$ is equivariant. With obvious modifications the same argument works in the 2-elementary case as well. □

Let us fix a quasi-invariant probability measure μ on ∂G such that the diagonal action of G on $(\partial^2 G, \mu \otimes \mu)$ is ergodic (for instance, the harmonic measure of any non-degenerate symmetric random walk on G has this property, see [**Kai94**]). On the other hand, Proposition 1.16 implies that the action of G on $\partial^3 G$ is proper (cf. [**Gro87**, 8.2.K]). Therefore, this action is dissipative with respect to any quasi-invariant measure.

EXAMPLE 2.24. *A hyperbolic equivalence relation with a unique 1-elementary system of boundary measures.* This is the orbit equivalence relation of the action of G on $(\partial G, \mu)$.

Proposition 2.23 applied to the identity map on ∂G at once provides us with a 1-elementary system. In order to prove its uniqueness we have, in view of Theorem 2.22, to exclude existence of another 1-elementary system of boundary measures disjoint with the one we have just constructed. By Proposition 2.23 existence of such a system is equivalent to existence of an equivariant map $\pi : \partial G \to \partial G$ such that a.e. $\pi(\gamma) \neq \gamma$. Such a map π would allow one to equivariantly embed $\partial^2 G$ into $\partial^3 G$ by the formula $(\gamma_1, \gamma_2) \mapsto (\gamma_1, \pi(\gamma_1), \gamma_2)$, which is impossible because the action of G on the space $(\partial^2 G, \mu \otimes \mu)$ is ergodic, whereas its action on $\partial^3 G$ (with respect to the image of the measure $\mu \otimes \mu$) must be dissipative.

EXAMPLE 2.25. *A hyperbolic equivalence relation with a unique system of boundary measures which is 2-elementary.* This is the orbit equivalence relation of the action of G on $\widetilde{\partial^2 G}$ (endowed with the image of the measure $\mu \otimes \mu$).

The system in question by Proposition 2.23 corresponds to the identity map on $\widetilde{\partial^2 G}$. By Theorem 2.22 in order to prove its uniqueness we have to exclude the possibility that this system is the half-sum of two 1-elementary systems, i.e., in view of Proposition 2.23, that there exists an equivariant map $\pi : \widetilde{\partial^2 G} \to \partial G$ with $\pi(\{\gamma_1, \gamma_2\}) \subset \{\gamma_1, \gamma_2\}$. Such a map π would give a G-equivariant measurable section of the projection $\partial^2 G \to \widetilde{\partial^2 G}$, which is impossible by the ergodicity of the action of G on $(\partial^2 G, \mu \otimes \mu)$.

EXAMPLE 2.26. *A hyperbolic equivalence relation with two 1-elementary systems of boundary measures.* This is the orbit equivalence relation of the action of G on $(\partial^2 G, \mu \otimes \mu)$.

2.6. Foliations and equivalence relations. Recall that a *foliation* $\mathcal{F}$ is a smooth manifold decomposed into smooth immersed submanifolds *(leaves)* organized in a local product structure (we shall use the same notation $\mathcal{F}$ both for the foliation and for its underlying manifold). In other words, $\mathcal{F}$ can be covered by *flow boxes* $U \cong B \times T$ (where B and T are open Euclidean balls) endowed with an additional product structure preserved by the transition maps. The sets $\{x\} \times T$, $x \in B$ and $B \times \{t\}$, $t \in T$ are called *transversals* and *plaques*, respectively. By $\mathcal{F}(x)$ we denote the *leaf* of the foliation $\mathcal{F}$ passing through a point x. *Laminations* are defined in precisely the same way except for allowing the transverse structure to be just continuous (rather than smooth), so that for laminations the transversals T are just topological spaces (without any smooth structure). However, the difference between foliations and laminations is insignificant for us, as we shall always (unless otherwise specified) be working in the *measure category*, and only require measurability of the transverse structure of $\mathcal{F}$ with respect to a certain *quasi-invariant transverse measure* μ. We shall refer to the couple $(\mathcal{F}, \mu)$ as a *measurable foliation* (see, for instance, [**MS88**] for more details).

If a measurable foliation $(\mathcal{F}, \mu)$ is endowed with a transversely measurable leafwise Riemannian structure, we shall call it a (leafwise) *Riemannian measurable foliation.* We shall say that $(\mathcal{F}, \mu)$ is *hyperbolic* if μ-a.e. leaf is δ-hyperbolic (with respect to its Riemannian metric). A particular case is the situation when a.e. leaf of $\mathcal{F}$ is a simply connected manifold with pinched negative sectional curvatures. In this case measurable systems of boundary measures can be defined as measurable systems of measures on leafwise unit tangent spheres (by identifying the latter with the leafwise hyperbolic≡visibility boundaries via the leafwise exponential maps). In the general situation one could apply the approach used for hyperbolic equivalence relations in Section 2.2 and Section 2.3. However, since we are only interested in *invariant* systems of boundary measures, we shall spare tedious technical details by reducing foliations to equivalence relations on appropriate systems of transversals (such a reduction goes back to Plante [**Pla75**] and Bowen [**Bow77**]).

DEFINITION 2.27. A system $\mathcal{T} = \{T_i\}$ of transversals of a Riemannian measurable foliation $\mathcal{F}$ is *uniform* if

(i) For any $r > 0$ the number of points from $\mathcal{T}$ in a.e. leafwise ball of radius r is uniformly bounded by a constant $C = C(r)$;
(ii) The intersection of $\mathcal{T}$ with a.e. leaf of $\mathcal{F}$ is non-empty, and there exists a constant $A > 0$ such that for any point x of a.e. leaf its leafwise distance to $\mathcal{F}(x) \cap \mathcal{T}$ does not exceed A.

A foliation is *uniform* if it admits a uniform system of transversals.

Compactness considerations immediately imply

PROPOSITION 2.28. *Any Riemannian measurable foliation admitting a compact underlying space is uniform.*

Let now $(\mathcal{F}, \mu)$ be a uniform Riemannian measurable foliation with a uniform system of transversals $\mathcal{T}$. Denote by $R_{\mathcal{F}}$ the induced equivalence relation on $\mathcal{T}$ (two points are $R_{\mathcal{F}}$-equivalent iff they belong to the same leaf of $\mathcal{F}$). Then the restriction $\mu_{\mathcal{T}}$ of the measure μ to $\mathcal{T}$ is $R_{\mathcal{F}}$-quasi-invariant in the sense of Section 2.1. Consider on $R_{\mathcal{F}}$ the graph structure

$$K = \{(x, y) \in R_{\mathcal{F}} : d(x, y) \leq 3A\} ,$$

where d denotes the leafwise Riemannian metric, and $A = A(\mathcal{T})$ is the constant from Definition 2.27. Then one can easily see that the leafwise graph metrics are roughly isometric to the leafwise Riemannian metrics, so that, in particular, if $\mathcal{F}$ is hyperbolic, then the graphed equivalence relation $(\mathcal{T}, \mu_{\mathcal{T}}, R_{\mathcal{F}}, K)$ is also hyperbolic, and the leafwise hyperbolic boundaries with respect to the Riemannian and the graph metrics coincide (e.g., see [**GdlH90**, Proposition 7.14]).

PROPOSITION 2.29. *An invariant system of boundary measures on a hyperbolic foliation which is measurable with respect to one uniform system of transversals is also measurable with respect to any other uniform system of transversals.*

PROOF. Let $\mathcal{T}$ and $\overline{\mathcal{T}}$ be two uniform systems of transversals, and let λ be a measurable invariant system of boundary measures of the equivalence relation $(\mathcal{T}, R_{\mathcal{F}})$. By Theorem 2.7 we may realize λ as the weak* leafwise limit of measurable systems $\lambda^n = \{\lambda^n_x\}$ of interior measures of $(\mathcal{T}, R_{\mathcal{F}})$. We shall need the following "transition maps" allowing one to pass from measures on $\mathcal{T}$ to measures on $\overline{\mathcal{T}}$ and back in a measurable way: denote by α_x, $x \in \mathcal{T}$ the probability measure uniformly distributed on the set of points from $\overline{\mathcal{T}}$ nearest to x in the leafwise metric of the leaf $\mathcal{F}(x)$, and by $\overline{\alpha}_{\overline{x}}$, $\overline{x} \in \overline{\mathcal{T}}$ the probability measure uniformly distributed on the set of points from $\mathcal{T}$ nearest to $\overline{x}$. By the definition of a uniform system of transversals the support of any measure α_x lies within a uniformly bounded distance from x, and the same property holds for the measures $\overline{\alpha}_{\overline{x}}$ as well. Obviously, both maps $x \mapsto \alpha_x$ and $\overline{\boldsymbol{x}} \mapsto \overline{\alpha}_{\overline{x}}$ are measurable. By using the measures α_x and $\overline{\alpha}_{\overline{x}}$ one can pass from the sequence λ^n to the sequence $\overline{\lambda}^n$ of measurable systems of interior measures of the equivalence relation $(\overline{\mathcal{T}}, R_{\mathcal{F}})$ defined as

$$\overline{\lambda}^n_{\overline{x}}(\overline{y}) = \sum_{x,y} \overline{\alpha}_{\overline{x}}(x)\, \lambda^n_x(y)\, \alpha_y(\overline{y}) \ .$$

Since the measures λ^n_x converge in the weak* topology to a leafwise constant limit, the measures $\overline{\lambda}^n_{\overline{x}}$ also converge to the same limit, which is therefore measurable with respect to the system of transversals $\overline{\mathcal{T}}$. □

By using the above Proposition, we can now define an invariant measurable system of boundary measures of a uniform hyperbolic foliation as an invariant measurable system of boundary measures of any uniform system of transversals. Theorem 2.20 and Theorem 2.22 then imply

THEOREM 2.30. *If $\lambda = \{\lambda_x\}$ is an invariant measurable system of boundary measures on a uniform conservative hyperbolic foliation $(\mathcal{F}, \mu)$, then a.e. measure λ_x is supported by at most 2 points.*

THEOREM 2.31. *Let $(\mathcal{F}, \mu)$ be a uniform ergodic hyperbolic foliation which has more than one leaf. Then it admits an invariant measurable system of boundary probability measures if and only if it belongs to one of the following 3 classes:*

(i) *$(\mathcal{F}, \mu)$ admits a unique invariant measurable system of boundary measures which is 1-elementary;*
(ii) *$(\mathcal{F}, \mu)$ admits a unique invariant measurable system of boundary measures which is 2-elementary;*
(iii) *$(\mathcal{F}, \mu)$ admits two disjoint 1-elementary invariant measurable systems of boundary measures, and any invariant measurable system of boundary measures is their convex combination.*

Examples of foliations from each of the classes described in Theorem 2.31 can be constructed along the same lines as in Section 2.5.

3. Amenability

3.1. Groups. The class of *amenable groups* is a natural generalization of the class of finite groups. We shall remind just two definitions of amenability of a locally compact group G (among a host of other definitions, see [**Gre69**], [**Pat88**], [**Pie84**]). The *Reiter condition*

(3.1) There exists an *approximatively invariant* sequence of probability measures θ_n on G, i.e., such that $\|g\theta_n - \theta_n\| \to 0$ for any $g \in G$.

by its constructiveness is very useful for verifying amenability of a given group, whereas, on the contrary, the *fixed point condition*

(3.2) Any continuous affine action of G on a compact space has a fixed point.

is the one which is most useful for applications of amenability. For instance, the fixed point condition implies at once that any continuous action of an amenable group on a compact space has an invariant measure.

We shall illustrate how these two definitions work together on the example of the following result. Recall that a group of isometries of a hyperbolic space $\mathcal{X}$ is called *elementary* if it either is compact or fixes a finite subset of $\partial\mathcal{X}$ (actually, if in the latter case the fixed set contains more than 2 points, then the group fixes also a compact subset of $\mathcal{X}$, and therefore is compact, see Proposition 1.18).

THEOREM 3.3. *A closed subgroup G of the group of isometries of an exponentially bounded hyperbolic space $\mathcal{X}$ is amenable if and only if it is elementary.*

PROOF. If G is amenable, then by the fixed point property (3.2) it has an invariant probability measure λ on $\partial\mathcal{X}$. If λ does not have atoms of weight at least $\frac{1}{2}$, then by Proposition 1.18 its quasi-barycenter set is compact and fixed by G together with the measure λ. Otherwise, G fixes the set of maximal weight atoms of the measure λ which consists either of 1 or of 2 points.

Conversely, if G is compact, then it is obviously amenable. Otherwise, if G fixes a point $\gamma \in \partial\mathcal{X}$, take a point $o \in \mathcal{X}$ and put $\theta_n = \lambda_n(o,\gamma)$, where λ_n are the maps from Theorem 1.33. Then

$$\|g\theta_n - \theta_n\| = \|g\lambda_n(o,\gamma) - \lambda_n(o,\gamma)\| = \|\lambda_n(go,\gamma) - \lambda_n(o,\gamma)\| \to 0 ,$$

so that Reiter's condition (3.1) is satisfied. If the fixed set of G consists of two points $\{\gamma_1,\gamma_2\}$, then just take $\theta_n = \big[\lambda_n(o,\gamma_1) + \lambda_n(o,\gamma_2)\big]/2$. □

This result is classical for the constant curvature spaces (i.e., for Kleinian groups), and, of course, should be of no surprise in the stated generality either, although I could not find it in the literature in an explicit form. For the case of a group with a finite critical exponent acting on a CAT(-1) space it is proved in [**BM96**, Corollary 2.5] (also see [**BS87**] and [**AB98**] for properties of amenable groups of isometries of CAT(0) spaces); note that the assumption of exponential boundedness of the space $\mathcal{X}$ in Theorem 3.3 is essential (see Section 3.5 for the corresponding counterexamples).

However, the main reason for giving this proof is that it contains the key geometric ingredients of the relationship between amenability and hyperbolicity, namely, boundary convergence of geodesics (essentially equivalent to Theorem 1.33) and existence of quasi-barycenter sets, as illustrated by the following diagram:

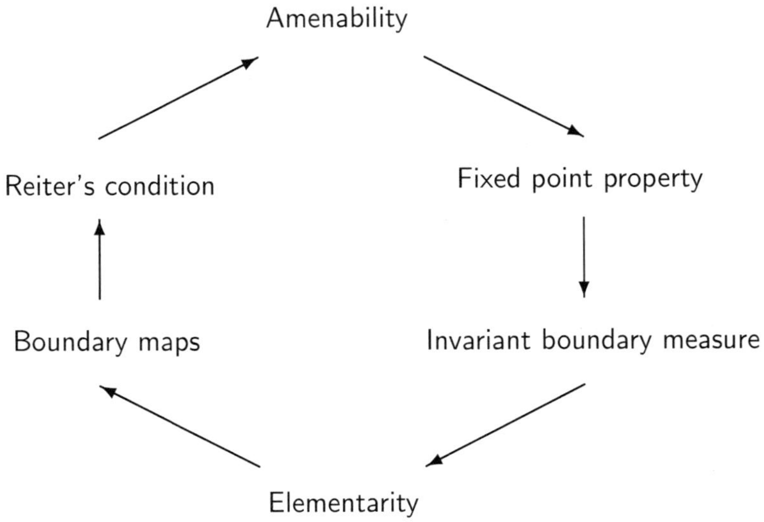

The notion of amenability can also be defined for a number of objects other than groups: discrete equivalence relations, foliations, groups actions, see [**Zim78**], [**CFW81**], [**HK87**], [**Ada92**], [**AEG94**], [**Kai97**] for various definitions. All these definitions can be put into the general framework of amenability for *measured groupoids*, see [**ADR00**], [**CHLI02**]. We shall refer the reader to the aforementioned references for precise definitions; note that in all these cases it is possible to give (equivalent) definitions of amenability both in terms of approximatively invariant sequences of probability measures (analogous to the Reiter condition (3.1); we shall briefly mention these definitions when dealing with the corresponding classes of objects) and in terms of an appropriate fixed point property (analogous to condition (3.2)).

We shall now proceed to establish analogues of Theorem 3.3 for these objects by taking stock of the results from Section 1 and Section 2, and following precisely the same scheme: amenability by the fixed point property implies existence of an invariant system of boundary measures, and therefore elementarity; whereas elementarity by virtue of Theorem 1.33 leads to existence of an approximatively invariant sequence of probability measures, i.e., to amenability.

3.2. Equivalence relations.

THEOREM 3.4. *A conservative hyperbolic equivalence relation with uniformly bounded vertex degrees is amenable if and only if it is elementary.*

PROOF. If the equivalence relation is amenable, then the fixed point property applied to the Banach bundle of leafwise boundary measures implies existence of an invariant measurable system of boundary measures, see [**ADR00**, Theorem 4.2.7] (based on the original argument from [**Zim77**]), which by Theorem 2.20 implies that the equivalence relation is elementary.

For proving the converse implication we have to construct a sequence of measurable maps λ_n assigning to a.e. point x from the state space of the equivalence relation a probability measure on the equivalence class $[x]$, and such that for a.e.

pair of equivalent points (x, y)

$$\|\lambda_n(x) - \lambda_n(y)\| \to 0 \tag{3.5}$$

(this is the analogue of the Reiter condition for equivalence relations).

The only point which needs an explanation here is the measurability of the approximatively invariant sequences of measures obtained in Theorem 1.33. It would follow from the existence of a measurable family of leafwise tempered measures. Under the assumption of uniform boundedness of vertex degrees the leafwise counting measures are obviously tempered. □

REMARK 3.6. S. Adams [**Ada90**] proved that a treed equivalence relation with a *finite invariant measure* is amenable if and only if a.e. tree has at most 2 ends. It is plausible that the same is true for hyperbolic equivalence relations as well, namely, that for a hyperbolic equivalence relation with a finite invariant measure the hyperbolic boundary of a.e. leaf has at most 2 points.

REMARK 3.7. It is well-known that if a graphed equivalence relation has *subexponential growth* (i.e., $[\operatorname{card} B(x,n)]^{1/n} \to 1$ for a.e. $x \in X$, where $B(x,n)$ are the balls in the leafwise graph metrics), then it is amenable, e.g., see [**AL91**, Proposition 3.3]. Since the sequences of leafwise balls centered at any two equivalent points are obviously sandwiched, Lemma 1.35 provides a direct "constructive" way of proving this fact, and this Lemma can also be used in numerous other arguments deducing amenability from subexponentiality in various contexts, see [**AVS57**], [**Gre69**], [**Ser79**], [**Sam79**], [**ADR00**], [**dlHGC99**]. However, in Theorem 1.33 Lemma 1.35 is used in a somewhat different situation, where the sandwiched sequences of sets are not metric balls.

REMARK 3.8. The boundary maps provided by Theorem 1.33 do not depend on the measure on the state space of the equivalence relation, so that in fact the argument in the proof of Theorem 3.4 shows amenability of elementary hyperbolic equivalence relations even in the *Borel* (rather than measure) category (cf. the discussion of definitions of amenable action in various categories in Section 3.4).

3.3. Foliations. In view of the discussion in Section 2.6, measurable foliations can be dealt with in precisely the same way as equivalence relations (by using Theorem 1.33 and Theorem 2.31). Once again, one of equivalent definitions of amenability amounts to existence of an approximatively invariant sequences of measures (3.5) (note that there is also a somewhat different notion of amenable foliation meaning that its *fundamental groupoid* is amenable, see the instructive discussion in [**CHLI02**]).

THEOREM 3.9. *A uniform conservative hyperbolic foliation is amenable if and only if it is elementary.*

COROLLARY 3.10. *A hyperbolic foliation admitting a compact underlying space is amenable if and only if it is elementary.*

EXAMPLE 3.11. Let $\mathcal{F}$ be a foliation with simply connected leaves of pinched negative sectional curvature. Denote by $S\mathcal{F}$ the associated *stable extension*. This is the foliation whose underlying space consists of leafwise unit tangent vectors of $\mathcal{F}$, and the leaves consist of tangent vectors from the same leaf of $\mathcal{F}$ pointing at the same point at infinity. Then the foliation $S\mathcal{F}$ is obviously elementary, so that it is amenable with respect to any quasi-invariant transverse measure. This result is a generalization of the well-known amenability of the stable foliation of the geodesic flow, see [**Bow77**]. Yet another example of elementary hyperbolic foliations is provided by foliations and laminations with pointed at infinity leaves arising in conformal dynamics [**KL01**].

REMARK 3.12. The construction of the stable extension can also be applied to an arbitrary hyperbolic foliation (not necessarily with Cartan–Hadamard leaves), in which case we shall call it the *boundary extension*. The underlying space of the boundary extension consists of the pairs (x,γ), where x is a point of the underlying space of the original foliation, and γ is a point of the hyperbolic boundary of the leaf of x. The leaves of the boundary extension consist of all pairs (x,γ) with the same γ. The boundary extension is elementary and therefore amenable.

3.4. Group actions. Let G be a locally compact group with a measure type preserving action on a measure space (X,μ). The "Reiter type" definition of amenability of this action requires existence of a sequence of measurable maps $\lambda_n : X \to \mathcal{P}(G)$ which are *approximatively equivariant* in the sense that

$$\|\lambda_n(gx) - g\lambda_n(x)\| \to 0 \tag{3.13}$$

for all $g \in G$ and μ-a.e. $x \in X$. There are several modifications of this definition. In particular, a Borel action of a locally compact group G on a Borel space X is called *universally amenable* [**Ada96**] (other terms: *measurewise amenable* [**ADR00**], *measure-amenable* [**JKL02**]) if it is amenable for any quasi-invariant measure μ on X. If, moreover, X is a topological space, and the action of G on X is continuous, then the action is called *topologically amenable* if the convergence in (3.13) is uniform on compact subsets of $G \times X$.

THEOREM 3.14. *An ergodic action of a word hyperbolic group G on a measure space (X,μ) is amenable if and only if it factorizes through the action of G either on the hyperbolic boundary ∂G, or on its symmetric square $\widetilde{\partial^2 G}$.*

PROOF. The amenability of the action of G on ∂G or on $\widetilde{\partial^2 G}$ is well-known (see Remark 3.16 below) and follows at once from Theorem 1.33. Therefore, any action which factorizes through ∂G or $\widetilde{\partial^2 G}$ is also amenable.

Conversely, amenability of the action on the space (X,μ) by the fixed point property applied to the Banach bundle of boundary measures implies existence of an equivariant measurable map from X to the space of probability measures on ∂G (in precisely the same way as in the proof of Theorem 3.4). By ergodicity we may assume that either these measures are a.e. purely non-atomic, or that almost all of them are equidistributed on a set of atoms whose cardinality is a.e. the same. In the first case, or if the number of atoms exceeds 2, we obtain by Proposition 1.18 an equivariant map from X to the space of finite subsets of G, which means that the action on X is dissipative, and therefore is confined to a single orbit. Otherwise we have the sought for equivariant projection from X to ∂G or $\widetilde{\partial^2 G}$. □

THEOREM 3.15. *Let G be a closed subgroup of the group of isometries of an exponentially bounded hyperbolic space $\mathcal{X}$. Then the action of G on the hyperbolic boundary $\partial\mathcal{X}$ is topologically amenable.*

PROOF. A sequence of approximatively invariant measures for this action is provided by Theorem 1.33. Indeed, let $\lambda_n : \mathcal{X} \times \partial\mathcal{X} \to \mathcal{P}(\mathcal{X})$ be the Iso($\mathcal{X}$)-equivariant maps constructed in Theorem 1.33. Choose a reference point $o \in \mathcal{X}$, and put $\widetilde{\lambda}_n(\gamma) = \lambda_n(o,\gamma)$ for any $g \in G$ and $\gamma \in \partial\mathcal{X}$. Then the maps $\widetilde{\lambda}_n : \partial\mathcal{X} \to \mathcal{P}(\mathcal{X})$ are approximatively equivariant, because

$$\|\widetilde{\lambda}_n(g\gamma) - g\widetilde{\lambda}_n(\gamma)\| = \|\lambda_n(o, g\gamma) - \lambda_n(go, g\gamma)\| \to 0$$

for any $g \in G$ and $\gamma \in \partial\mathcal{X}$ uniformly on compacts, see Remark 1.41. Since the action of G on $\mathcal{X}$ is proper, there exists an equivariant Borel lift $\mathcal{P}(\mathcal{X}) \to \mathcal{P}(G)$ (e.g., see [**ADR00**, Proposition 2.1.10]). Applying it to the sequence $\widetilde{\lambda}_n$ gives an approximatively equivariant sequence of maps from $\partial\mathcal{X}$ to $\mathcal{P}(G)$. □

REMARK 3.16. The action of the group $SL(2,\mathbb{Z})$ on the circle at infinity of the hyperbolic plane was the first example of an amenable action of a non-amenable group [**Bow77**], [**Ver78**]. Since then amenability of boundary actions was established in a number of other situations [**Spa87**], [**SZ91**], [**Ada94**]; the most general result being universal amenability of the boundary action of the group of isometries of an exponentially bounded hyperbolic space [**Ada96**]. In the particular case of the boundary action of a word hyperbolic group the arguments of Adams were recently simplified by Germain [**Ger00**], who proved topological amenability of this action. However, the approach based on using Theorem 1.33 and Lemma 1.35 is more streamlined even in this case.

In the CAT(-1) case replacing Theorem 1.33 with Theorem 1.38 in the proof of Theorem 3.15 yields

THEOREM 3.17. *Let G be a closed subgroup of the group of isometries of a hyperbolic space $\mathcal{X}$ with finite critical exponent. Then the action of G on the hyperbolic boundary $\partial\mathcal{X}$ is topologically amenable.*

REMARK 3.18. Burger and Mozes in [**BM96**, Corollary 1.4] stopped short of saying that under the assumptions of Theorem 3.17 the action of G on $\partial\mathcal{X}$ is amenable with respect to any Patterson measure (although they basically proved that). Moreover, they also made the first step to establishing universal amenability of the boundary action by showing that the stabilizer of *any* boundary point is amenable (which is a necessary condition for universal amenability of the action, see [**Ada92**]).

REMARK 3.19. As we shall see in Section 3.5 below, without any assumptions on the growth of the space $\mathcal{X}$ Theorem 3.15 and Theorem 3.17 may fail even in the weakest form of amenability of the action with respect to a given quasi-invariant boundary measure, which answers a question formulated by Adams in [**Ada96**].

3.5. Counterexamples to boundary amenability. As it was pointed out by Burger and Mozes in [**BM96**] (Remark after Proposition 1.6; in our notations $X(\infty) \equiv \partial X$),

> The product $X = [0,\infty) \times \mathbf{H}^2$ with Riemannian metric $(dt)^2 + e^{-2t}\frac{|dz|^2}{y^2}$, where $\mathbf{H}^2$ denotes the upper half-plane, is a CAT(-1) space on which $SL(2,\mathbb{R})$ acts isometrically, fixing a point in $X(\infty)$.

Note that the metric on X is obtained from the hyperbolic metric on $\mathbf{H}^2$ in precisely the same way as the hyperbolic metric on (a horoball in) $\mathbf{H}^2$ is obtained from the Euclidean metric on $\mathbb{R}$, i.e., by exponential rescaling in the vertical direction. In fact, this example is a particular case of the following general *hyperbolization construction* (cf. [**KL01**]). This construction resembles that of a *κ-cone over a metric space* (e.g., see [**BH99**]).

PROPOSITION 3.20. *For a complete* CAT(0) *space Y put $X = \mathbb{R}\times Y$, and define the metric d_X on X by putting for any two points $(t_1, y_1), (t_2, y_2) \in X$*

$$d_X\big((t_1, y_1), (t_2, y_2)\big) = d_{\mathbf{H}^2}\big((t_1, 0), (t_2, d_Y(y_1, y_2))\big) , \tag{3.21}$$

where d_Y is the metric on Y, and $d_{\mathbf{H}^2}$ is the metric

$$ds^2 = dt^2 + e^{-2t} dl^2$$

on the hyperbolic plane $\mathbf{H}^2 \cong \mathbb{R} \times \mathbb{R}$ realized in the upper half-plane model with logarithmic rescaling of the vertical coordinate. Then d_X is a complete metric on X, the space X is CAT(−1) *with respect to this metric, and the boundary ∂X of the hyperbolic compactification of X is the union $Y \cup \{\omega\}$, where ω is the common limit point of vertical geodesic rays $[0, \infty) \times \{y\}$, $y \in Y$ in X directed upwards. The point of ∂X corresponding to a point $y \in Y$ is the limit point of the directed downwards vertical geodesic ray $[0, -\infty) \times \{y\}$. If Y is non-compact, then ∂X is homeomorphic to the one-point compactification of Y, and if Y is compact, then ∂Y is homeomorphic to the disjoint union of Y and ω.*

DEFINITION 3.22. We shall call the CAT(−1) space $(X, d_X) = (\mathcal{H}Y, d_{\mathcal{H}Y})$ constructed in Proposition 3.20 the *hyperbolization* of the CAT(0) space (Y, d_Y).

PROOF OF PROPOSITION 3.20. The verification of the triangle inequality for d_X using formula (3.21) is straightforward. Completeness of the metric d_X is obvious. Further, the space (X, d_X) is geodesic, and the (unique) geodesic $\overline{\xi}$ joining two points (t_1, y_1) and (t_2, y_2) is the image of the geodesic joining the points $(t_1, 0)$ and $(t_2, d_Y(y_1, y_2))$ in $\mathbf{H}^2$ under the natural isometry between the strip $\mathbb{R}\times[0, d_y(y_1, y_2)]$ in $\mathbf{H}^2$ and the "curtain" $\mathbb{R} \times \xi$ in X over the geodesic ξ joining y_1 and y_2 in Y.

In order to check the CAT(−1) property let us consider a geodesic triangle $\Delta_X = \big[(t_0, y_0), (t_1, y_1), (t_2, y_2)\big]$ in X. It is more convenient to construct a comparison triangle for Δ_X in the 3-dimensional hyperbolic space $\mathbf{H}^3$ rather than in the 2-dimensional hyperbolic plane $\mathbf{H}^2$, and to use the fact that $\mathbf{H}^3$ is obtained from the Euclidean plane $\mathbf{E}^2$ in precisely the same way as X is obtained from Y. Let $\Delta_{\mathbf{E}^2} = [z_0, z_1, z_2]$ be the comparison triangle in $\mathbf{E}^2$ for the triangle $\Delta_Y = [y_0, y_1, y_2]$ in Y. Then the definition of the distance d_X implies that $\Delta_{\mathbf{H}^3} = \big[(t_0, z_0), (t_1, z_1), (t_2, z_2)\big]$ is a comparison triangle for Δ_X. Now take points (t_i', y_i'), $i = 1, 2$, on the sides $[(t_0, y_0), (t_i, y_i)]$ of the triangle Δ_X, and denote by (t_i', z_i') the corresponding points on the sides of the comparison triangle $\Delta_{\mathbf{H}^3}$. By the above description of the geodesics in X the "heights" t_i' remain the same, and $d_Y(y_0, y_i') = d_{\mathbf{E}^2}(z_0, z_i')$. Therefore the CAT(0) property of the space Y and the definition of the metric d_X imply that

$$d_X\big((t_1', y_1'), (t_2', y_2')\big) \le d_{\mathbf{H}^3}\big((t_1', z_1'), (t_2', z_2')\big) .$$

The description of the hyperbolic boundary ∂X easily follows from the above description of the geodesics in X. □

Proposition 3.20 now allows one to obtain examples showing that Theorem 3.3, Theorem 3.15 and Theorem 3.17 may well fail without any bounded geometry assumptions on the space or on the group.

PROPOSITION 3.23. *Let Y be any CAT(0) space with a non-amenable closed group of isometries G, and let $X = \mathcal{H}Y$ be its hyperbolization as described in Proposition 3.20. Then the group G acting on X as $g(t,y) = (t, gy)$ is also closed in* Iso(X)*, and it fixes the point $\omega \in \partial X$. Therefore,*

- *The group G is elementary with respect to the space X and non-amenable;*
- *The action of the group G on ∂X is non-amenable with respect to the invariant measure δ_ω concentrated at the fixed point ω.*

In order to obtain more concrete examples one can take Y to be, for instance, the hyperbolic plane (as did Burger and Mozes; actually they took just the horoball in $\mathcal{H}\mathbf{H}^2$ centered at the point ω rather than the whole space $\mathcal{H}\mathbf{H}^2$), or the Cayley 1-complex of the free group. Still, it would be interesting to have examples of non-amenable actions with respect to a non-atomic boundary measure.

References

[AB98] Scot Adams and Werner Ballmann, *Amenable isometry groups of Hadamard spaces*, Math. Ann. **312** (1998), no. 1, 183–195. MR **99i:**53032

[Ada90] Scot Adams, *Trees and amenable equivalence relations*, Ergodic Theory Dynam. Systems **10** (1990), no. 1, 1–14. MR **91d:**28041

[Ada92] ———, *Generalities on amenable actions*, Unpublished notes.

[Ada94] ———, *Boundary amenability for word hyperbolic groups and an application to smooth dynamics of simple groups*, Topology **33** (1994), no. 4, 765–783. MR **96g:**58104

[Ada96] ———, *Reduction of cocycles with hyperbolic targets*, Ergodic Theory Dynam. Systems **16** (1996), no. 6, 1111–1145. MR **98i:**58135

[ADR00] C. Anantharaman-Delaroche and J. Renault, *Amenable groupoids*, Monographies de L'Enseignement Mathématique [Monographs of L'Enseignement Mathématique], vol. 36, L'Enseignement Mathématique, Geneva, 2000, With a foreword by Georges Skandalis and Appendix B by E. Germain. MR **2001m:**22005

[AEG94] Scot Adams, George A. Elliott, and Thierry Giordano, *Amenable actions of groups*, Trans. Amer. Math. Soc. **344** (1994), no. 2, 803–822. MR **94k:**22010

[AL91] Scot Adams and Russell Lyons, *Amenability, Kazhdan's property and percolation for trees, groups and equivalence relations*, Israel J. Math. **75** (1991), no. 2-3, 341–370. MR **93j:**43001

[AVS57] G. M. Adelson-Velskii and Yu. A. Shreider, *The Banach mean on groups*, Uspehi Mat. Nauk (N.S.) **12** (1957), no. 6(78), 131–136. MR 20 #1238

[BCG96] Gérard Besson, Gilles Courtois, and Sylvestre Gallot, *Minimal entropy and Mostow's rigidity theorems*, Ergodic Theory Dynam. Systems **16** (1996), no. 4, 623–649. MR **97e:**58177

[BH99] Martin R. Bridson and André Haefliger, *Metric spaces of non-positive curvature*, Grundlehren der Mathematischen Wissenschaften [Fundamental Principles of Mathematical Sciences], vol. 319, Springer-Verlag, Berlin, 1999. MR **2000k:**53038

[BM96] M. Burger and S. Mozes, CAT(-1)*-spaces, divergence groups and their commensurators*, J. Amer. Math. Soc. **9** (1996), no. 1, 57–93. MR **96c:**20065

[Bow77] Rufus Bowen, *Anosov foliations are hyperfinite*, Ann. of Math. (2) **106** (1977), no. 3, 549–565. MR 57 #1569

[BS87] M. Burger and V. Schroeder, *Amenable groups and stabilizers of measures on the boundary of a Hadamard manifold*, Math. Ann. **276** (1987), no. 3, 505–514. MR **88b:**53049

[CFW81] A. Connes, J. Feldman, and B. Weiss, *An amenable equivalence relation is generated by a single transformation*, Ergodic Theory Dynamical Systems **1** (1981), no. 4, 431–450 (1982). MR **84h:**46090

[CHLI02] K. Corlette, L. Hernández Lamoneda, and A. Iozzi, *A vanishing theorem for the tangential de Rham cohomology of a foliation with amenable fundamental groupoid*, Preprint.

[DE86] Adrien Douady and Clifford J. Earle, *Conformally natural extension of homeomorphisms of the circle*, Acta Math. **157** (1986), no. 1-2, 23–48. MR **87j**:30041

[dlHGC99] P. de la Harpe, R. I. Grigorchuk, and T. Ceccherini, *Amenability and paradoxical decompositions for pseudogroups and discrete metric spaces*, Tr. Mat. Inst. Steklova **224** (1999), 68–111. MR **2001h**:43001

[FD88] J. M. G. Fell and R. S. Doran, *Representations of *-algebras, locally compact groups, and Banach *-algebraic bundles. Vol. 1*, Pure and Applied Mathematics, vol. 125, Academic Press Inc., Boston, MA, 1988, Basic representation theory of groups and algebras. MR **90c**:46001

[FM77] Jacob Feldman and Calvin C. Moore, *Ergodic equivalence relations, cohomology, and von Neumann algebras. I*, Trans. Amer. Math. Soc. **234** (1977), no. 2, 289–324. MR 58 #28261a

[GdlH90] É. Ghys and P. de la Harpe (eds.), *Sur les groupes hyperboliques d'après Mikhael Gromov*, Progress in Mathematics, vol. 83, Birkhäuser Boston Inc., Boston, MA, 1990, Papers from the Swiss Seminar on Hyperbolic Groups held in Bern, 1988. MR **92f**:53050

[Ger00] E. Germain, *Approximate invariant means for boundary actions of hyperbolic groups*, 2000, Appendix B in the book [**ADR00**].

[Gre69] Frederick P. Greenleaf, *Invariant means on topological groups and their applications*, Van Nostrand Mathematical Studies, No. 16, Van Nostrand Reinhold Co., New York, 1969. MR 40 #4776

[Gro87] M. Gromov, *Hyperbolic groups*, Essays in group theory, Math. Sci. Res. Inst. Publ., vol. 8, Springer, New York, 1987, pp. 75–263. MR **89e**:20070

[HK87] S. Hurder and A. Katok, *Ergodic theory and Weil measures for foliations*, Ann. of Math. (2) **126** (1987), no. 2, 221–275. MR **89d**:57042

[JKL02] S. Jackson, A. S. Kechris, and A. Louveau, *Countable Borel equivalence relations*, J. Math. Log. **2** (2002), no. 1, 1–80. MR **2003f**:03066

[Kai94] Vadim A. Kaimanovich, *Ergodicity of harmonic invariant measures for the geodesic flow on hyperbolic spaces*, J. Reine Angew. Math. **455** (1994), 57–103. MR **95g**:58130

[Kai97] ———, *Amenability, hyperfiniteness, and isoperimetric inequalities*, C. R. Acad. Sci. Paris Sér. I Math. **325** (1997), no. 9, 999–1004. MR **98j**:28014

[Kai02] ———, *Random walks on Sierpiński graphs: hyperbolicity and stochastic homogenization*, Preprint.

[KL01] Vadim A. Kaimanovich and Mikhail Lyubich, *Conformal and harmonic measures on laminations associated with rational maps*, 2001, Preprint.

[MS88] Calvin C. Moore and Claude Schochet, *Global analysis on foliated spaces*, Mathematical Sciences Research Institute Publications, vol. 9, Springer-Verlag, New York, 1988, With appendices by S. Hurder, Moore, Schochet and Robert J. Zimmer. MR **89h**:58184

[Pat76] S. J. Patterson, *The limit set of a Fuchsian group*, Acta Math. **136** (1976), no. 3-4, 241–273. MR 56 #8841

[Pat88] Alan L. T. Paterson, *Amenability*, Mathematical Surveys and Monographs, vol. 29, American Mathematical Society, Providence, RI, 1988. MR **90e**:43001

[Pie84] Jean-Paul Pier, *Amenable locally compact groups*, Pure and Applied Mathematics, John Wiley & Sons Inc., New York, 1984, A Wiley-Interscience Publication. MR **86a**:43001

[Pla75] J. F. Plante, *Foliations with measure preserving holonomy*, Ann. of Math. (2) **102** (1975), no. 2, 327–361. MR 52 #11947

[Sam79] Manuel Samuélidès, *Tout feuilletage à croissance polynomiale est hyperfini*, J. Funct. Anal. **34** (1979), no. 3, 363–369. MR **81m**:57021

[Ser79] Caroline Series, *Foliations of polynomial growth are hyperfinite*, Israel J. Math. **34** (1979), no. 3, 245–258 (1980). MR **82i**:28019

[Spa87] R. J. Spatzier, *An example of an amenable action from geometry*, Ergodic Theory Dynam. Systems **7** (1987), no. 2, 289–293. MR **88j**:58100

[SZ91] Ralf J. Spatzier and Robert J. Zimmer, *Fundamental groups of negatively curved manifolds and actions of semisimple groups*, Topology **30** (1991), no. 4, 591–601. MR **92m**:57047

[Ver78] A. M. Vershik, *The action of* PSL(2, **Z**) *in* $\mathbf{R}^1$ *is approximable*, Uspehi Mat. Nauk **33** (1978), no. 1(199), 209–210. MR 58 #13201

[Woe00] Wolfgang Woess, *Random walks on infinite graphs and groups*, Cambridge Tracts in Mathematics, vol. 138, Cambridge University Press, Cambridge, 2000. MR **2001k:**60006

[Zim77] Robert J. Zimmer, *Hyperfinite factors and amenable ergodic actions*, Invent. Math. **41** (1977), no. 1, 23–31. MR 57 #10438

[Zim78] ———, *Amenable ergodic group actions and an application to Poisson boundaries of random walks*, J. Functional Analysis **27** (1978), no. 3, 350–372. MR 57 #12775

CNRS UMR 6625, IRMAR, Université Rennes-1, Campus Beaulieu, 35042 Rennes, France

E-mail address: kaimanov@univ-rennes1.fr

Contemporary Mathematics
Volume **347**, 2004

Spectral Analysis on Tree Like Spaces from Gauge Theoretic View Points

Tsuyoshi Kato

ABSTRACT. In this paper we propose an approach to measure complexity of smooth structure on four manifolds by constructing Yang Mills moduli spaces over Casson handles.

0. Introduction

There are several classes of open Riemannian manifolds whose spectral geometry have been deeply studied, the universal covering spaces of compact manifolds, cusp manifolds, etc. In many situations, one will not restrict dimension for these classes.

Here we will study the bottom of spactra (near zero) for a new class of open manifolds restricted to the four dimension's case. This is so-called *Casson handles* (CH). A Casson handle is an open smooth four manifold.

Let us describe how Casson handles appear inside smooth four manifolds. Let λ be a non degenerate quadratic form over a free $\mathbf{Z}$ module Λ. We denote by $|\lambda|$, a simply connected topological closed orientable four manifold with the intersection form λ ([Fr]). More precisely this means that there exists an isomorphism:

$$(H^2(|\lambda|;\mathbf{Z}), (\quad,\quad)) \cong (\Lambda, \lambda).$$

When Λ is a direct sum of $\mathbf{Z}$, $\Lambda = \oplus \mathbf{Z}$, then one says that this isomorphism is a marking of $|\lambda|$.

Let M be a simply connected smooth closed oriented four manifold of even type. Then by Freedman, there exists a homeomorphism $M \cong k|-E_8| \oplus l(S^2 \times S^2)$ where $k|-E_8|$ is k times connected sums of $|-E_8|$. For simplicity of the notation, one restricts to the K3 surface M. However the process will be applied to the general case. Let us fix a marking:

$$(H^2(M;\mathbf{Z}), (\quad,\quad)) \mapsto (\mathbf{Z}^{16} \oplus \mathbf{Z}^6, -2E_8 \oplus 3H).$$

There exists an open subset $S \subset M$ homeomorphic to $3(S^2 \times S^2)\backslash$ pt, which is compatible with the decomposition of the forms. In particular S admits an induced smooth structure. This smoothing on the end is far from the standard

2000 *Mathematics Subject Classification.* 14J80, 57M30, 57R10, 57R57, 57R65.
Key words and phrases. Yang Mills theory, Casson handles.

product $S^3 \times [0, \infty)$. By the Casson-Freedman construction, the end of S admits a structure of *Casson handles*, $S \cong D^4 \cup_{i=1}^6 CH(T_i)$ where each $CH(T_i)$ is a Casson handle represented by T_i, an infinite tree with one end point, whose edges are assigned with signs $\pm$. We will call T_i an infinite signed tree.

There are sufficiently many Casson handles. In fact for each infinite signed tree, one can associate a Casson handle. It is known that there are uncountably many Casson handles which are mutually non diffeomorphic ([G], or see refs in [K]). On the other hand, they are all homeomorphic to the standard 2 handle $(D^2 \times D^2, S^1 \times D^2)$.

Question: Can one find a complete Riemannian metrics on CH which allow 'nice' spectral analysis ?

A Casson handle is expressed by an infinite tree with one end point and with signs $\pm$ on every edge of it. We say that such a tree is a signed tree.

In this paper we will define a subclass of infinite trees which is called homogeneous trees of bounded type. These are constructed by iterating to attatch infinitely many half periodic trees. A connected infinite subtree of a homogeneous tree of bounded type is called a tree of bounded type. We will call shortly a Casson handle of bounded type for a Casson handle constructed from a signed homogeneous tree of bounded type. Notice that a tree of bounded type grows polynomially. On the other hand homogeneous trees grow faster than exponentially in general, and these can be approximated by a family of homogeneous trees of bounded type.

We say that an open four manifold S has *tree-like end* if there are signed trees $T_1, \dots, T_l$, such that S is diffeomorphic to $D^4 \cup_j^l CH(T_j)$, where $CH(T_j)$ is the Casson handle corresponding to T_j, and every $CH(T_j)$ is attatched to the zero handle D^4 along the attatching S^1 of the first stage kinky handle (which corresponds to the end point in T_j). Notice that there is a natural smooth structure on it induced from the handle body.

Let S be of even type. Then we have the following:

THEOREM 0.1 ([**K**]). *Let $S = D^4 \cup_j^{2l} CH(T_j)$ have a tree-like end whose trees are homogeneous of bounded type. Then there are complete Riemannian metrics g of bounded geometry and weight functions w on S so that a bounded complex:*

$$0 \longrightarrow W_w^{k+2}(S; \Lambda^0) \xrightarrow{d} W_w^{k+1}(S; \Lambda^1) \xrightarrow{d_+} W_w^k(S; \Lambda_+^2) \longrightarrow 0$$

is Fredholm with $H^0 = 0$, $H^1 = 0$ and $l \leq \dim H_+^2(S; \mathbf{R}) \leq \dim H^2 \leq 2l = \dim H^2(S; \mathbf{R})$.

One will choose Riemannian metrics explicitly on each building block, kinky handle, and will attach them isometrically. Thus the construction of the Riemannian metrics and weight functions works for arbitrary Casson handles.

It is a linearized complex of the *Yang Mills equation* whose solution space is one of the most important subjects in four dimensional geometry. By the above theorem, one can construct Yang Mills moduli spaces over such S, generically as smooth and finite dimensional manifolds. In order to verify the above, one will use the analytic tool developed by [T1]. In the paper, it is shown that one can construct a gauge theory over periodic open 4 manifolds, using the Fourier-Laplace transform between spaces. Here in order to obtain Fredholmness for AHS complex over Casson handles, one uses the Fourier-Laplace transform iteratedly.

We found some difficulty to verify Fredholm property over S of unbounded type with respect to the constructed Riemannianm metrics and weight function. Here we propose:

Conjecture: For homogeneous Casson handles of unbounded type, there are no such metrics and weight functions as above.

In particular, this includes a class of Casson handles consisting of signed trees of exponential degree. In this note, we will explain one possible way to approach to this, which is still under progress.

By use of the above metrics and weight functions, one can construct Yang-Mills moduli spaces over S with tree-like end of bounded type. Here one will obtain some results on S using these moduli spaces. We have included outlines of these results in section 1.

The author would like to thank to the organizers of the JAMS symposium for their hospitarity at Matsushima, Sendai. Part of this work was done when the author visited the Max Planck Institut. He would like to express his gratitude to MPI for their hospitality.

1. Results and outline

1.A How to describe wildness of the ends; a basic approach: Let M be an oriented closed smooth four manifold of even type with $\pi_1(M) = 1$. Let us take a marking (an isomorphism from a unimodular form to the standard one):

$$\varphi : (H_2(M;\mathbf{Z}), <,>) \cong (\oplus^{8k}\mathbf{Z} \oplus^{2l} \mathbf{Z}, \quad k(-E_8) \oplus lH).$$

Then by Casson-Freedman, one finds an open four manifold S with tree-like end, homeomorphic to the interior of $l(S^2 \times S^2)\backslash D^4$, and finds a smooth embedding $S \hookrightarrow M$ which induces an embedding of the form:

$$(H^2(S;\mathbf{Z}), < \quad , \quad >) \cong (\oplus^{2l}\mathbf{Z}, lH) \hookrightarrow (H^2(M;\mathbf{Z}), < \quad , \quad >).$$

In order to construct the moduli space of $SO(3)$ ASD connections over M, one takes $\hat{\alpha} \in H^2(M;\mathbf{Z}_2)$ which is mod 2 reduction of $\alpha \in H^2(M;\mathbf{Z})$. Let us take $\hat{\alpha} \in H^2(M;\mathbf{Z}_2)$. Suppose the intersection form of M is isomorphic to $L = (-E_8) \oplus (k-1)(-E_8) \oplus lH$ for $k \geq 2$ and $l \geq 3$. Then one has the corresponding decomposition $\hat{\alpha} = \hat{\alpha}_1 \oplus \hat{\alpha}_2 \oplus \hat{\alpha}_3$. Let $\Sigma_1, \ldots, \Sigma_d \subset S$ be embedded surfaces which represent homology classes $[\Sigma_i] \in H_2(S;\mathbf{Z}) \hookrightarrow H_2(M;\mathbf{Z})$.

THEOREM 1.1. *Suppose $\hat{\alpha}_1, \hat{\alpha}_2, \hat{\alpha}_3 \neq 0$, and $S \subset M$ consists of signed homogeneous trees. Then if the $SO(3)$ Donaldson's invariant $Q_{k,\hat{\alpha}}(M; [\Sigma_1], \ldots, [\Sigma_d])$ does not vanish, then (1) at least one of the trees is of unbounded type, or (2) the L^2 ASD moduli space over S is generically w- separated.*

Here w is a weight function over S. The assumptions in the theorem are realized for some algebraic surfaces. This will be studied in section 4 (see 1.C).

Our main construction, when S is of bounded type, is to equip a complete Riemannian metric on S and construct the Atiyah, Hitchin and Singer complex of Fredholm type between w weighted Sobolev spaces. One can use any L^2 ASD connection over S as a base point, and can construct the L^2_w ASD moduli space. w-separated property is some discontinuity of the L^2_w ASD moduli space in the larger space, L^2 ASD moduli space over S, under generic deformation of metrics (see 2.D). The condition (2) states that if one gets an L^2 ASD connection as a limit element in $\mathfrak{M}(P,(M,h_i);[\Sigma_1],\ldots,[\Sigma_m])$, then it cannot be a base of any generic L^2_w ASD moduli space over S. Here h_i are Riemannian metrics on M converging to a complete Riemannian metric g on S on each compact subset.

Vanishing of Q is well known when M decomposes non trivially as a connected sum $M = W_1 \sharp W_2$ diffeomorphically with $b^2_+(W_i) > 0$ ([D2]). In a parallel way, one may expect that the vanishing result would also hold if the end of S is not so complicated smoothly.

The outline of the proof is as follows. Suppose the invariant is non zero and all trees are of bounded type. Let $L_2 = lH$ which is the intersection form of $l(S^2 \times S^2)$. Let us remove a small disk D^4 from this, and denote $S' = l(S^2 \times S^2)\backslash D^4$. Then one has a smooth immersion $S' \to M$ which is compatible with the intersection form. By Casson and Freedman's tower construction, one finds an open subset $S \subset S'$ and a smooth embedding $S \hookrightarrow M$ which realizes the form L_2. S has tree-like end. Notice that one may assume that all Σ_i are embedded in S. Our main construction is to introduce a complete Riemannian metric g and a weight function w on S so that one can do Fredholm part in gauge theory over (S,g), when S is of bounded type. Let us choose a family of Riemannian metrics g_i on M so that $g_i|S$ converges to g on every compact subset of S. Let $\mathfrak{M}_{k,\hat{\alpha}}((M,g_i);[\Sigma_1],\ldots,[\Sigma_d])$ be an oriented zero dimensional moduli space of ASD connections whose algebraic number of the points is the Donaldson invariant for any i. Its number is invariant under generic deformations of Riemannian metrics and is a smooth invariant. In particular for all i, the above spaces are all non empty. Let us take a sequence $[A_i] \in \mathfrak{M}_{k,\hat{\alpha}}((M,g_i);[\Sigma_1],\ldots,[\Sigma_d])$. Then one finds a subsequence such that their restrictions on S converges on every compact subset, to $[A_\infty] \in \mathfrak{M}_{k',\hat{\alpha}'}((S,g,w);[\Sigma_1],\ldots,[\Sigma_{d'}])$. If (2) could not be satisfied, then the latter moduli space should be a smooth manifold. Since $\hat{\alpha}_i \neq 0$, this space has a negative formal dimension. This is a contradiction.

1.B On asymptotics of bottom of spectra: The complex is of Fredholm if and only if the corresponding self adjoint elliptic operator has a discrete spectrum near zero. The proof of the main theorem shows that one can obtain some estimates on spectra of the AHS complex for CH of unbounded type, by approximating the tree by a family of trees of bounded type.

Let $T_1,\ldots,T_k$ be signed homogeneous trees defining open four manifolds S with tree-like end, not necessarily of bounded type. In this general case, one will not obtain moduli spaces as smooth manifolds over S. However the proof of the main theorem will give some estimates on the decay of the ASD connections $\{A_i\}$ over S. This can be obtained by approximating S by a family of open manifolds such that ASD moduli spaces are constructed over the approximations.

Let S be an open four manifold with tree-like end, smoothly embedded into a compact manifold M. Suppose the assumptions in the above theorem are satisfied. Notice that any tree with one end point admits a natural distance from the fixed point. Let us use e to denote the edges in T_j. Let $F(\{T_j\}) : \mathbf{N} \to \mathbf{N}$ be a function:

$$F(\{T_j\})(n) = \operatorname{Sup}_{v,j} \#\{e| \text{ one end of } e \text{ is } v, \text{ and } v \in T_j, |v| = n\}.$$

For any tree T, let us put $T(n) = \{x \in T : |x| \leq n\}$. For any unbounded homogeneous tree, one gets a family of homogeneous trees of bounded type $\{T^n\}_n$

such that $T(n)$ and $T^n(n)$ are isomorphic. Let us put:

$$Y_n = D^4 \cup_{j=1}^k CH(T_j^n), \quad W_n = D^4 \cup_{j=1}^k CH(T_j(n)).$$

Each Y_n admits a compatible Riemannian metric g_n, $g_n|W_n = g_n$, and a weight function w_n such that one can do Fredholm part in gauge theory over (Y_n, g_n, w_n).

Let us choose a sequence $A_i \in \mathfrak{M}((M, g_i), [\Sigma_1], \ldots, [\Sigma_m]) \neq \phi$. Then a subsequence $\{A_i\}$ converges to an L^2 ASD connection A_∞ over S on each compact subset. Let us define $G(\{A_i\}) : \mathbf{N} \times \mathbf{N} \mapsto (0, C]$ and λ_i by:

$$G(\{A_i\})(n, n') = |F_{A_n}|L^2_{w_n}(W_n \backslash W_{n-n'}, g; \Lambda^2_+),$$

(1)
$$\lambda_n = \inf \frac{|(d^+_{A_n})^*_{w_n}(u)|L^2_{w_n}(W_n)}{|u|L^2_{w_n}(W_n)}.$$

Let $\mathbf{F} = \{F : \mathbf{N} \mapsto \mathbf{N}\}$, $\mathbf{G} = \{h : \mathbf{N} \times \mathbf{N} \mapsto (0, C(h)]\}$ and $\mathbf{H} = \{h : \mathbf{N} \mapsto (0, C(h)]\}$ be the set of functions.

Let P be an $SO(3)$ bundle over M with $\hat{\alpha}_i \neq 0$, $i = 1, 2, 3$ as before. Suppose the restriction of the Donaldson invariant $Q|H_2(S; \mathbf{Z})$ is non zero as before. Then one can choose a sequence of ASD connections $A_i \in \mathfrak{M}((M, h_i), P)$ which converges to A_∞ as before.

THEOREM 1.2. *There exists a universally defined map $\mathfrak{F} : \mathbf{F} \times \mathbf{H} \mapsto \mathbf{G}$ and a constant C such that one has the pointwise estimate at (i, j) with $j \geq C\lambda_i^{-1}$:*

$$\mathfrak{F}(F(\{T_j\}), \{\lambda_i\}) \leq G(\{A_i\}).$$

The estimate is independent of any data on $M \backslash S$ and embedded surfaces. One will use the fact that the Donaldson invariant is polynomial. Notice that all Y_n, Riemannian metrics and weight functions are determined by the asymptotics of $\{T_j\}_j$. Thus the above statement saids that $\mathfrak{F}$ above is given by the growth function $F(\{T_j\}_j)$ and $\{\lambda_i\}$.

In order to verify this, one can use the above argument. If $G(\{A_i\})(n, n')$ is sufficiently small, then one gets A'_n on Y_n by cutting off A_n around $W_n \backslash W_{n-n'}$. Since A'_n has a sufficiently small $L^2_{w_n}$ norm on its curvature, one can find an ASD connection A''_n on Y_n around A'_n. Since Y_n has tree-like end of bounded type, one can construct ASD moduli space $\mathfrak{M}(Y_n, [\Sigma_1], \ldots, [\Sigma_{m'}])$. Its formal dimension is negative. This is a contradiction.

1.C Kähler forms and Picard lattice: The situation in the main result (theorem 1.1) is realized by some algebraic surfaces, using some relation between Kähler forms and Picard lattices.

Let M be a simply connected complex algebraic surface of even type with $p_g > 0$. Let us take a marking:

$$\psi : H^2(M; \mathbf{Z}) \cong \oplus^8 \mathbf{Z} \oplus^{8(k-1)} \mathbf{Z} \oplus^{2l} \mathbf{Z}$$

with respect to the decomposition of the form as $(-E_8) \oplus_{k-1} (-E_8) \oplus^l H$, $k \geq 2$. Then one has an open manifold with tree-like end $S \subset M$, which corresponds to lH. Let h be a hyperplane class of M, and decompose $h = h_1 \oplus h_2 \oplus h_3$ with respect to above. Suppose there is an integral $(1,1)$ form $\alpha = \alpha_1 \oplus \alpha_2 \oplus \alpha_3 \in H^{1,1}(M: \mathbf{R}) \cap H^2(M; \mathbf{Z})$, such that the mod 2 reduction $\hat{\alpha_1}, \hat{\alpha_2}, \hat{\alpha_3} \neq 0 \in H^2(M; \mathbf{Z}_2)$.

COROLLARY 1.3. *Let M be as above, and α satisfy the above condition. Then if there is a hyperplane class $h \in H^2(M; \mathbf{Z})$ with $h_1 = h_2 = 0$, then at least one* *of T_j for S have unbounded type, or L^2 ASD moduli space over S is generically w-separated.*

This follows from the Theorem 1.1, and the non vanishing result:

THEOREM 1.4 ([**D1**]). *Let M be as above and take any hyperplane class h. Then there is $k_0 = k(M, h)$ such that for any $k \geq k_0$, and any $\hat{\alpha}$ which is mod 2 reduction of an integral $(1,1)$ class, the Donaldson invariant $Q_{\hat{\alpha},k}(h)$ is positive.*

The non realizable result above, gives some restrictions on the corresponding trees, from a relation between the Kähler cone and the Picard lattice.

In section 4, we will introduce a notion, for the Picard group $\text{Pic}(M)$ in $H^2(M; \mathbf{Z})$ to be *large*. For example the Picard groups of all projective Kummer surfaces are large (more than 17).

Let M be as above, and suppose the Picard group of M is large.

COROLLARY 1.5. *Suppose a complex algebraic surface M of even type has a large Picard group. Then there is a marking, and the corresponding $S \subset M$ represented by signed homogeneous trees $T_1, \ldots, T_{2l}$ such that at least one of them must be of unbounded type, or L^2 ASD moduli space over S is generically w-separated.*

1.D An approach to the conjecture: Let M be a simply connected oriented four manifold of even type. Let us take a topological decomposition $M \cong \sharp^k | -E_8 | \sharp^l S^2 \times S^2$. Then one has an open manifold with tree-like end $S \subset M$, which corresponds to $\sharp^l S^2 \times S^2$ as before.

Let us recall the conjecture proposed in the introduction. Here we will indicate a possible way to approach this. The idea has already appeared in the proof of the main theorem.

Step 1: Let us choose some M (e.g. algebraic surfaces) so that $\mathfrak{M}(E, (M, h)) \neq \phi$ for all generic h. By Freedman, one may express $S \cong D^4 \cup \cup_{i=1}^{2l} CH(T_i) \subset M$. In general these Casson handles will be of unbounded type.

Step 2: Suppose S could admit an admissible pair (g, w). Then choose an exaustion:

$$
\begin{aligned}
&K_0 \Subset K_1 \Subset \cdots \subset S \subset M, \\
&h_i : \text{ family of Riemannian metrics on } M \text{ with } h_i | K_i = g | K_i.
\end{aligned}
\tag{2}
$$

Let us choose an $SO(3)$ bundles $E \to M$ so that $w_2(E|S) \neq 0$, $w_2(E|\sharp^k| - E_8|) \neq 0$.

Now we take any element $[A_i] \in \mathfrak{M}(E, (M, h_i)) \neq \phi$. For simplicity we assume here dim $\mathfrak{M}(E, (M, h_i)) = 0$ (In general, we use the Donaldson polynomial). A subsequence $\{A_{k_i}\}_i$ converges to an ASD connection A_∞ over $E' \to (S, g)$ (see 1.C). In particular $\mathfrak{M}(E', (S, g, w)) \neq \phi$.

Step 3: The last is dimension counting:

$$
\begin{aligned}
&\dim \mathfrak{M}(E, (M, h_i)) = -2p_1(E) - 3(1 + b_+^2(M)) = 0, \\
&\dim \mathfrak{M}(E', (S, g, w)) = -2p_1(E') - 3b_+^2(S)
\end{aligned}
\tag{3}
$$

where $-p_1(E') + 2 \leq -p_1(E)$ and $b_+^2(M) \leq b_+^2(S)$ holds. Thus we have a negative formal dimension of $\mathfrak{M}(E', (S, g, w))$. Since this is non empty, this would be a contradiction if these moduli spaces should have structure of smooth manifolds.

More detail can be found in the proof of theorem 3.1. In order to accomplish this method, one would need a more elaborated transversality argument.

2. Moduli space

2.A Admissible four manifolds: Let Y be a non compact smooth four manifold. Throughout this section, one always assumes that Y is simply connected, and simply connected at infinity. In this section, one will construct Yang Mills moduli spaces over Y. For details, we refer to [K].

In [K], we have introduced a notion on Y, *admissible* for a pair (g, w) of a complete Riemannian metric g of bounded geometry and a weight function w over a non compact four manifold. Essentially this requires Fredholmness of the corresponding Atiyah, Hitchin and Singer complex and computability of the cohomology. There one has verified the following:

THEOREM 2.1 ([**K**]). *Let Y be diffeomorphic to a Casson handle of homogeneously bounded type. Then one can equip an admissible pair on it.*

Theorem 0.1 follows from this by definition of admissibility. In the following, we fix an admissible pair (g, w) over Y. We will use weighted Sobolev spaces $W^k_w(Y)$ with the norms:

$$|u|W^k_w = (\Sigma_{l\leq k}\int \exp(w)|\nabla^l(u)|^2 vol)^{\frac{1}{2}}.$$

2.B ASD moduli space: Let $E \to Y$ be a $G = SO(3)$ vector bundle such that except some compact subset $K \subset Y$, $E|Y\backslash K$ is trivial. One denotes the corresponding principal G bundle by P. For $SO(3)$ bundles, E is determined by $w_2(E) \in H^2(Y; \mathbf{Z}_2)$. In later sections, we fix a trivialization of $E|Y\backslash K$. Thus we fix an $SO(3)$ bundle with w_2 and $p_1(E)$. Let A be a smooth connection over E such that except some compact subset on Y, it satisfies the ASD equation:

$$F_A + *F_A = 0,$$

where F_A is the curvature form of A. Let us denote by $\tilde{R}(Y)$ the set of smooth connections as above satisfying $|F_A|L^2(Y) < \infty$. By changing a trivialization, one may assume $|a|L^2(Y\backslash K) < \infty$ where $A = d + a$ on $Y\backslash K$. Later we always assume this property. If A is the trivial connection except some compact subset, then it is an element in $\tilde{R}(Y)$.

Let us put $Ad(P) = P \times_G \mathfrak{G}$, where $\mathfrak{G}$ is the Lie algebra of G. $(P \times_G \mathbf{R}^3 = E)$. Let us fix $A_0 \in \tilde{R}(Y)$. One defines an affine Hilbert space as:

$$\mathfrak{A}_k(P)_0 = \{A_0 + a | a \in W^k_w(Y; AdP \otimes T^*Y)\}, \quad k \geq 3.$$

Let us take $g \in C^1_{\text{loc}}(Y; AutE)$. By embedding as $g \in C^1_{\text{loc}}(Y; Hom(E, E))$, one may consider $\nabla_{A_0} g \in C^0(Y; Hom(E, E) \otimes T^*Y)$. Notice that if g is locally W^4, then it is of C^1 class. Now one defines the weighted Sobolev gauge group $(l \geq 4)$:

$$(4)\quad \begin{aligned}\mathfrak{G}_l(P) &= \{h \in W^l_{loc}(Y; Aut(E)) | \nabla_{A_0} h \in W^{l-1}_w(Y; Hom(E, E) \otimes T^*Y)\},\\ \mathfrak{G}_l(P)_0 &= \{h \in W^l_{loc}(Y; Aut(E)) | h - id \in W^l_w(Y; Hom(E, E) \otimes T^*Y)\}.\end{aligned}$$

The Lie algebras of $\mathfrak{G}_l(P)$ and $\mathfrak{G}_l(P)_0$ are correspondingly as follows:

$$(5)\quad \begin{aligned}\mathfrak{g}_l(P) &= \{h \in W^l_{loc}(Y; Ad(P)) | \nabla_{A_0} h \in W^{l-1}_w(Y; Ad(P) \otimes T^*Y)\},\\ \mathfrak{g}_l(P)_0 &= W^l_w(Y; Ad(P)).\end{aligned}$$

These spaces all admit structure of Banach manifolds. $\mathfrak{G}_{k+1}(P)_0$ acts on $\mathfrak{A}_k(P)_0$ by $g^*(A_0 + a) = g^{-1}\nabla_{A_0} g + g^{-1} a g$. Notice that $\mathfrak{G}_{k+1}(P)$ may not act on $\mathfrak{A}_k(P)_0$, since A_0 may live only in $L^2(Y)$.

Let us take any $A \in \hat{\mathfrak{M}}_k(P)$. Then one has the AHS complex coefficient in A:

$$(6)\quad \begin{aligned} 0 \longrightarrow W_w^{k+1}((Y,g');\Lambda^0 \otimes ad\mathfrak{g}) &\xrightarrow{d_A} W_w^k((Y,g');\Lambda^1 \otimes ad\mathfrak{g}) \\ &\xrightarrow{d_A^+} W_w^{k-1}((Y,g');\Lambda_+^2 \otimes ad\mathfrak{g}) \longrightarrow 0. \end{aligned}$$

PROPOSITION 2.2. *Suppose* $d_A^+ : W_w^k(Y; AdP \otimes T^*Y) \mapsto W_w^{k-1}(Y; AdP \otimes \Lambda_+^2)$ *is surjection for* $A \in \hat{\mathfrak{M}}_k(P)$. *Then* $\mathfrak{M}_k(P) = \hat{\mathfrak{M}}_k(P)/\mathfrak{G}_{k+1}(P)_0$ *is a finite dimensional smooth manifold at* $[A]$. *Its tangent space is naturally isomorphic to* $\text{Ker } d_A^+/\text{im } d_A$.

In order to construct smooth moduli spaces, one uses K.Uhlenbeck's generic metric theorem. Let (g, w) be an admissible pair for Y. Let us choose a smooth map $h : Y \mapsto [0, \infty)$ with $h(x) \geq \frac{w}{2}(x)$ and a small $\epsilon > 0$. Then one introduces the following Banach manifold:

$$\mathfrak{C} = \{\phi \in C^l(Gl(TM)) : Sup_K(\Sigma_{j=0}^l e^h|\nabla^j(\phi^* g - g)|K) < \epsilon, K : \text{ compact}\}.$$

Let us take $\phi \in \mathfrak{C}$, and put $g' = \phi^* g$. Then one has the AHS complex with respect to g':

$$0 \longrightarrow W_w^{k+1}((Y,g');\Lambda^0) \xrightarrow{d} W_w^k((Y,g');\Lambda^1) \xrightarrow{d^+} W_w^{k-1}((Y,g');\Lambda_+^2) \longrightarrow 0.$$

One sees that this is a Fredholm complex with the same index as the unpurturbed one. The following result is by [FU] (also see [**K**]):

PROPOSITION 2.3. *Suppose* Y *is indefinite. Let us fix* A_0, *an* L^2 *ASD connection with respect to* g. *Then for a Baire set of* $\phi \in \mathfrak{C}$, *there are no reducible connections in* $\mathfrak{A}_k(P)_0$ *with respect to* $\phi^*(g)$. *In particular* $\mathfrak{M}_k(P, \phi^*(g))$ *is a smooth manifold.*

2.C Large deformation of metrics: Let M be an oriented smooth closed four manifold, and take an open subset $Y \subset M$. Suppose Y can admit (g, w), an admissible pair over Y. Here one aims to obtain ASD connections over Y, from sequences of ASD connections over M.

Let us take an exaustion $K_0 \Subset K_1 \Subset \cdots \subset Y$ by compact subsets, and choose a generic family of Riemannian metrics $\{h_i\}_i$ on M such that $|h_i|K_i - g|K_i|C^l(K_i) \to 0$ as $i \to \infty$.

Let us take an $SO(3)$ bundle $P \mapsto M$, and take a sequence of ASD connections A_i over P with respect to (M, h_i).

PROPOSITION 2.4 ([**U1**], [**U2**]). *After gauge transformation* $g_i^*(A_i)$ *and taking a subsequnce (which we also denote* A_i), $\{A_i\}$ *converges to an ASD connection* A_∞ *with respect to* (Y, g), *on every compact subsets of* Y *except for a finite set of points.*

Notice that one knows only that F_{A_∞} lies in $L^2((Y,g);\Lambda_+^2)$. Recall that in order to construct an ASD moduli space $\mathfrak{M}_k(Y, g)$, one has to choose an L^2 'base connection'. Here we will take A_∞ as the base. After changing a trivialization on the end, one may assume $a \in L^2(Y \backslash K)$ where $A_\infty|Y \backslash K = d + a$.

In order to avoid the case when A_∞ is a flat connection, one will require some conditions on $w_2(P) \in H^2(M; \mathbf{Z})$. This will be considered later.

2.D Discontinuity of L^2 ASD connection under small deformation of metrics: In order to obtain a moduli space as a smooth manifold, one needs to make a small deformation of Riemannian metrics. Suppose for any generic metric h on M, the ASD moduli space $\mathfrak{M}_k(M,h)$ is non empty. Let us take a family of irreducible ASD connections $\{A_i \in \mathfrak{M}_k(M,h_i)\}$ such that $h_i|K_i$ converges to g, and A_i converges to A_∞ as above. Let us choose a generic small purturbation $g' = \varphi^*(g)$ of g on Y, where $\varphi \in \mathfrak{C}$. Then one considers the following continuity problem for the limit ASD connection:

Continuity problem: Let g and $g' = \varphi^*(g)$ be as above. Then for any sequence $\{A_i\}$ as above, does there exist generic small perturbations h_i' of h_i with $h_i' \to \varphi^*(g)$ on compact subsets, and another family $\{A_i' \in \mathfrak{M}_k(M,h_i')\}$ satisfying the following properties ? A_i' converges to $A_\infty' \in \mathfrak{M}_k(Y,g')$ on every compact subset except finite points, where:

$$A_\infty' \in \mathfrak{A}(P)_0 = A_\infty + W_w^k(Y;Ad(P)\otimes\Lambda^1). \quad (**)$$

Let us formulate a w-separation property which states some discontinuity, and does not satisfy $(**)$. Let (g,w) be an admissible pair on Y, and take an $SO(3)$ bundle $P \to Y$. Then one defines the L^2 ASD moduli space $\mathfrak{M}(g)$ as the set of smooth ASD connections on P with $\int_Y |F_A|^2 < \infty$. By changing a trivialization near the end, one may assume $A \in W^k(Y;Ad(P)\otimes\Lambda^1)$.

In section 3, one will introduce $Q(M,P;[\Sigma_1],\dots,[\Sigma_m]) \in \mathbf{Z}$, the Donaldson invariant. For this, one will choose generic sections of line bundles $\Lambda^1,\dots,\Lambda^m$ over $\Sigma_j \subset Y$.

Now let us take any $A \in \mathfrak{M}(g)$. We say that A is *generically w-separated* under small perturbation, if there exists a Baire set $\mathfrak{B} \subset \mathfrak{C}$ such that for any $\varphi \in \mathfrak{B}$ and any generic sections $s_1,\dots,s_m$, one has the following property; let us choose another families of generic sections $s_1^i,\dots,s_m^i$ and h_i so that (1) each $s_j^i \to s_j$ as $i \to \infty$, (2) $\mathfrak{M}(P,(M,h_i);s_1^i,\dots,s_m^i)$ is generically defined, and (3) $h_i \to \varphi^*(g)$.

Then for any convergent subsequence $A_i \in \mathfrak{M}(P,(M,h_i);s_1^i,\dots,s_m^i)$ as 2.C with the limit A_∞, A and A_∞ has w- separated property:

$$A - A_\infty \notin L_w^2((Y,g);\Lambda^1).$$

If each $A \in \mathfrak{M}(g)$ is generically w-separated, then the L^2 moduli space $\mathfrak{M}(g)$ is called generically w-separated (with respect to $Q(M,P;[\Sigma_1],\dots,[\Sigma_m])$).

2.E Asymptotics of ASD connections: Here one formulates asymptotics of L^2 norms of F_A for ASD connections A over a complete Riemannian manifold (Y,g). One will get some lower bound of the sequence, if A is obtained as a limit under deformation of Riemannian metrics on closed manifold M. The estimate is independent of any data on $M\backslash Y$.

Let $Y \subset M$ be an open connected subset. Let us equip a complete Riemannanian metrtic g of bounded geometry. An admissible approximation of (M,Y,g) is a sequence of admissible triples $\{(Y_i,g_i,w_i)\}$ with the following property; there exists an exaustion of Y by compact subsets $L_0 \subset L_1 \subset \cdots \subset Y$ and diffeomorphisms $\varphi_i|L_i$ from L_i into Y_i such that (1) $g|L_i = g_i|L_i$,

(2) $d(\partial L_{i+1},L_i) \geq 1$ and

(3) $\pi_1(L_{l+1}\backslash L_l) \mapsto \pi_1(L_{l+2}\backslash L_l)$ has zero image.

We always assume that any admissible approximation consists of unformly bounded geometry.

Let us take a generic family of Riemannian metrics $\{h_i\}$ on M such that $|h_i - g|C^l(L_i) < \epsilon_i \to 0$. Then one chooses a sequence $\{A_i \in \mathfrak{M}_k(M, h_i)\}_i$. By taking a subsequence, one may assume that $\{A_i\}$ converges on each compact subset, to A_∞, an ASD connection on (Y, g). One defines an *asymptotics* of convergent sequence $\{A_i\}$ and a partial lowest spectrum:

$$\{a_{i,j} = |F_{A_i}|L^2_{w_i}(L_i \backslash L_{i-j}; \Lambda^2_+)\}_{i,j}, \quad \lambda_i = \lambda(L_i, A_i) = \inf \frac{|(d^+_{A_i})^*_{w_i}(u)|L^2_{w_i}(L_i)}{|u|L^2_{w_i}(L_i)}.$$

Let us take an admissible approximation $\{(Y_i, g_i, w_i)\}$ of (M, Y, g) as above. Let $\{A_i\}$, $\{\lambda_i\}$ and $\{a_{i,j}\}$ be as above. Let us take cut off functions $\varphi_{i,j}$ for each i, j on M with $\varphi_{i,j}|M \backslash L_i \equiv 0$ and $\varphi_{i,j}|L_{i-j} \equiv 1$.

PROPOSITION 2.5. *There is a function $G : \mathbf{N} \times \mathbf{N} \times (0,1] \mapsto (0,1]$ which only depends on the admissible approximation with the following property; if one has bound $G(i,j,\lambda_i) > a_{i,j}$ for (i,j) with $j \geq C\lambda_i^{-1}$, then one gets an ASD connection A'_i over Y_i with resepct to g_i such that* (1) $|A'_i|W^k_{w_i}(Y_i, g_i) < \infty$, *and* (2) *its norm satisfies $|A'_i - \varphi_{i,j}A_i|W^k_{w_i} < C(Y_i)a_{i,j}$.*

The continuity method is a well known process to obtain an ASD connection by slightly perturbing an 'almost' ASD connection. The proof follows by applying the continuity method over each (Y_i, g_i). Notice the inequality:

(7)

$$\begin{aligned}
|u|L^2_{w_i}(Y_i) &\leq |(1-\varphi_{i,j})u|L^2_{w_i} + |\varphi_{i,j}u|L^2_{w_i} \\
&\leq \lambda_i^{-1}(|d(1-\varphi_{i,j})u|L^2_{w_i} + |d_{A_i}\varphi_{i,j}u|L^2_{w_i}) \\
&\leq \lambda_i^{-1}(|(1-\varphi_{i,j})d_{A'_i}u|L^2_{w_i} + |\varphi_{i,j}d_{A'_i}u|L^2_{w_i}) + \frac{C}{2}j^{-1}\lambda_i^{-1}|u|L^2_{w_i}(L_i \backslash L_{i-j}).
\end{aligned}$$

3. Casson handles and the Donaldson invariant

3.A The Donaldson invariant: Let M be an oriented simply connected smooth closed four manifold. Let h be any element in $H_2(M; \mathbf{Z})$. One may represent h by a smoothly embedded Riemann surface. Let $\Sigma_1, \ldots, \Sigma_m \subset M$ be embedded surfaces representing homology classes $h_i = [\Sigma_i] \in H_2(M; \mathbf{Z})$. Let $\phi : H_2(M; \mathbf{Z}) \cong \oplus\mathbf{Z}$ be a marking.

Suppose M is of even type. Thus the intersection form is isomorphic to $(-E_8) \oplus (k-1)(-E_8) \oplus lH$, $k \geq 2$. Associated to the decomposition of this form, one may express:

$$\oplus\mathbf{Z} = \oplus_8 \mathbf{Z} \oplus_{8(k-1)} \mathbf{Z} \oplus_{2l} \mathbf{Z}.$$

Through ϕ and this decomposition, one has decompositions of $[\Sigma_i]$ as $[\Sigma_i] = h_i^1 \oplus h_i^2 \oplus h_i^3$, where $h_i^1 \in \oplus_8\mathbf{Z}$, $h_i^2 \in \oplus_{8(k-1)}\mathbf{Z}$, and $h_i^3 \in \oplus_{2l}\mathbf{Z}$.

Now suppose $Y \subset M$ is homeomorphic to $l(S^2 \times S^2) \backslash D^4$ which induces an isomorphism:

$$i : H^2(Y; \mathbf{Z})(\hookrightarrow H^2(M; \mathbf{Z})) \quad \cong \oplus_{2l}\mathbf{Z}$$

through ϕ. Moreover suppose the induced smooth structure on Y admits a Casson handle of homogeneously bounded type. Then one has an admissible pair (g, w) on Y (see [K]). Under the above assumption, one can construct the Donaldson invariant $Q(P; (Y, g, w); h_1, \ldots, h_m)$. One can construct Q for all oriented closed smooth four manifold. The constructions for Y are parallel to the closed case. Here is a review for the latter.

We follows [FS2]. Let E be an $SO(3)$ bundle over M, and take an embedded Riemann surface $\Sigma \subset M$. Let $\mathfrak{B}_k^*(\Sigma; E|\Sigma) = \mathfrak{A}(E|\Sigma)_k^*/\mathfrak{G}_{k+1}(E|\Sigma)$ be the gauge equivalence classes of irreducible Sobolev k connections. Similarly one has $\tilde{\mathfrak{B}}_k^*(\Sigma; E|\Sigma) = \mathfrak{A}(E|\Sigma)_k^*/\mathfrak{G}_{k+1}(E|\Sigma)_0$, where $\mathfrak{G}_{k+1}(E|\Sigma)_0 \subset \mathfrak{G}_{k+1}(E|\Sigma)$ is a subgroup with $g(x_0)$ to be the identity for a fixed point $x_0 \in \Sigma$. Thus one has a fibration:

$$SO(3) \hookrightarrow \tilde{\mathfrak{B}}_k^*(\Sigma; E|\Sigma) \mapsto \mathfrak{B}_k^*(\Sigma; E|\Sigma),$$

which gives an $SO(3)$ principal bundle β over $\mathfrak{B}_k^*(\Sigma; E|\Sigma)$. One may assume that restriction of any irreducible connection A over E to $E|\Sigma$ is also irreducible ([FS]). Thus let us choose a lift $\tilde{\beta}$ over $\mathfrak{B}_k^*(\Sigma; E|\Sigma)$. One has $c_1(\tilde{\beta}) = w_2(\beta)$ mod 2. Let us choose a complex line bundle Λ over $\mathfrak{B}_k^*(\Sigma; E|\Sigma)$ with $c_1(\Lambda) = c_1(\tilde{\beta})$. Now one has the restriction map:

$$r : \mathfrak{M}_k^*(P) \mapsto \mathfrak{B}_k^*(\Sigma; E|\Sigma).$$

One chooses a generic section s of Λ. Then by the transversality argument ([DK] p192), one has a smooth manifold $\mathfrak{M}_k^*(P) \cap r^{-1}(\text{ Ker} s)$ of codimension 2. One may generalize this construction. Suppose $\dim\mathfrak{M}_*(P) = 2m$, and choose m embedded surfaces $\Sigma_1, \dots, \Sigma_m \subset M$. As before, one has the determinant bundles over $\mathfrak{B}_k^*(\Sigma_j; E|\Sigma_j)$. Then for generic sections $s_1, \dots, s_m$, one has a smooth submanifold of dimension 0 as:

$$\mathfrak{M}_*(P) \cap r_1^{-1}(\text{ Ker} s_1) \cap \dots \cap r_m^{-1}(\text{ Ker} s_m).$$

By choosing these embeddings with general positions, one may assume that this 0 dimensional manifold consists of finite oriented points. Thus one may count the number of points so that one obtains an integer. This number is, by definition, $Q(M, P; [\Sigma_1], \dots, [\Sigma_m])$ (modulo constant).

3.B Main theorem: Our main theorem (theorem 1.1) is as follows:

THEOREM 3.1. *Let $\pi_1(M) = 1$, all $h_i^1 = h_i^2 = 0$, $1 \le i \le m$ and $\hat{\alpha}_i \neq 0$, $i = 1, 2, 3$. Then if the Donaldson invariant does not vanish $Q(M, P; h_1, \dots, h_m) \neq 0$, then the end of Y is tree like of unbounded type, or the L^2 ASD moduli space over (Y, g) is generically w-separated.*

Proof : Let $T_1, \dots, T_{2l}$ be $2l$ signed trees, and put $Y = D^4 \cup CH(T_1) \cup \dots \cup CH(T_{2l})$. Let us fix a marking of $H^2(M; \mathbf{Z})$, and take a smooth embedding $Y \hookrightarrow M$, which is compatible with the forms.

On the other hand, there exists a subset $\bar{Y} \subset Y$ whose interior is homeomorphic to Y, such that $M \backslash \bar{Y}$ is homeomorphic to $M_1 \sharp M_2 \backslash$ pt, where $M_1 = |-E_8|$ and $M_2 = |(k-1)(-E_8)|$ respectively.

By taking any integral lift $\alpha \in H^2(M; \mathbf{Z})$, it gives a lift of the $SO(3)$ bundle to another $U(2)$ bundle $\bar{P}$. Then for any $h \in H_2(M; \mathbf{Z})$ and its mod 2 reduction $[h]_2$, one has the equality $< w_2(P), [h]_2 >=< c_1(\bar{P}), h >$ mod 2.

Now suppose both conclusions would not hold. In particular Y could have the end of bounded type. Then by Theorem 0.1, one can equip an admissible pair, a complete Riemannian metric g of bounded geometry and a smooth function $w : Y \mapsto [0, \infty)$, so that one gets weighted Sobolev spaces $W_w^l(Y; \Lambda^*)$. Let us choose arbitrary complete Riemannian metrics of bounded geometry g_1 and g_2 on the interiors of M_1 and M_2.

Let $E \to Y$ be an $SO(3)$ bundle with $p_1(E)$ and $w_2(E) = \hat{\alpha}$, which satisfies the conditions in the theorem. Let us take an L^2 ASD connection A_0 over (E, Y, g). Then by a generic choice of g, one sees that for all A_0, d_{A_0} is an injection and $d^+_{A_0}$ is a surjection. Thus one has a moduli space $\mathfrak{M}_*(E)$ of ASD connections over E, which is smooth and of dimension $2m$. Let $\Sigma_1, \dots, \Sigma_{m'} \subset Y$ be embedded surfaces as above. Then one has a smooth manifold $\mathfrak{M}(E; [\Sigma_1], \dots, [\Sigma_{m'}])$ defined similarly as before. This space has a formal dimension $2(m - m')$.

Now let $\Sigma_1, \dots, \Sigma_m \subset M$ be as in 3.A. By the assumption, one may assume that all these surfaces are included in Y. Moreover one may assume that these are in general position; for any $\Sigma_i, \Sigma_j, \Sigma_k$, one has empty intersection $\Sigma_i \cap \Sigma_j \cap \Sigma_k = \phi$.

Suppose one has $Q(P, M; [\Sigma_1], \dots, [\Sigma_m]) \neq 0$. This shows that for a generic choice of Riemannian metric g on M, and also generic choice of these surfaces, one has non empty 0 dimensional moduli space $\mathfrak{M}(P, (M, g); [\Sigma_1], \dots, [\Sigma_m])$. The algebraic number of the space is equal to the above invariant Q.

Let us choose an exausting family of compact subsets as:

$$K_0 \subset K_1 \subset \cdots \subset Y, \quad K'_0 \subset K'_1 \subset \cdots \subset M_1, \quad K''_0 \subset K''_1 \subset \cdots \subset M_2.$$

Let us choose a family of metrics g'_i on M with $g'_i|K_i = g|K_i$, $g'_i|K'_i = g|K'_i$ and $g'_i|K''_i = g|K''_i$. Let us choose a family of small constants $\epsilon_i > 0$, $i = 0, 1, \dots$, so that $\epsilon_i \to 0$. In order to construct a smooth moduli space $\mathfrak{M}(E, M; [\Sigma_1], \dots, [\Sigma_m])$, one needs to perturbe g'_i slightly. Let us choose a family of generic Riemannian metrics g_i on M with $|g - g_i|W^l_w(K_i \cup K'_i \cup K''_i, g \cup g_1 \cup g_2) \leq \epsilon_i$. One may assume that the family $\{g_i\}$ consists of uniformly bounded geometry.

Then one has a family of smooth manifold:

$$\{\mathfrak{M}(P, (M, g_i); [\Sigma_1], \dots, [\Sigma_m])\}_{i \geq 0}.$$

Let us choose any sequence $[A_i] \in \mathfrak{M}(P, (M, g_i); [\Sigma_1], \dots, [\Sigma_m])$, $i = 0, 1, \dots$. Then there is a subsequence which we also denote by $[A_i]$, and gauge transformations $h_i \in Aut(P)$ so that $h^*_i(A_i)$ converges to $[A_\infty]$, an L^2 ASD connection with respect to (Y, g) in the C^∞ topology on every compact subset of $Y \backslash$ points. More precisely, for any compact subset $K_n \subset Y$ and any small $\epsilon > 0$, there exists $i_0 \gg 0$ so that for any $i \geq i_0$, one has $|h^*_i(A_i) - A_\infty|C^l(K_n) < \epsilon$.

Similarly for $l = 1, 2$, $h^*_i(A_i)|M_l \backslash$ pts converge to L^2 ASD connections A^l_∞ over (M_l, g_l) respectively.

Now let $\{p_1, \dots, p_l\} \subset Y$ be the bubbling points for the sequence $\{A_i\}$. Then at most $2l$ surfaces can contain these points. Thus by reordering surfaces, one obtains a non empty moduli space as a smooth finite dimensional manifold by Proposition 2.3:

$$A_\infty \in \mathfrak{M}(E', (Y, g); [\Sigma_1], \dots, [\Sigma_{m-2l}]).$$

Notice that here we have used both assumptions of the contradiction argument; Y is equipped with an admissible pair (g, w), and one might choose a generic A_∞ so that the conclusion in Proposition 2.3 holds with respect to A_∞.

Let us see that A_∞ defines an $SO(3)$ bundle E' on Y with strictly smaller $-p_1(E')$ compared to $-p_1(E)$ on M. For this, one will see $|F_{A^l_\infty}|L^2 > 0$ and A^l_∞ define non trivial bundles on M_l, $l = 1$ or 2. One has chosen E so that each $\hat{\alpha}_j \neq 0$ where $w_2(E) = \hat{\alpha} = \Sigma^3_{j=1} \hat{\alpha}_j \in \mathbf{Z}^8_2 \oplus_{k-1} \mathbf{Z}^8_2 \oplus \oplus_{2l} \mathbf{Z}_2 \cong H^2(M; \mathbf{Z}_2)$.

Let $\alpha_j \in H^2(M; \mathbf{Z})$ be the corresponding lift. Let us take homology classes $h = h_1 + h_2 + h_3 \in H_2(M; \mathbf{Z})$ such that the mod 2 reduction of $< \alpha_j, h_j >_2 \in \mathbf{Z}_2$ are all non zero $(= 1)$.

First one sees $w_2(E') = \hat{\alpha}_3 \in H^2(Y; \mathbf{Z}_2)$. Let $\Sigma_j \subset M$ be embedded surfaces which represent h_j. As below, one has $w_2(E') \neq 0$. If the family $\{F_{A_i}\}_i$ concentrates at some points $p \in M_1$ and $p' \in M_2$, then clearly one has $-p_1(E') \leq -p_1(E) - 4 - 4l$. Now suppose it does not concentrate on M_1. Let $\Sigma_1 \subset M_1$ be as above. Then one has the equality $< w_2(E), \Sigma_1 >=< c_1(\bar{P}), \Sigma_1 >=< c_1(A_i), \Sigma_1 >\neq 0$ mod 2. Since there are no concentrating points, one has also $< c_1(A^1_\infty), \Sigma_1 >\neq 0$. Then clearly lim $\sup_i |F_{A_i}|L^2(M_1, g_i|M_1) \geq 4\pi^2$. Thus in any situation, one has $-p_1(E') \leq -p_1(E) - 2 - 4l$.

Then using A_∞ as the base point, one has the non empty moduli space of ASD connections $\mathfrak{M}(E', (Y, g); [\Sigma_1], \ldots, [\Sigma_{m-2l}])$. One may assume that this space contains no reducible orbits. Then one has

$$\dim \mathfrak{M} = \dim \operatorname{Ker} d^+_{A_\infty} / \operatorname{im} d_{A_\infty} - 2(m - 2l).$$

By Theorem 0.1, one obtains an inequality:

$$\begin{aligned} \text{(8)} \qquad \dim \operatorname{Ker} d^+_{A_\infty} / \operatorname{im} d_{A_\infty} &\leq -2p_1(E') + 2p_1(E) + 2m + 3 - 2(m - 2l) \\ &\leq -4 - 8l + 2m + 3 - 2(m - 2l) = -1 - 4l < 0, \end{aligned}$$

where 3 comes from dimension of the constant gauge transformation. In particular, the formal dimension of this space is negative. This is a contradiction. This completes the proof.

3.C Asymptotics of ASD connections: Let T be any infinite tree and $|\ \ |$ be the norm which is the distance from the base point. Every integer value corresponds to some vertex of T. Let us use notations e and v to denote an edge and a vertex respectively. If v is one end point of e, then one denotes $v \in e$. One defines a finite subtree $E(v) = \{e \in T | v \in e\}$. Let us define the growth function of T by $F(T) : \mathbf{N} \mapsto \mathbf{N}$:

$$F(T)(n) = \sup_v \sharp\{e \in T | v \in e, |v| = n\}.$$

For a family of trees $\{T_j\}_j$, one puts $F(\{T_j\}_j)(n) = \max_j F(T_j)(n)$. Let us denote $T(n) = \{x \in T : |x| \leq n\}$. Any homogeneous tree T admits a family of homogeneous trees of bounded type $\{T^n\}$ such that there are isomorphisms $\varphi_n : T(n) \subset T \cong T^n(n) \subset T^n$. It is called an *asymptotics* of T.

Let M be a simply connected oriented smooth closed manifold of even type. One obtains an embedding $Y = D^4 \cup \cup_{j=1}^l CH(T_j) \subset M$ as before. Suppose the defining signed trees $T_1, \ldots, T_l$ are homogeneous trees of unbounded type. Construction of complete Riemannian metrics g and weight functions w on Y also works for the unbounded case ([K]). By taking asymptotics $\{T_j^n\}$ for all T_j, one gets a family of open manifolds by $Y_n = D^4 \cup \cup_{j=1}^l CH(T_j^n)$. Let us put $W_n = D^4 \cup \cup_{j=1}^l CH(T_j^n(n))$. Recall that each kinky handle of $CH(T)$ is determined by each $E(v) \subset T$. Thus the family $\{Y_n\}_n$ can be regarded as an approximation of $CH(T)$. Let us take a family of admissible pairs (g_n, w_n) over Y_n. Let g be a complete Riemannian metric of bounded geometry on Y with $g_n|W_n = g|W_n$.

Let us choose a sequence $A_i \in \mathfrak{M}((M, g_i), [\Sigma_1], \ldots, [\Sigma_m])$. Then a subsequence $\{A_i\}$ converges to an L^2 ASD connection A_∞ over (Y, g) on each compact subset except finite set of points. Then one can assign an asymptotics of $\{A_i\}_i$ as a function $G(\{A_i\}) : \mathbf{N} \times \mathbf{N} \mapsto (0, C]$ by:

$$G(\{A_i\})(i, j) = |F_{A_i}| L^2_{w_i}(W_i \backslash W_{i-j}, g).$$

Let $\mathbf{F} = \{F : \mathbf{N} \mapsto \mathbf{N}\}$, $\mathbf{G} = \{h : \mathbf{N} \mapsto (0, C(h)]\}$ and $\mathbf{H} = \{h : \mathbf{N} \mapsto (0, C(h)]\}$ be the set of functions.

Let P be an $SO(3)$ bundle over M with $\hat{\alpha}_i \neq 0$, $i = 1, 2, 3$ as before. Suppose the restriction of the Donaldson's invariant $Q|H_2(Y; \mathbf{Z})$ is non zero. Then there exists a family of ASD connections A_i over P which converges to A_∞ with the following property. Let $\{\lambda_i\}_i$ be the family of partial lower spectrums with respect to $\{W_i\}$ (see 1.B).

THEOREM 3.2. *There exists a universally defined map* $\mathfrak{F} : \mathbf{F} \times \mathbf{H} \mapsto \mathbf{G}$ *such that one has the pointwise estimate at* (m, n) *with* $n \geq C\lambda_m^{-1}$*:*

$$\mathfrak{F}(F(\{T_j\}_j), \{\lambda_i\}) \leq G(\{A_i\}).$$

Proof : This follows from Proposition 2.5. Recall the properties $g|W_n = g_n|W_n$ and $w_{n+1}|W_n = w_n|W_n$. We show that for a small $\epsilon = \epsilon(F(\{T_j\}_j)(n), \lambda_n)$, if $G(\{A_i\})(n, n') < \epsilon$, then one can find $A'_n \in \mathfrak{M}((Y_n, g_n), [\Sigma_1], \dots, [\Sigma_{m'}])$, $m' \leq m$, which should be a smooth manifold of negative formal dimension. This is enough for the proof.

Let $H_2(Y; \mathbf{Z}) \cong \oplus^l \mathbf{Z}$ be an isomorphism with a generating set $a_1, \dots, a_l$. Since Q is polynomial, there exist some choice of $a_{i(1)}, \dots, a_{i(m)}$ from the generating set with $Q(M, P; a_{i(1)}, \dots, a_{i(m)}) \neq 0$. Then one chooses embedded surfaces $\Sigma_i \subset Y$ with $[\Sigma_1] = a_{i(1)}, \dots, [\Sigma_m] = a_{i(m)}$, and $A_i \in \mathfrak{M}((M, h_i); a_{i(1)}, \dots, a_{i(m)})$ so that the above ϵ can be chosen, independently of any choices from $H^2(Y; \mathbf{Z})$. In particular ϵ depends only on Y, but is independent of any data on $M \backslash Y$.

Let us take a subsequence of $\{A_i\}$ which converges to $A_\infty = \lim_i A_i$ on each compact subset except for finite points. If the above condition is satisfied, then applying Proposition 2.5, one finds an ASD connection A'_n over (Y_n, g_n) such that $|F^+_{A'_n}|L^2_{w_n}(W_n) \leq C(g_n)G(A_\infty)(n, n')$. Both of ϵ and C are determined by $F(\{T_j\})$ and $\{\lambda_j\}$. Let us take a cut off function φ on M with $\varphi|W_{n-n'} \equiv 1$ and $\varphi|(W_n)^c \equiv 0$. Then if $G(\{A_i\})(n, n')$ is sufficiently small, one can get a smooth connection A'_n over (Y_n, g_n) such that $A'_n|W_{n-n'}$ is ASD and $|F^+_{A'_n}|L^2_{w_n}((Y_n, g_n), \Lambda^2_+)$ is sufficiently small. Then by the continuity method, one gets $A''_n \in \mathfrak{M}((Y_n, g_n), [\Sigma_1], \dots, [\Sigma_{m'}])$. The argument that this space has negative formal dimension, is the same as the proof of Theorem 3.1. This completes the proof.

4. The position of Kähler forms and Picard lattice

In this section we show that the situation in the main result (Theorem 1.1) is realized by some algebraic surfaces, using some relation between Kähler forms and Picard lattices.

Let M be a simply connected complex algebraic surface with $p_g > 0$. Suppose M is of even type, and take a marking $\psi : H^2(M; \mathbf{Z}) \cong \oplus^8 \mathbf{Z} \oplus^{8k} \mathbf{Z} \oplus^{2l} \mathbf{Z}$ with respect to the decomposition of the form as $(-E_8) \oplus (k-1)(-E_8) \oplus lH$. Then one has an open manifold $S \subset M$ which corresponds to lH. The diffeomorphism type of the end of S is determined by $T_1, \dots, T_{2l}$, where each T_j is a homogeneous signed tree, possibly of unbounded type.

M can be holomorphically embeddable into a projective space $\mathbf{C}P^N$. Let us take such embedding $i : M \hookrightarrow \mathbf{C}P^N$. Then one has a natural homology class $h \in H_2(M; \mathbf{Z})$, the hyperplane class. This is defined as the intersection of M with the hyperplane $\mathbf{C}P^{N-1} \subset \mathbf{C}P^N$. Let ω be the Fubini Study form over $\mathbf{C}P^N$. Then the Poincaré dual of h is the restriction of $\omega = \omega|M$. This form gives a

Kähler (Hodge) structure on M. There are some restriction of ω on the position in $H^2(M;\mathbf{R})$. For example, let y be a closed analytic subvariety in M. Then one has a positivity $<\omega, y>> 0$.

Let h be a hyperplane class of M, and decompose $h = h_1 \oplus h_2 \oplus h_3$ with respect to above.

Now suppose there is an integral $(1,1)$ form $\alpha = \alpha_1 \oplus \alpha_2 \ominus \alpha_3 \in H^{1,1}(M;\mathbf{R}) \cap H^2(M;\mathbf{Z})$, such that the mod 2 reduction $\hat{\alpha_i} \neq 0 \in H^2(M;\mathbf{Z}_2)$, $i = 1,2,3$.

THEOREM 4.1. *Let M be as above and α satisfy the above condition. Then if there exists a hyperplane class $h \in H^2(M;\mathbf{Z})$ with $h_1 = h_2 = 0$, then one of the following holds; (1) at least one of T_j is of unbounded type, or (2) the L^2 ASD moduli space over (S,g) is generically w-separated.*

This theorem follows from Theorem 3.1 and Donaldson's non vanishing theorem (Theorem 1.4).

LEMMA 4.2. *Let M be a complex algebraic surface of even type. Let ω be a Hodge form. Then there exists a marking $\phi : H^2(M;\mathbf{Z}) \cong \oplus_{8k}\mathbf{Z} \oplus_{2l}\mathbf{Z}$ corresponding to the decomposition $k(-E_8) \oplus lH$, such that $\phi(\omega) \in \oplus_{2l}\mathbf{Z}$.*

This follows easily from the result in [J] for unimodular forms over $\mathbf{Z}$.

Recall that the Picard number is the rank of the image of the Picard group $\mathrm{Pic}(M)$ in $H^2(M;\mathbf{Z})$. Let M be a complex algebraic surface with even type of the form, $k(-E_8) \oplus lH$. Let us say that the Picard group of M is large if the Picard number is grater than $\max\{8k+1, 2l+1\}$. Thus the Picard numbers of all projective Kummer surfaces are large (more than 17).

Then one has the following statement; let M be as above, and suppose the Picard group of M is large. Then there exists a marking ϕ as above, such that $\phi(\omega) \in \oplus_{2l}\mathbf{Z}$, and for some $\alpha = \alpha_1 + \alpha_2 + \alpha_3 \in \mathrm{im}\ \mathrm{Pic}(M)$, one has $[\alpha_j]_2 \neq 0 \in \oplus \mathbf{Z}_2$.

In order to see this, one uses the following fact. Let M be an oriented smooth closed 4 manifold with $\pi_1(M) = 1$. Then $H^2(M;\mathbf{Z})$ is a free abelian group. Let us take a marking $\varphi : (H^2(M;\mathbf{Z}), <\ ,\ >) \mapsto (\oplus \mathbf{Z}, L)$ where L is a unimodular form on a free $\mathbf{Z}$ module, and $<\ ,\ >$ is the intersection form. In general there are many choices of markings. For example suppose L is of even type which contains at least two hyperbolic matrices. Then for any isomorphic primitive sublattices $l, l' \subset L$ of rank 1, there is an automorphism on L which sends l to l'. In particular there exists a marking ϕ as above, such that $\phi(h) \in \oplus_{2l}\mathbf{Z}$, and for some $\alpha = \alpha_1 + \alpha_2 + \alpha_3 \in \mathrm{im}$ $\mathrm{Pic}(M)$, one has $[\alpha_1]_2, [\alpha_2]_2, [\alpha_3]_2 \neq 0 \in \oplus \mathbf{Z}_2$.

COROLLARY 4.3. *Let M and ϕ be as above with the corresponding $S \subset M$ with tree-like end. Then for the defining homogeneous signed trees $T_1, \dots, T_{2l}$, at least one of them is of unbounded type, or the L^2 ASD moduli space over (S,g) is generically w-separated.*

For a $K3$ surface M, the Kähler cone is described as $\mathfrak{C} = \{y \in H^{1,1}(M;\mathbf{C}) | < y, y >> 0, < y, d >> 0, d \in \Delta\}$, where Δ consists of the effective classes with self intersection -2 in the Picard lattice.

There is a 1 dimensional space of non-vanishing holomorphic 2 forms whose generator x satisfies (1) $< x, x >= 0$ and $< x, \bar{x} >> 0$. For $x \in H^{2,0}(M;\mathbf{C})$, let $E(x) \subset H^2(M;\mathbf{R})$ be the linear 2 plane spanned by $\mathrm{Re}x$ and $\mathrm{Im}x$.

Then $y \in H^2(M;\mathbf{R})$ can be a Kähler form on some K3 surface M if and only if the following conditions are satisfied:

(2) $< y, E(x) >= 0$, $< y, y >> 0$,
(3) $< y, d > \neq 0$ for any $d \in H^2(M; \mathbf{Z})$ with $< d, d >= -2$ and $< x, d >= 0$.

PROPOSITION 4.4. *Let (M, ϕ) be a marked Kähler K3 surface with Kähler form y and $x \neq 0 \in H^{2,0}(M; \mathbf{C})$. Suppose $H^2(M; \mathbf{Z}) \cap E(x)$ consists of a primitive rank 2 lattice in $H^2(M; \mathbf{Z})$. Then there is an automorphism τ on $H^2(M; \mathbf{Z})$ which is independent of y such that one has $(\tau \circ \phi(y))_1 \neq 0$, where $\tau \circ \phi(y) = (\tau \circ \phi(y))_1 \oplus (\tau \circ \phi(y))_2 \in \oplus^{16}\mathbf{Z} \oplus^6 \mathbf{Z}$.*

Recall that a sublattice $L' \subset L$ is primitive if L/L' is free as a $\mathbf{Z}$ module.

Proof : Let us put $L = H^2(M; \mathbf{Z}) \cap E(x)$. Then there is an automorphism τ of $H^2(M; \mathbf{Z})$ such that $\tau(L)_1 = 0 \in \oplus^{16}\mathbf{Z}$ (see [BPV] p15 2.9(2)). Since $\tau(L)$ is primitive, one has an orthogonal decomposition $H^2(M; \mathbf{Z}) = (\tau(L))^\perp \oplus \tau(L)$. Let us take $d \in (\tau(L))^\perp$ with $< d, d >= -2$. Then by the condition of y above, one has $< \tau(y), d > \neq 0$. This shows the result.

References

[AHS] M. ATIYAH, N. HITCHIN AND I. SINGER, *Self-duality in four dimensional Riemannian geometry*, Proc. R. Soc. London, **362**(1978), 425 - 461 .

[BPV] W.BARTH, C.PETERS AND A.VAN DE VEN, *Compact complex surfaces*, Springer, Berlin New York Heidelberg Tokyo, 1984.

[C] A. CASSON, *Three lectures on new infinite constructions in 4-dimensional manifodls.*

[Co] A. CONNES, *Non commutative geometry*, Academic press Inc., San Diego, CA, 1994.

[D1] S. K. DONALDSON, *Poynomial invariants for smooth 4-manifolds*, Topology **29** (1990), 257 - 315.

[D2] S. K. DONALDSON, *An application of gauge theory to four dimensional topology*, Journal of Differential Geometry **18** (1983), 279 - 315.

[DK] S. K. DONALDSON AND P. B. KRONHEIMER, *The geometry of four-manifolds*, Clarendon Press , Oxford, 1990.

[FS1] R. FINTUSHEL AND R. J. STERN, *SO(3)-connections and the topology of 4-manifolds*, Journal of Gifferential Geometry **20**(1984), 523-539.

[FS2] R. FINTUSHEL AND R. J. STERN, *Definite 4-manifolds*, Journal of Differential Geometry **28**(1988), 133-141.

[F] A. FLOER, *An instanton invariant for 3 manifolds*, Comm. Math. Phy. **118**(1988), 215 - 240.

[FU] S. FREED AND K. UHLENBECK *Instantons and four manifolds*, Springer, 2nd edition, New York, 1991.

[Fr] M. H. FREEDMAN, *The topology of four-dimensional manifolds*, Journal of Differential Geometry **17** (1982), 357 - 454.

[FQ] M. FREEDMAN AND F. QUINN, *The topology of four dimensional manifolds*, Princeton Univ. press, 1990.

[G] R.GOMPF, *Periodic ends and knot concordance*, Topology and its applications **32**(1989), 141 - 148.

[GL] M. GROMOV AND H. B. LAWSON, *Positive scalar curvature on complete Riemannian manifolds*, I. H. E. S Publ. Math. **58**(1983), 295-408.

[J] I. D. JAMES, *On Witt's theorem for unimodular quadratic froms*, Pacific Journal of Math. **26**(1968), 303-316.

[K] T. KATO, *Four manifolds with tree like ends*, MPI preprint, 2000.

[Ki] R. C. KIRBY, *The topology of 4 manifolds*, Springer L. N. **1374**.

[M] J. MILNOR *On simply connected 4-manifolds*, Symp. Int. top. alg., Mexico, 1958 122 - 128.

[Q] F. QUINN, *Ends of maps III in dimension 4 and 5*, Journal of Differential Geometry **17** (1982), 503-521.

[T1] H. TAUBES, *Gauge theory on asymptotically periodic 4 manifolds*, Journal of Differential Geoetry **25** (1987), 363 - 430.

[T2] H. TAUBES, *Casson's invariant and gauge theory*, Journal of Differential Geoetry **31**(1990), 547 - 599.

[U1] K. Uhlenbeck, *Connections with L^p bounds on curvature*, Comm. Math. Phys. **83**(1982), 31 - 42.

[U2] K. Uhlenbeck, *Removable singularities in Yang-Mills fields*, C. M. P. **83**(1982), 11 - 29.

Department of Mathematics, Faculty of Science, Kyoto University, Kyoto 606-8502, Japan

E-mail address: tkato@math.kyoto-u.ac.jp

Contemporary Mathematics
Volume **347**, 2004

The Dehn filling space of a certain hyperbolic 3-orbifold

Sadayoshi Kojima and Shigeru Mizushima

ABSTRACT. We construct the first example of a "one-cusped" hyperbolic 3-orbifold for which we see the true shape of the space of hyperbolic Dehn fillings.

1. Introduction

The hyperbolic Dehn filling theory was established by W. Thurston in his original note [**8**]. Besides the general theory, he extensively analyzed the space of Dehn fillings of the figure eight knot complement in terms of the decomposition by two ideal tetrahedra. The analysis leaded us to extremely fascinating research activities on Dehn fillings in the last 25 years.

One of unsolved problems in his analysis was to determine the global shape of the Dehn filling space. Since then, there have been many deep researches such as [**4, 2**] discussing this problem in fact, however, it is still mysterious. Though W. Neumann and A. Reid [**7**] have determined the true shape for the Whitehead link complement with one cusp component being complete, there even seems to be no concrete examples of "one-cusped" hyperbolic manifolds or orbifolds for which we see the true shape of the Dehn filling space. In this paper, we construct hopefully the first such example of an orbifold, regretfully rather than a manifold.

The example we present here is topologically $\mathbb{Z}_2 \times \mathbb{Z}_2$-covered as an orbifold by the complement of a seven component link in the connected sum of three copies of $\mathbf{S}^2 \times \mathbf{S}^1$. Based on the study of one-circle packings on complex affine tori by the second author in [**6**], we construct the example, which we will denote by N, and its all possible Dehn filling deformations in the following three sections. Then using careful analysis of deformations of such tori in [**6**] again, we determine the global shape of the Dehn filling space of N in subsequent two sections.

2. Construction

We start with the *hexagonal packing* $\mathcal{P}$ on the complex plane $\mathbb{C}$ by equi-radii circles, see Figure 1. The set of euclidean translations which leave $\mathcal{P}$ invariant forms a group Γ acting freely on $\mathbb{C}$. It is isomorphic to $\mathbb{Z} \times \mathbb{Z}$ and the quotient is a particular euclidean torus admitting a cyclic symmetry of order 6, which we call

2000 *Mathematics Subject Classification.* Primary 57M50; Secondary 52C15, 30F99.
Key words and phrases. hyperbolic 3-manifold, orbifold, cone-manifold, Dehn filling.
The first author was supported in part by JSPS Grant 12440015.

the *hexagonal torus*. $\mathcal{P}$ then descends to what we call a one-circle packing on the hexagonal torus.

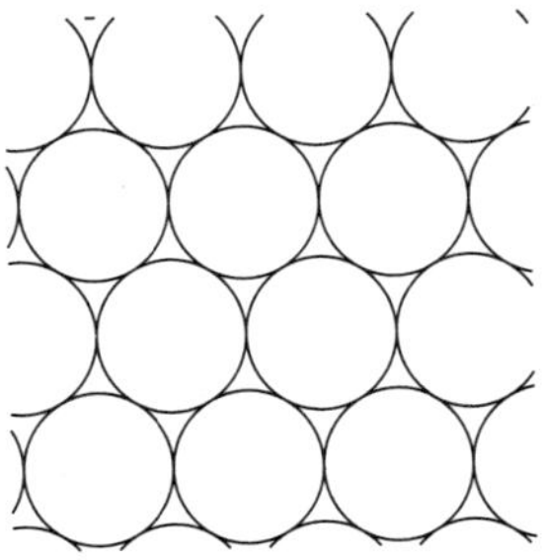

FIGURE 1. Hexagonal packing

Let us regard the complex plane $\mathbb{C}$ as the boundary of the upper half space model $\mathbb{H}^3$ of the hyperbolic 3-space. $\mathcal{P}$ together with the dual packing $\mathcal{P}^*$ which consists of circumscribed circles of interstices in $\mathcal{P}$ fills $\mathbb{C}$. Each member of both $\mathcal{P}$ and $\mathcal{P}^*$ bounds a hemisphere in $\mathbb{H}^3$ and such a hemisphere bounds a hemiball facing $\mathbb{C}$. Cutting off those hemiballs from $\mathbb{H}^3$, we obtain a region L with ideal polygonal boundary in $\mathbb{H}^3$. The intersections of $\mathcal{P}$ and $\mathcal{P}^*$ correspond to ideal vertices of L with rectangular section.

The group Γ acts properly discontinuously on $\mathbb{H}^3$ by the Poincaré extensions and the region L is invariant under the action. Hence taking quotient, we obtain a hyperbolic 3-manifold N with ideal polygonal boundary and a cusp. The polygonal boundary consists of two ideal triangles and one ideal hexagon, where they intersect orthogonally.

Take the double of N along triangular faces, and then take again the double of the result along the boundary which now consists of two ideal hexagons. This double doubling construction gives us a hyperbolic 3-manifold M with seven cusps among which 4 are old and 3 new. It is obvious by the construction that M admits a $\mathbb{Z}_2 \times \mathbb{Z}_2$-symmetry generated by two reflections associated with the doubling. N is the quotient of M by this symmetry and therefore we regard N as an orbifold from now on, and study its Dehn filling deformations as an orbifold.

Topologically N is a solid torus with a core loop and three points on the boundary removed. To see the effect on the first double, stretch the removed vertices on ∂N to arcs appropriately so that they become circles after doubling. Then by first doubling, we obtain a genus 3 handle body with two core loops and three loops on the boundary removed. This is depicted in Figure 2, where the shaded disks represent triangular faces along which we took the double.

A framed link representation of M, the double of Figure 2, can be immediately described by Figure 3.

3. Circle Packings on Complex Affine Tori

We will construct Dehn filling deformations of N using deformations of the hexagonal packing on $\mathbb{C}$ with the same combinatorial type. The hexagonal packing descends to a one-circle packing on the hexagonal torus. Conversely, the universal cover of a one-circle packing on the torus combinatorially equivalent to one on the hexagonal torus is a packing on the plane equivalent to the hexagonal packing.

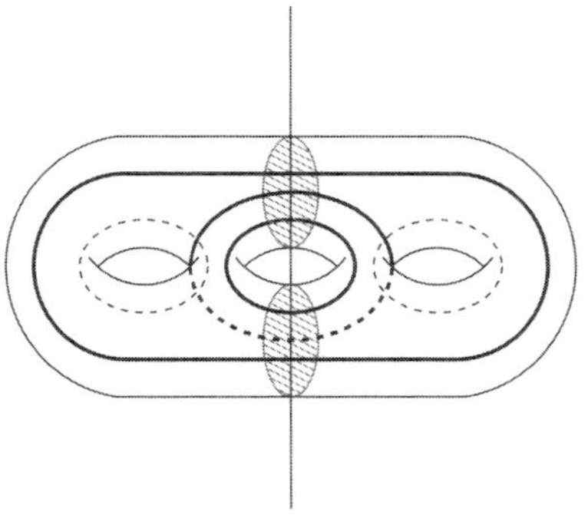

FIGURE 2. First double

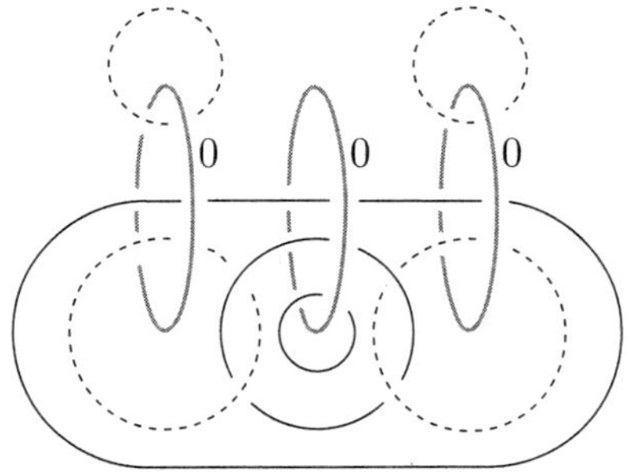

FIGURE 3. Framed link representation of M

However since the hexagonal packing is rigid in euclidean geometry, there are no other one-circle packings on euclidean tori than one on the hexagonal torus. Hence to get deformations, we will work with one-circle packings not on euclidean but rather on complex affine tori. This section is to review what has been known about complex affine tori and circle packings on them according to [**3, 6**].

A *complex affine structure* on the torus is by definition a collection of local charts on $\mathbb{C}$ whose transition function is a restriction of a complex affine transformation, so in particular it defines a complex structure. The uniformization assigning the conformal class to each affine torus defines a map of the space $\mathcal{A}$ of marked complex affine tori to the Teichmüller space $\mathcal{H}$ of the torus. It is a natural complex line bundle projection, where the fiber is parameterized by explicit description of developing maps of affine structures under some normalization as we will see below. $\mathcal{H}$ is identified with the upper half plane $\mathbb{H}^2$ in $\mathbb{C}$ and hence $\mathcal{A}$ is homeomorphic to the product $\mathbb{H}^2 \times \mathbb{C}$. We call the first factor a *Teichmüller parameter* $\omega \in \mathbb{H}^2$ and the second an *affine parameter* $c \in \mathbb{C}$. Complex affine tori sharing the same ω are conformally equivalent. We denote by $T_{\omega,c}$ the marked complex affine torus associated with $(\omega, c) \in \mathbb{H}^2 \times \mathbb{C}$. Also we denote by α, β specified simple closed curves generating the fundamental group of the torus for marking.

The zero section of the line bundle $\mathbb{H}^2 \times \mathbb{C} \to \mathbb{H}^2$ precisely consists of euclidean structures. In this case, the developing map

$$\mathrm{dev} : \widetilde{T}_{\omega,0} \to \mathbb{C},$$

is a homeomorphism, and the images of α and β by the holonomy ρ are translations defined for each $z \in \mathbb{C}$ by

$$\begin{cases} \rho(\alpha) : z \mapsto z + 1 \\ \rho(\beta) : z \mapsto z + \omega \end{cases}$$

after appropriate normalization.

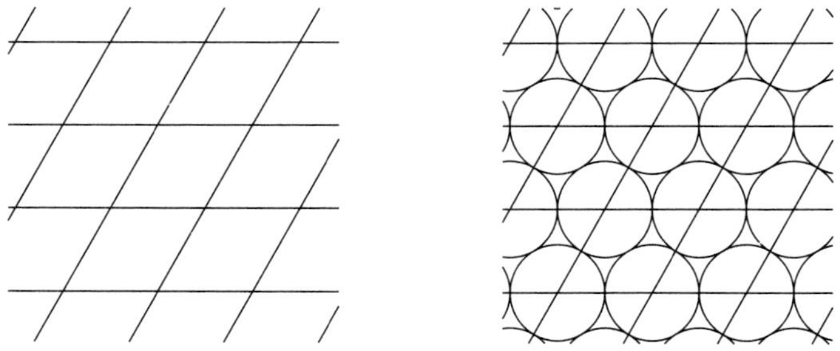

Images of α and β Image of a circle packing

FIGURE 4. Developed image of torus with euclidean structure

In affine but non-euclidean case, identifying the universal cover $\widetilde{T}_{\omega,c}$ with $\widetilde{T}_{\omega,0} \cong \mathbb{C}$ by the uniformization, the developing map is described by an exponential map

$$\mathrm{dev} : z \mapsto e^{cz}$$

under the identification of $\mathcal{A}$ with $\mathbb{H}^2 \times \mathbb{C}$. Here ω and c are the Teichmüller and affine parameters of the complex affine torus in question. The images of α, β by the holonomy ρ are obviously similarities

$$\begin{cases} \rho(\alpha) : z \mapsto e^{c} z \\ \rho(\beta) : z \mapsto e^{c\omega} z \end{cases}$$

which fix the origin.

Images of α and β Image of a circle packing

FIGURE 5. Developed image of torus with complex affine structure

Since a complex affine transformation is a similarity, it sends a circle on $\mathbb{C}$ to a circle. Hence we can still discuss about circle packings on complex affine tori. The second author showed in [**6**] that the set of complex affine tori which admit a one-circle packing forms a real two-dimensional moduli space $\mathcal{M}$ in $\mathbb{H}^2 \times \mathbb{C}$. In constructing the moduli space, the marking for the packing was specified so that the edges of the nerve of the packing are homotopic to α, β and $\beta\alpha^{-1}$. The moduli space $\mathcal{M}$ contains only one euclidean torus associated with $\omega_0 = e^{\pi i/3}$, which is the hexagonal one. Moreover, the restriction of the uniformization $\mathbb{H}^2 \times \mathbb{C} \to \mathbb{H}^2$ to $\mathcal{M}$ is onto and two to one except at the hexagonal torus. The affine parameters of a generic fiber consist of nonzero c and its negative $-c$.

4. Deformations

To each pair (ω, c) on the moduli space $\mathcal{M}' = \mathcal{M} - \{(\omega_0, 0)\}$, we construct a Dehn filling deformation of N as an orbifold based on the construction of N. To see this, we start with a complex affine torus $T_{\omega,c}$ packed by one circle combinatorially equivalent to the one-circle packing on $T_{\omega_0,0}$. The developing map of $T_{\omega,c}$ is an exponential map and $\mathbb{Z}$-covers $\mathbb{C} - \{0\}$. Thus if we let X be the universal cover of $\mathbb{C} - \{0\}$, it canonically lifts to a homeomorphism : $\widetilde{T}_{\omega,c} \to X$. By mapping a circle packing on $\widetilde{T}_{\omega,c}$ to X, we obtain a circle packing $\mathcal{P}_{\omega,c}$ on X combinatorially equivalent to the hexagonal packing on $\mathbb{C}$. Also the holonomy lifts uniquely to an injective homomorphism of $\pi_1(T_{\omega,c}) \cong \mathbb{Z} \times \mathbb{Z}$ to the group of transformations of X. We denote its image by $\Gamma_{\omega,c}$.

Then we let ℓ denote the geodesic in $\mathbb{H}^3$ joining 0 and ∞. The universal cover Y of $\mathbb{H}^3 - \ell$ will be a substitution of $\mathbb{H}^3$ in the euclidean case. In fact, cutting off hemiballs in Y bounded by the members of a deformed hexagonal packing $\mathcal{P}_{\omega,c}$ and its dual $\mathcal{P}^*_{\omega,c}$ on X, we obtain a region $L_{\omega,c}$ with ideal polygonal boundary in Y. Also since the action of $\Gamma_{\omega,c}$ on X extends to a properly discontinuous action on Y which leaves $L_{\omega,c}$ invariant, by taking a quotient we obtain a hyperbolic 3-manifold with ideal polygonal boundary and incomplete end near ℓ. We denote its completion by $N_{\omega,c}$. It carries a Dehn surgery type singularity by definition, see [**8**], and we thus obtain to each (ω, c) in $\mathcal{M}'$ a Dehn filling deformation of $N = N_{\omega_0,0}$.

On the other hand, let N_* be a Dehn filled deformation of N as an orbifold and R an orbifold obtained from N_* by removing the points attached by the completion. R is homeomorphic to the torus times $[0, 1)$ with three points on the boundary removed. The boundary of R consists of two ideal triangles and one ideal hexagon, where they intersect orthogonally. Choose a developing map D of R so that a neighborhood of an incomplete end winds ℓ in $\mathbb{H}^3$. By this normalization, the holonomy of R maps $\pi_1(R) \cong \mathbb{Z} \times \mathbb{Z}$ to transformations fixing 0 and ∞, the image of the holonomy contains an element of infinite order by the definition of the Dehn filling deformation.

R is not complete, but not so wild. Let U be an equidistant neighborhood of an incomplete end and B its complement. It is obvious that any point in R admits the unique shortest path to $\partial U = \partial B$. Thus U is developed to an equidistant neighborhood of ℓ and B its complement. In particular, the image of the developing map D misses ℓ, and D is in fact a map of the universal cover $\widetilde{R}$ into $\mathbb{H}^3 - \ell$. Then D lifts to a map $\widetilde{D}$ of the universal cover $\widetilde{R}$ to Y, the universal cover of $\mathbb{H}^3 - \ell$. Accordingly, the holonomy of R lifts to a homomorphism to the group of transformations of Y. We denote its image by Γ_*.

Let Z be the preimage of the boundary of the equidistant neighborhood of ℓ in Y. Γ_* acts properly discontinuously and freely on Z by the definition of the Dehn filling deformation, and Z bijectively corresponds to the preimage of Z by $\widetilde{D}$. This shows that $\widetilde{D}$ is injective at least on a neighborhood of the preimage of Z by $\widetilde{D}$. We will show that $\widetilde{D}$ is globally injective. Suppose we have two points in $\widetilde{R}$ mapped by $\widetilde{D}$ to a point $y \in Y$. y admits the unique shortest path to the point y_0 on Z. Since the preimage of Z by $\widetilde{D}$ bijectively corresponds to the preimage of $\partial U = \partial B$ in $\widetilde{R}$, the preimage of y_0 consists of a single point n_0. Hence two points in question lie on the orthogonal path to the preimage of $\partial U = \partial B$ through n_0 on the same side with the same distance to n_0. Thus they must be the same.

Then the ideal hexagonal faces on the boundary of $\widetilde{R}$ in the developed image define a packing on X, the universal cover of $\mathbb{C} - \{0\}$, combinatorially equivalent to the hexagonal one. Moreover the packing is invariant under the induced action of Γ_* on X. Hence it descends to a one-circle packing on some affine torus $T_{\omega,c}$ parameterized by some pair $(\omega, c) \in \mathcal{M}'$, and N_* is isometric to $N_{\omega,c}$.

5. Dehn Filling Coefficients

Following [**6**], we review how to find a pair (ω, c) lying on the moduli space $\mathcal{M}'$. If (ω, c) lies on $\mathcal{M}'$, a simple observation shows that ω and c must satisfy the identities,

$$\frac{\cos \operatorname{Im}(c/2)}{\cosh \operatorname{Re}(c/2)} = \frac{\cos \operatorname{Im}(c\omega/2)}{\cosh \operatorname{Re}(c\omega/2)} = \frac{\cos \operatorname{Im}(c(\omega-1)/2)}{\cosh \operatorname{Re}(c(\omega-1)/2)}. \tag{5.1}$$

Moreover, the inequalities

$$|\operatorname{Im}(c)| < \pi, \quad |\operatorname{Im}(c\omega)| < \pi, \quad |\operatorname{Im}(c(\omega-1))| < \pi$$

must be satisfied by obvious reason. Let $f : \mathcal{C} \to [0, 1)$ be a smooth function defined by

$$f(z) = 1 - \frac{\cos \operatorname{Im}(z/2)}{\cosh \operatorname{Re}(z/2)},$$

where

$$\mathcal{C} = \{z \in \mathbb{C} \mid |\operatorname{Im}(z)| < \pi\}$$

is the largest possible region for c to vary in $\mathcal{M}$. The identity (5.1) shows that c, $c\omega$ and $c(\omega - 1)$ lie on the same level set $L_s = \{z \in \mathcal{C} \mid f(z) = s\}$. When $s = 0$, L_0 consists of a single point corresponding to the hexagonal torus. For each $0 < s < 1$, L_s is topologically a simple loop around the origin, has symmetries about the real and imaginary axis, and bounds a strictly convex interior.

For $c \in \mathcal{C}' = \mathcal{C} - \{0\}$, the origin 0, c, $c\omega$ and $c(\omega - 1)$ form a parallelogram by obvious reason in plane geometry. Conversely, if $c \in \mathcal{C}'$ is given and let L_s be the level set on which c lies, then there are unique two points c_1, c_2 on L_s so that c, c_1, c_2 lie in counterclockwise order on L_s, and they together with the origin form a parallelogram. Then we obtain a pair (ω, c) in the moduli space $\mathcal{M}'$ by letting $\omega = c_1/c$. Hence $c \in \mathcal{C}'$ determines ω. In view of this fact, we identify the moduli space $\mathcal{M}'$ with the region $\mathcal{C}'$.

In the level set L_s for $0 < s < 1$, there are twelve special points depicted in Figure 6 which give rise to pairs on $\mathcal{M}'$ rather obviously. Here, p_1 and p_7 are intersections with the real axis. Similarly p_4, p_{10} lie on the imaginary axis. p_i for $i = 2, 3, 5, 6, 8, 9, 11, 12$ are points such that p_3p_{11}, p_5p_9, p_2p_6 and p_8p_{12} are the bisectors of the segments $0p_1$, $0p_4$, $0p_7$ and $0p_{10}$ where 0 is the origin in $\mathcal{C} \subset \mathbb{C}$. Three points p_i, p_{i+2}, p_{i+4} (mod 12) together with the origin 0 form a parallelogram and give rise to a pair on $\mathcal{M}'$. As s decreases from 1 toward 0, an ellipselike closed curve L_s shrinks down to a point L_0, however the order of p_i's on the level set never changes. Hence the trace of p_j defines a locus l_j in the region $\mathcal{C}$, and $\mathcal{C}$ is divided by these loci into twelve regions denoted by $C_1, \cdots, C_{12}$ as in Figure 7.

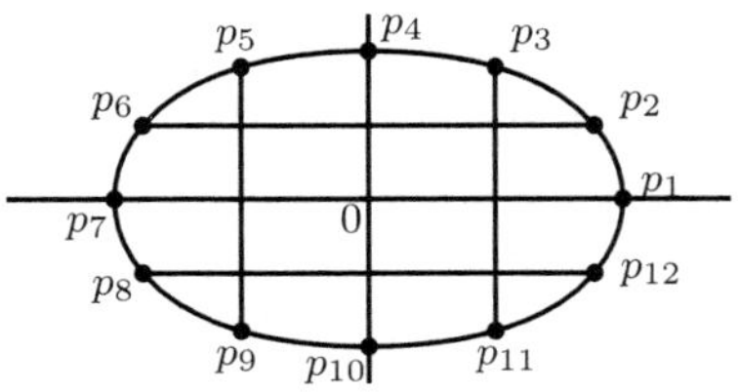

FIGURE 6.
12 points on L_s

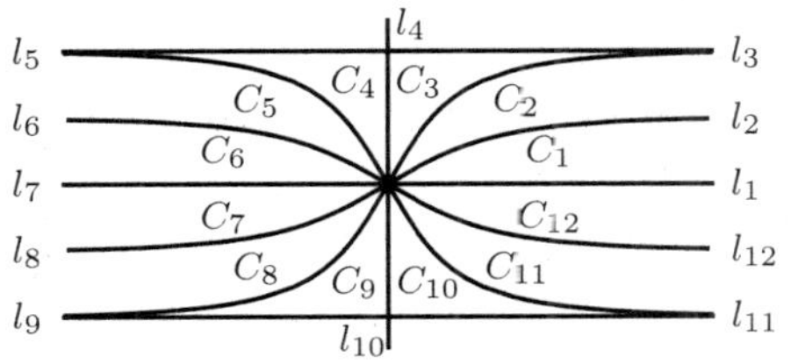

FIGURE 7.
12 regions on $\mathcal{C}$

To each pair (ω, c) on the moduli space $\mathcal{M}'$, we choose four points $0, c, c\omega, c(\omega - 1)$ which form a parallelogram. Choosing the origin and the third and fourth vertices $c\omega, c(\omega - 1)$, and adding the negative of the second one at the end. we obtain a next parallelogram spanned by $0, c\omega, c(\omega - 1), -c$. Doing the same shift twice, we obtain parallelograms spanned by $0, c(\omega - 1), -c, -c\omega$ and $0, -c, -c\omega, -c(\omega - 1)$. The last one is symmetric to the original parallelogram about the origin. This shows that the moduli space $\mathcal{M}' = \mathcal{C}'$ has a rotational symmetry of order 6. On the other hand, the complex conjugation gives an orientation reversing involutive symmetry. Thus by these all together, the moduli space $\mathcal{M}' = \mathcal{C}'$ admits a dihedral symmetry of order 12.

Regarding α, β as a meridian-longitude pair for N, we discuss the hyperbolic Dehn filling coefficients of $N_{\omega,c}$. Recall that $\rho(\alpha)$ is a similarity defined by $z \mapsto e^c z$ and $\rho(\beta)$ by $z \mapsto e^{c\omega} z$ for nonzero c. Then the Dehn filling coefficients $(\mu, \lambda) \in \mathbb{R}^2$ for $N_{\omega,c}$ is by definition, see [**8**], a continuous solution of the equation

$$\rho(\alpha)^\mu \rho(\beta)^\lambda = \text{rotation by } \pm 2\pi$$

under the normalization that $\rho(\alpha), \rho(\beta)$ go to the identity when the structure approaches the complete one. More precisely for our choice of c and ω, the Dehn filling coefficient (μ, λ) of $N_{\omega,c}$ is a solution of the linear equation

$$\mu c + \lambda c\omega = \pm 2\pi i. \tag{5.2}$$

Note that a pair (ω, c) corresponds to $\pm(\mu, \lambda)$ and conversely a pair (μ, λ) to $(\omega, \pm c)$. We define (μ, λ) for $c = 0$ as (∞, ∞).

To each finite Dehn filling coefficient (μ, λ), we assign a slopelike number $t = \mu/(\mu + \lambda)$, where the slope is μ/λ by definition. Though (ω, c) corresponds to coordinates $\pm(\mu, \lambda)$, the unique t is assigned for them. Dividing (5.2) on both sides by $\mu + \lambda$ and taking the real parts, we obtain an obvious identity,

$$\operatorname{Re}(tc + (1 - t)c\omega) = 0,$$

which shows how to read off t with respect to the Dehn filling coefficient of $N_{\omega,c}$ by an elementary plane geometry in $\mathbb{C}$ as follows. c and $c\omega$ lie on the same level set of f in $\mathcal{C}'$. Here we draw the straight line through c and $c\omega$. Then the intersection of the line with the imaginary axis is the point which divides the segment connecting c and $c\omega$ in $1 - t$ to t.

For our convenience, we define a function $T : \mathcal{C}' \to \mathbb{R} \cup \{\infty\}$ by assigning $t = \mu/(\mu + \lambda)$ to each $c \in \mathcal{C}' = \mathcal{M}'$. It is easy to see that the level set of T at

$t = -1, 0, 1/2, 1, 2, \infty$ coincide with the union of l_1 and l_7, the union of l_2 and l_8, $\cdots$, the union of l_6 and l_{12}. When c moves along the level set of f in counterclockwise direction, $c\omega$ does the same. Hence T is strictly increasing along a counterclockwise move on the level set of f except a gap at $t = \infty$.

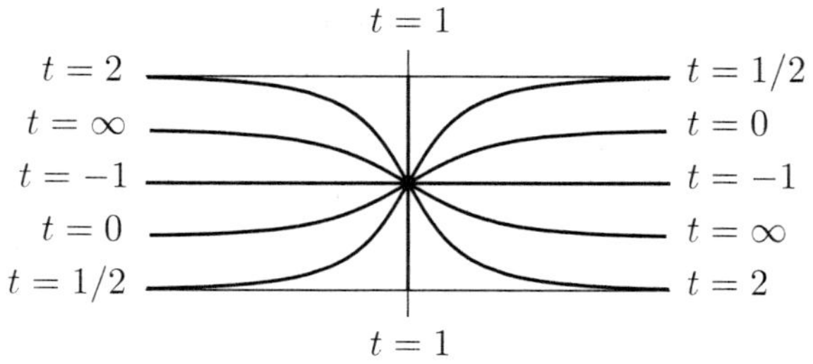

FIGURE 8

Choose a point c in the region C_3, then t assigned lies in $(1/2, 1)$. When c moves along the level set of T toward the boundary of $\mathcal{C}$, $c\omega$ must move along the level set of T in C_5. Since $\mathrm{Re}(c\omega)$ goes to $-\infty$ and $\mathrm{Re}(tc+(1-t)c\omega)$ remains zero, $\mathrm{Re}(c)$ must go to ∞. The limit of the curve l_3 is $\infty + \pi i$, and this gives $\mathrm{Im}(c) \to \pi$. On the other hand, by (5.1), $\cos \mathrm{Im}(c\omega/2) = \cos \mathrm{Im}(c(\omega - 1)/2) \cosh \mathrm{Re}(c\omega/2)/ \cosh \mathrm{Re}(c(\omega - 1)/2)$. Since $\cos \mathrm{Im}(c(\omega - 1)/2)$ is bounded and $\cosh \mathrm{Re}(c\omega/2)/ \cosh \mathrm{Re}(c(\omega - 1)/2)$ is asymptotically $e^{-\mathrm{Re} c/2}$, $\cos \mathrm{Im}(c\omega/2)$ converges to 0 and the limit of $\mathrm{Im}(c\omega)$ is π. Thus, $tc + (1 - t)c\omega \to \pi i$, and we have

$$2tc + 2(1 - t)c\omega \to 2\pi i.$$

This shows that $(\mu, \lambda) \to \pm(2t, 2(1 - t))$ and $\mu + \lambda \to \pm 2$.

By the dihedral symmetry of the moduli space, we can calculate the limit for other t. The image of the Dehn filling coefficients (μ, λ) parameterized by the moduli space $\mathcal{M} = \mathcal{C}$ then covers the shaded region $\mathcal{D}$ in Figure 9, and the boundary consists of a hexagon as follows:

$$\begin{aligned} &\{(\mu, \lambda) \mid \mu + \lambda = 2, \mu \geq 0, \lambda \geq 0\} \cup \{(\mu, \lambda) \mid \mu = 2, -2 \leq \lambda \leq 0\} \\ &\quad \cup \{(\mu, \lambda) \mid \lambda = -2, 0 \leq \mu \leq 2\} \cup \{(\mu, \lambda) \mid \mu + \lambda = -2, \mu \leq 0, \lambda \leq 0\} \\ &\qquad \cup \{(\mu, \lambda) \mid \mu = -2, 0 \leq \lambda \leq 2\} \cup \{(\mu, \lambda) \mid \lambda = 2, -2 \leq \mu \leq 0\}. \end{aligned}$$

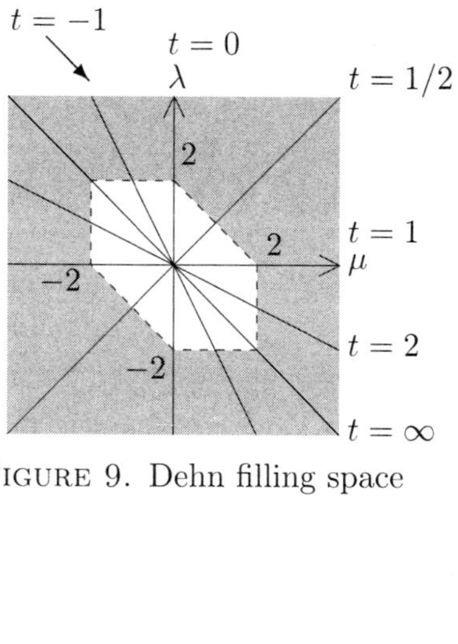

FIGURE 9. Dehn filling space

6. Degenerations

We check how the degeneration occurs when they approach the boundary.

We first think of the deformations along the path defined by the level set of T at $t = 1$. Then the deformations are realized by hyperbolic cone-manifolds and the meridian loop α bounds a disk hitting the cone singularity. See [**1**] for details about cone-manifolds. Since

$$|\mathrm{Im}(c)| \to \pi \quad \text{and} \quad |\mathrm{Re}(c\omega)| \to \infty,$$

the cone angle converges to π and the length of the singular locus diverges. In fact, the Dehn filled manifold splits along an euclidean sub cone-manifold, a disk with two corner of angle $\pi/2$ and a cone point of cone angle π, at the limit. The same result holds when $t = 0$ and $t = \infty$ because of the dihedral symmetry.

More generally, choose coprime integers p and q and think of the deformations of N along the path on the level set of T at $t = p/(p+q)$. Then again the deformations are realized by cone-manifolds and a loop homotopic to $\alpha^p\beta^q$ bounds a disk hitting the cone singularity. For $t \in (0,1)$, the level set of T lies in $C_2 \cup l_3 \cup C_3$. When the deformations approach the boundary along this path, $\mathrm{Re}(c) \to \infty$, $\mathrm{Im}(c) \to \pi$, $\mathrm{Re}(c\omega) \to -\infty$ and $\mathrm{Im}(c\omega) \to \pi$. The calculation similar to one in the previous section shows that

$$|\mathrm{Im}(pc + qc\omega)| \to |(p+q)\pi|,$$

and hence the cone angle converges to $(p+q)\pi$. To see the length of the singular locus, we choose integers r and s such that $ps - qr = 1$. Then, $\alpha^r\beta^s$ represents a curve which intersects $\alpha^p\beta^q$ once and hence the translation factor of $\rho(\alpha^r\beta^s)$, or equivalently $\mathrm{Re}(rc + sc\omega)$ represents the length of the singular locus. Making use of the property $\mathrm{Re}(pc + qc\omega) = 0$, we have

$$|\mathrm{Re}(rc + sc\omega)| = |\frac{rq - ps}{q}\mathrm{Re}(c)| \to \infty,$$

which implies that the length of the singular locus diverges and a corresponding cone-manifold splits as before. We can see similar facts for other rational t by the dihedral symmetry. In fact, the cone angle approaches $|p|\pi$ if $t > 1$ and $|q|\pi$ if $t < 0$. In either case, the length of the singular locus diverges.

7. Remark

If we relax the complex affine structure modeled on $\mathbb{C}$ to the complex projective structure modeled on the Riemann sphere, the relaxation defines the involution τ on $\mathbb{H}^2 \times \mathbb{C}$ which sends (ω, c) to $(\omega, -c)$. The moduli space $\mathcal{M}_*$ of complex projective tori admitting a one-circle packing is the quotient of $\mathcal{M}$ by τ. This space is studied in [**5**] by the cross ratio parameters and represented as a semi algebraic set by

$$\mathcal{M}_* = \{(x,y,z) \,|\, xyz - x - y - z = 0, x, y, z > 0\},$$

see the appendix in [**5**].

Assigning the holonomy representation to each projective structure, we obtain a map

$$\mathcal{M}_* \to \mathbb{H}^2 \times \mathbb{C}/\tau \to X(\pi_1),$$

of $\mathcal{M}_*$ to the space $X(\pi_1)$ of representations of the fundamental group of the torus in $\mathrm{PGL}(2,\mathbb{C})$ up to conjugation. It is a regular map with respect to the algebraic structures on the source and target, and easily seen to be a proper injective homeomorphism by concrete computation in [**5**]. This in particular implies also that the boundary we described here is the true one since holonomy representations diverge at the boundary.

References

[1] D. Cooper, C. Hodgson and S. Kerckhoff: *Three-dimensional orbifolds and cone-manifolds*, MSJ Memoirs, **5** (2000).

[2] J. Dowty: *Orthlength and hyperbolic Dehn surgery*, Ph.D. thesis, Melbourne Univ., 2001.

[3] R. C. Gunning: *Affine and projective structures on Riemann surfaces*, Riemann surfaces and related topics: Ann. of Math. Stud., **97** (1981), 225-244.

[4] C. Hodgson: *Degeneration and regeneration of geometric structures on three-manifolds*, Ph.D. thesis, Princeton Univ., 1986.

[5] S. Kojima, S. Mizushima and S. P. Tan: *Circle packings on surfaces with projective structures*, J. Differential Geom., **63** (2003), no. 3, 349–397.

[6] S. Mizushima: *Circle packings on complex affine tori*, Osaka J.Math., **37** (2000), 873-881.

[7] W. D. Neumann and A. Reid: *Arithmetic of hyperbolic manifolds*, Topology '90, ed. by B. Apanasov et al, de Gruyter, (1992), 273-310.

[8] W. P. Thurston: The geometry and topology of 3-manifolds, Lecture Notes, Princeton Univ., 1977/78.

DEPARTMENT OF MATHEMATICAL AND COMPUTING SCIENCES, TOKYO INSTITUTE OF TECHNOLOGY, MEGURO, TOKYO 152-8552 JAPAN

E-mail address: sadayosi@is.titech.ac.jp

DEPARTMENT OF MATHEMATICAL AND COMPUTING SCIENCES, TOKYO INSTITUTE OF TECHNOLOGY, MEGURO, TOKYO 152-8552 JAPAN

E-mail address: mizusima@is.titech.ac.jp

Contemporary Mathematics
Volume **347**, 2004

An Asymptotic of the Large Deviation for Random Walks on a Crystal Lattice

Motoko Kotani

Dedicated to Professor Takushiro Ochiai on the occasion of his sixtieth birthday

ABSTRACT. A crystal lattice is an abelian cover of a finite graph. Typical examples are the integer lattices, the triangular lattice and the hexagonal lattice. Large deviation holds for random walks on a crystal lattice. We will discuss how the rate function appearing in the large deviation relates with geometry of the crystal lattice. Because of the periodic structure of the crystal lattice, by reducing the scale smaller and smaller, it converges in the Gromov-Hausdorff topology. The limit space plays an role in the large deviation theorem.

1. Introduction

A *crystal lattice* X is an infinite graph on which abelian group acts freely with a finite graph X_0 as its quotient graph. In other words, X is the abelian cover of a finite graph X_0 with the covering transformation group Γ. For symplicity, we assume Γ is torsion free, therefore isomorphic to $\mathbb{Z}^d$. The integer lattice $\mathbb{Z}^d$, the triangular lattice and the hexagonal lattice are typical examples of crystal lattices.

We consider random walks on a crystal lattice X whose transition probability p is Γ-invariant. Since X is an abelian cover of a finite graph X_0, random walks on X are described as a Markov chain x_n valued in the 1-chain group $C_1(X_0)$ of X_0. With this idea, long time behaviors of random walks on crystal lattices are analyzed in details. For examples, the central limit theorem, the asymptotic expansion of the n-step transition probability [**4**], [**5**], [**6**], large deviations [**7**] are obtained In those explicit formula, geometrical interpretations are emphasized.

It has been explained in [**6**] that the euclidean metric (Albanese metric) appearing in the central limit theorem is related with the standard realization (energy minimizing Γ-equivariant map) of a crystal lattice. The rate function (to measure the decay rate) appearing in the large deviation theorem is related to the tangent cone at the infinity of a crystal lattice. To be more precise, let us consider

2000 *Mathematics Subject Classification.* 60G50, 53C21, 58J35, 39A12.
Key words and phrases. transition probability, large deviation, crystal lattice, entropy.
Supported by Grant-in-Aid for Scientific Research C (2)14540057.

a given crystal lattice X as a metric space (X, d) with the graph metric d. Fix a base point $x_0 \in X$ and then change scale. Namely, we consider a family of metric spaces $X_\epsilon = (X, \epsilon d, x_0)$ and let ϵ go to zero. Then it converges in the pointed Gromov-Hausdorff topology (cf.[**3**]) to the d-dimensional vector space $\Gamma \otimes \mathbb{R}$ with a non-euclidean metric d_∞. We found in [**7**], the closed unit ball $\overline{\mathcal{D}}$ in $\Gamma \otimes \mathbb{R}$ with respect to d_∞ is exactly the essential domain where the rate function is defined to take finite value.

In §2, we study random walks on a finite graph and how its geometry is related with long time asymptotics of random walks. Then by using these results, we see the rate function I is expressed as the relative entropy with respect to the given transition probability p:

THEOREM 1.1. *For ξ which belongs to $\mathcal{D}_0$ or the interior of a face of $\overline{\mathcal{D}_0}$ of the maximal dimension, we have*

$$I(\xi) = -H(p_\xi, p),$$

where

$$H(p_\xi, p) = \sum_{e \in E_0} m_\xi(e) \log\Big(\frac{p_\xi(e)}{p(e)}\Big)$$

is the relative entropy of the transition probability p_ξ with respect to p.

In §4, we study a precise asymptotic of the n-step transition probability with large deviation scaling in the senses of S. Lalley [**8**] and M. Babillot–F. Ledrappier [**1**].

COROLLARY 1.2. *Let $A \subset \Gamma \otimes \mathbb{R}$ and $\xi \in \mathcal{D}$.*

$$P_x(\xi_n - n\xi \in A) \sim \Big(\frac{2\pi}{n}\Big)^{d/2} e^{-nI(\xi)} m(x) \sqrt{\det(\mathrm{Hess}\, I)(\xi)} \int_A e^{-\langle u_\xi, \zeta\rangle} d\lambda(\zeta).$$

2. Random walks on a finite graph and their homological directions

In this section we study random walks on a finite graph and their homological directions. Later we use results in this section to analyze random walks on a crystal lattice.

Let X_0 be a finite graph. The set of all oriented edges is denoted by E_0. For an oriented edge e, its origin and terminus are denoted by $o(e)$ and $t(e)$, respectively, and the inverse edge is denoted by $\overline{e}$. Through out this article, we assume that if $e \in E_0$ then $\overline{e} \in E_0$. A subset E_0^+ of E_0 is called an *orientation* of X_0 if $E_0 = E_0^+ \sqcup E_0^-$ with $E_0^- = \{\overline{e} \mid e \in E_0^+\}$. Let $E_x := \{e \in E_0 \mid o(e) = x\}$ be the set of edges with $o(e) = x \in X_0$. A *path* c of X_0 of length n is a sequence $c = (e_1, \ldots, e_n)$ of oriented edges e_i with $o(e_{i+1}) = t(e_i)$.

A *transition probability* p on X_0 is a non-negative function $p : E_0 \to \mathbb{R}_{\geq 0}$ of E_0 satisfying $\sum_{e \in E_x} p(e) = 1$. It gives a random walk on X_0 and the n-step transiton probability $p_n(x, y)$ is given by $p_n(x, y) = \sum_c p(c)$, where sum is taken over all paths $c = (e_1, \ldots, e_n)$ of length n with the origin $o(c) = x$ and the terminus $t(c) = y$ and $p(c) = p(e_1) \cdots p(e_n)$. A transition probability p is called *strongly connected* (irreducible) when, for every x and y, there is an $n \in \mathbb{N}$ such that $p_n(x, y) > 0$. It is known that when p is strongly connected, there exists the unique invariant positive probability measure $m : X_0 \to \mathbb{R}_{>0}$ on X_0;

$$m(x) = \sum_{e \in E_x} p(\overline{e}) m(t(e)), \qquad \sum_{x \in X_0} m(x) = 1.$$

Put $m(e) = m(o(e))p(e)$ to define $m : E_0 \to \mathbb{R}_{>0}$ satisfying

$$m(x) = \sum_{e \in E_x} m(e) = \sum_{e \in E_x} m(\bar{e}), \quad \sum_{e \in E_0} m(e) = 1.$$

A strongly connected transition probability p induces the probability measure P_x on the set Ω_x of infinite paths whose origin $o(c) = x$, and $P = \sum_{x \in X_0} m(x) P_x$ of the set Ω of paths of infinite length, respectively, in the usual manner. Let $C_1(X_0, \mathbb{R})$ be the 1-chain group of X_0, i.e. the free abelian group generated by E_0 over $\mathbb{R}$ with relations $\bar{e} = -e$ for all oriented edge e and its inverse edge $\bar{e}$. Consider the random variables $x_i(c) = e_i$, $i = 1, 2, \ldots$ valued in $C_1(X_0, \mathbb{R})$ for $c = (e_1, e_2, \ldots) \in \Omega$ and put $\xi_n(c) = \sum_{i=1}^n x_i(c) \in C_1(X_0, \mathbb{R})$. From the law of large numbers, we see the empirical mean of this process

$$\frac{1}{n}\xi_n(c) \to \gamma_p \quad (n \to \infty, P - a.e.\ c \in \Omega),$$

and γ_p is given by $\sum_{e \in E_0} m(e)e$. Because m is an invariant measure, γ_p belongs to the kernel of the boundary map, hence induces an element of the first homology group $H_1(X_0, \mathbb{R})$. We call it the *homological direction* of the given random walk and denote by the same symbol γ_p.

Put $\mathcal{D}_0 := \{\gamma_p \mid \text{non-degenerate transition probability } p\}$, where by *non-degenerate* we mean that p is a positive valued transition probability over E_0. In [**7**], it is shown that the closure $\overline{\mathcal{D}_0}$ is the unit ball in $H_1(X_0, \mathbb{R})$ with respect to the ℓ^1-norm:

$$\|\xi\|_1 = \sum_{e \in E_0^+} |a(e)|,$$

where $\xi = \sum_{e \in E_0^+} a(e)e$. Note that, with a given orientation E_0^+, every element $\xi \in H_1(X_0, \mathbb{R})$ is uniquely written as $\xi = \sum_{e \in E_0^+} a(e)e$ and the definition of $\|\cdot\|_1$ does not depend on the choice of an orientation E_0^+.

Since $\mathcal{D}_0$ is a metric ball, it is symmetric around the origin (the origin corresponds to symmetric random walks) and moreover, we have

THEOREM 2.1 ([**7**]).

(1) *$\overline{\mathcal{D}_0}$ is "rational" in the sense that all extreme points of $\overline{\mathcal{D}_0}$ are in $H_1(X_0, \mathbb{Q})$. Actually, $\gamma \in H_1(X_0, \mathbb{Q})$ is an extreme point of $\mathcal{D}_0$ if and only if $\xi = c/\|c\|_1$ for a circuit (simple closed path) c in X_0.*

(2) *A face of $\overline{\mathcal{D}_0}$ of the maximal dimension corresponds to an orientation E_0^+ of X_0 which is strongly connected. A face of lower dimensions corresponds to an oriented subgraph contained in a strongly connected orientation (with a compatible orientation).*

(3) *Let F_1 and F_2 be faces of $\overline{\mathcal{D}_0}$ and X_{F_1} and X_{F_2} are the corresponding oriented subgraphs. $F_1 \subset F_2$ if and only if $X_1 \subset X_2$ as oriented graphs (namely they have compatible orientations).*

The corrspondence is given in the following way. For ξ in the interior of a face F of maximal dimension, we find a strongly connected transition probability p_ξ whose homological direction γ_{p_ξ} is equal to ξ. The support $\mathrm{sppt}(p_\xi) = \{e \mid p(e) \neq 0\}$ coincides with an orientation E_0^+. An element ξ in a subface of F is written as $\xi = \sum_{e \in E_0^+} a(e)e$ uniquely, with this orientation E_0^+. The support $\mathrm{sppt}(a) = \{e \mid a(e) \neq 0\} \subset E_0^+$ gives an oriented subgraph of E_0^+.

For $\xi \in \mathcal{D}_0$, from the definition, there is a transition probability p such that $\xi = \gamma_p$. Such transition probability is not unique, but there is a canonical one. We now explain how this canonical p_ξ is actually given. Let us take a reference transition probability p and a cohomology class $u \in H^1(X_0, \mathbb{R})$ represented by a 1-form, i.e. $u : E_0 \to \mathbb{R}$ with $u(\overline{e}) = -u(e)$. For a 1-form u, we define the *twisted transition operator* $L_u : \ell^2(X_0, m) \to \ell^2(X_0, m)$ by

$$L_u\varphi(x) = \sum_{e \in E_x} p(e)\exp\big(u(e)\big)\varphi(t(e)), \quad (\varphi \in \ell^2(X_0, m)). \tag{2.1}$$

It has the unique simple positive eigenvalue $\alpha(u)$ (Perron-Frobenius eigenvalue) and the right eigenfunction ϕ_u i.e. $L_u\phi_u = \alpha(u)\phi_u$, and the left eigenfunction ψ_u, which is the eigenfunction of the transpose $L_u^t : \ell^2(X_0, m) \to \ell^2(X_0, m)$;

$$L_u^t\psi_u(x) := \sum_{e \in E_x} p(\overline{e})\exp\big(u(\overline{e})\big)\psi_u(t(e)) = \alpha(u)\psi_u.$$

Both ϕ_u and ψ_u are strictly positive valued, and normalized as $\sum_{e \in E_0} \phi_u(e)\psi_u(e) = 1$.

Put

$$p_u(e) = p(e)e^{u(e)}\frac{\phi_u(t(e))}{\phi_u(o(e))}\alpha(u)^{-1}, \tag{2.2}$$

$$m_u(e) = p(e)e^{u(e)}\phi_u(t(e))\psi_u(o(e))\alpha(u)^{-1}. \tag{2.3}$$

Then we see that p_u is a non-degenerate transition probability with the invariant measure $m_u(x) = \sum_{e \in E_x} m_u(e)$. Note that, neither p_u nor m_u depends on the choice of a representative u but only on its cohomology class. Actually, take another representative $u' = u + df$ with $f : X_0 \to \mathbb{R}$. Then we readily see that $e^{f(\cdot)}L_{u'} = L_u(e^{f(\cdot)})$ and so $\alpha(u') = \alpha(u)$, $\phi_{u'} = e^{-f(\cdot)}\phi_u$, $\psi_{u'} = e^{f(\cdot)}\psi_u$.

We see that $\beta := \log\alpha : H^1(X_0, \mathbb{R}) \to \mathbb{R}$ is real analytic and strictly convex. Thus the correspondence $u \mapsto \xi = (\nabla\beta)(u)$ is a diffeomorphism of $H^1(X_0, \mathbb{R})$ onto an open subset $\mathcal{D}_p$ in $\Gamma \otimes \mathbb{R}$. In [**7**], it is shown that $\mathcal{D}_p$ coincides with $\mathcal{D}_0$, hence does not depend on the reference transition probability p in particular.

Thus for $\xi \in \mathcal{D}_0$, there is a $u \in H^1(X_0, \mathbb{R})$ which attains the maximum of

$$\sup_{u \in H^1(X_0, \mathbb{R})} \langle u, \xi \rangle - \beta(u),$$

and $\xi = \gamma_{p_u} = \sum_{e \in E_0} m_u(e)e$. Define I as the Fenchel-Legendre transform of β

$$I(\xi) = \sup_{u \in H^1(X_0, \mathbb{R})} \langle u, \xi \rangle - \beta(u).$$

Then, the Gärtner-Ellis Theorem shows the following theorem:

THEOREM 2.2 (Large Deviation Principle, cf. [**2**]). *For the random walk x_n on a finite graph X_0 given by the transition probability p, there exists a lower semi-continuous function $I : H_1(X_0, \mathbb{R}) \to [0, \infty]$ satisfying*

(1) *The set $\{\xi \mid I(\xi) \le c\}$ is compact for every $c < \infty$.*
(2) *For every $A \subset H_1(X_0, \mathbb{R})$,*

$$\begin{aligned} -I(\operatorname{Int} A) &\le \liminf_{n \to \infty} \frac{1}{n} \log P_x(\frac{1}{n}\xi_n \in \operatorname{Int} A) \\ &\le \limsup_{n \to \infty} \frac{1}{n} \log P_x(\frac{1}{n}\xi_n \in \overline{A}) \le -I(\overline{A}), \end{aligned}$$

where $I(K) = \inf\{I(\mathbf{z}) \mid \mathbf{z} \in K\}$ *for* $K \subset H_1(X_0, \mathbb{R})$.

The function I is called the *rate function.*

Remark that I has the unique minimum inside of $\mathcal{D}_0$ and is lower semicontinuous. It takes finite value up to $\overline{\mathcal{D}_0}$ (for this, we use the fact that the extreme point of $\mathcal{D}_0$ corresponds to a simple closed path on X_0), and thus I is continuous over $\overline{\mathcal{D}_0}$.

The rate function I is the relative entropy with respect to the original transition probability p.

THEOREM 2.3. *For* $\xi \in \mathcal{D}_0$ *or the interior of a face of the maximal dimension (which corresponds to a strongly connected transition probability* p_ξ *with the invariant measure* m_ξ*), we have*

$$I(\xi) = -H(p_\xi, p),$$

where

$$H(p_\xi, p) = \sum_{e \in E_0} m_\xi(e) \log\Big(\frac{p_\xi(e)}{p(e)}\Big)$$

is the relative entropy of the transition probability p_ξ *with respect to* p.

PROOF. As we note in Theorem 2.1, $\mathcal{D}_0$ is a convex set and a face F of maximal dimension has extremal points $a_1, \dots, a_k$ which correspond to circuits $c_1, \dots, c_k$ in X_0. By denoting the homology class of c_i by the same symbol c_i, we write $a_i = c_i/|c_i|$. The orientations of all c_i's are compatible with the orientation E_0^+ of F.

For $\xi \in \mathcal{D}_0$, we write $\xi = \sum \mu_i a_i$ with $0 \le \mu_i$ and $\sum \mu_i < 1$. The corresponding transition probability p_ξ and its invariant measure m_ξ is given by (2.2). Since the homological direction of p_ξ is equal to ξ, we have

$$\xi = \sum_{e \in E_0} m_\xi(e) e = \sum_{e \in E_0^+} [m_\xi(e) - m_\xi(\overline{e})] e,$$
$$I(\xi) = \langle \xi, u \rangle - \log \alpha(u),$$

with the unique $u \in H^1(X_0, \mathbb{R})$.

We now compute $H(p_\xi, p)$ to show it coincides with $I(\xi)$. Since

$$\frac{p_\xi(e)}{p(e)} = e^{u(e)} \frac{\phi_u(t(e))}{\phi_u(o(e))} \alpha(u)^{-1},$$

$$\sum_{e \in E_0} m_\xi(e) \log\Big(\frac{p_\xi(e)}{p(e)}\Big) = \sum_{e \in E_0^+} m_\xi(e) \log\Big(\frac{p_\xi(e)}{p(e)}\Big) + \sum_{e \in E_0^-} m_\xi(e) \log\Big(\frac{p_\xi(e)}{p(e)}\Big)$$
$$= \sum_{e \in E_0^+} [m_u(e) - m_u(\overline{e})][u(e) + \log \frac{\phi_u(t(e))}{\phi_u(o(e))}] - [m_u(e) + m_u(\overline{e})] \log \alpha(u),$$

where we use $\sum_{e \in E_0^+} [m_u(e) - m_u(\overline{e})] u(e) = \langle \xi, u \rangle$, $\sum_{e \in E_0^+} [m_u(e) + m_u(\overline{e})] = 1$.

On the other hand, as $\xi = \sum \mu_i a_i$, we have $[m_u(e) - m_u(\bar{e})] = \sum \mu_i/|c_i|$, where the sum is taken over all circuits c_i's with $e \in c_i$. Thus we get

$$\sum_{e \in E_0^+} [m_u(e) - m_u(\bar{e})] \log \frac{\phi_u(t(e))}{\phi_u(o(e))}$$
$$= \sum_i \frac{\mu_i}{|c_i|} \sum_{e \in c_i} \log \frac{\phi_u(t(e))}{\phi_u(o(e))} = \sum_i \frac{\mu_i}{|c_i|} \log\Big(\prod_{e \in c_i} \frac{\phi_u(t(e))}{\phi_u(o(e))}\Big).$$

Because c_i is a closed curve , we notice $\prod_{e \in c_i} \frac{\phi_u(t(e))}{\phi_u(o(e))} = 1$.

Putting all these together, we obtain

$$\sum_{e \in E_0} m_\xi(e) \log\Big(\frac{p_\xi(e)}{p(e)}\Big) = \langle \xi, u \rangle - \log \alpha = I(\xi)$$

for $\xi \in \mathcal{D}_0$.

For $\xi \in \operatorname{Int} F$, there exists a strongly connected transition probability p_ξ whose support coincides with E_0^+. By using the invariant measure m_ξ, ξ is given by $\xi = \sum_{e \in E_0^+} m_\xi(e)e$, but p_ξ and m_ξ are not given by (2.2). Here we use continuity of I. Take a sequence $\xi_k \in \operatorname{Int} \mathcal{D}_0 \to \xi$ and let u_k be the corresponding 1-forms. Then we know

$$\xi_k = \sum_{e \in E_0^+} [m_{u_k}(e) - m_{u_k}(\bar{e})]e,$$
$$I(\xi_k) = -H(p_{\xi_k}, p).$$

As $\xi_k \to \xi$, we have $m_{u_k}(e) \to m_\xi(e)$ and $m_{u_k}(\bar{e}) = m(t(e))p_{u_n}(\bar{e}) \to 0$. Because $\lim_{\epsilon \to 0} \epsilon \log \epsilon = 0$, we obtain

$$I(\xi) = \lim_{k \to \infty} \sum_{e \in E_0} m_{u_k}(e) \log\Big(\frac{p_{u_k}(e)}{p(e)}\Big) = \sum_{e \in E_0^+} m_\xi(e) \log\Big(\frac{p_\xi(e)}{p(e)}\Big).$$

□

COROLLARY 2.4. *Let F be a face (not necessarily of maximal dimension) whose extreme points are $a_i = c_i/|c_i|$, and $\xi = \sum \mu_i \frac{c_i}{|c_i|} \in \operatorname{Int} F$ with $0 \leq \mu_i \leq 1$, $\sum \mu_i = 1$. Define $m_\xi(e) \geq 0$ so that $\xi = \sum_{e \in E_0^+} m_\xi(e)e$ (uniquely written in this way), where E_0^+ is the orientation corresponding to F and the corrsponding transition probability $p_\xi(e) = m_\xi(e)/\sum_{e' \in E_(o(e))} m_\xi(e')$. Then we have*

$$I(\xi) = I_{p_\xi} - \sum \frac{\mu_i}{|c_i|} \log P(c_i).$$

3. Large deviation for a crystal lattice

Let X be a crystal lattice, the Γ-cover of a finite graph X_0. To discuss large deviation, we realize X in $\Gamma \otimes \mathbb{R} \cong \mathbb{R}^d$. Namely we take a map $\Phi : X \to \Gamma \otimes \mathbb{R}$ satysfying $\Phi(\sigma x) = \Phi(x) + \sigma$ for every $\sigma \in \Gamma$, where the σ in the right hand side is the corresponding vector $\sigma \otimes 1 \in \Gamma \otimes \mathbb{R}$. For later use, we choose Φ which takes values in rational points with respect to the lattice Γ and $\Phi(x_0) = 0$ for a fixed base point $x_0 \in X$.

Now we consider a random walk on X whose transition probability p is Γ-invariant. We fix a base point $x_0 \in X$ and let Ω_{x_0} be the space of infinite paths

$c = (e_1, \dots)$ with $o(c) = x_0$, and let $\xi_n(c) = \Phi(t(e_n)) \in \Gamma \otimes \mathbb{R}$ be where the random walker is at the n-th step. Since p is Γ-invariant, it induces a random walk in X_0. We denote its homological direction by $\gamma_p \in H_1(X_0, \mathbb{R})$. It is known that an abelian cover of X_0 is a subcover of the universal abelian cover of X_0 whose covering transformation group is $H_1(X_0) = \pi_1(X_0)/[\pi_1(X_0), \pi_1(X_0)]$. The surjective homomorphism $\rho : H_1(X_0) \to \Gamma$ extends to $\rho : H_1(X_0, \mathbb{R}) \to \Gamma \otimes \mathbb{R}$. Now the law of large numbers is stated as

$$\frac{1}{n}\xi_n(c) \to \rho(\gamma_p) \in \Gamma \otimes \mathbb{R}, \quad (P - a.e.c).$$

Let Γ' be the dual lattice of Γ. For $u \in \mathrm{Hom}(\Gamma, \mathbb{R}) = \Gamma' \otimes \mathbb{R}$, we define a 1-form of X_0, denoted by the same symbol u, by

$$u : e \mapsto u(e) = \langle u, d\Phi(e) \rangle,$$

where $d\Phi(e) = \Phi(t(e)) - \Phi(o(e))$. In this way, we identify $\mathrm{Hom}(\Gamma, \mathbb{R}) \subset H^1(X_0, \mathbb{R})$. Let L_u be the twisted transition operator defined by (2.1), and $\alpha(u)$ be its Perron-Frobenius eigenvalue for $u \in \mathrm{Hom}(\Gamma, \mathbb{R})$.

LEMMA 3.1.

(1) *Let $\mathbb{E}_{x_0}$ be the expectation with respect to P_{x_0}. Then we have*

$$\lim_{n \to \infty} \frac{1}{n} \log \mathbb{E}_{x_0}(e^{\langle u, \xi_n \rangle}) = \log \alpha(u) =: \beta(u).$$

(2) *The function $\beta : \mathrm{Hom}(\Gamma, \mathbb{R}) \to \mathbb{R}$ is real analytic and strictly convex. Thus the correspondence $u \mapsto (\nabla\beta)(u)$ is a diffeomorphism of $\mathrm{Hom}(\Gamma, \mathbb{R})$ onto the $\mathcal{D} = \rho(\mathcal{D}_0)$ in $\Gamma \otimes \mathbb{R}$.*

Due to the discussions in §2, we have a large deviation theorem ([**7**]) as in Theorem 2.2 with $I : \Gamma \otimes \mathbb{R} \to [0, \infty]$ defined by the Fenchel-Legendre transform of β, i.e.

$$I(\xi) = \sup_{u \in \mathrm{Hom}(\Gamma, \mathbb{R})} \langle u, \xi \rangle - \beta(u).$$

The essential domain of I coincides with $\overline{\mathcal{D}}$, which is the unit ball with respect to the metric d_1 defined by the norm

$$\|\xi\|_1 := \inf\{\|\alpha\|_1 \mid \alpha \in H_1(X_0, \mathbb{R}) \text{ with } \rho(\alpha) = \xi \}.$$

As is shown in [**7**], it is the limit norm of the tangent cone of X at the infinity. Namely,

THEOREM 3.2. *Let (X, d) be a crystal lattice with the graph distance d. Then*

$$\lim_{\epsilon \to 0}(X, \epsilon d, x_0) = (\Gamma \otimes \mathbb{R}, d_1, 0)$$

in the pointed Gromov-Hausdorff topology.

Remark: In [**9**], [**11**], the "heat flow" is shown to be the gradient flow of the relative entropy as a function of the space of probability measures (on a Riemannian manifold) with the L^2-Wasserstein distance. It would be interesting how the rate function I, which turns out to be the relative entropy, is related with the geometry of the tangent cone at the infinity of the crystal lattice.

4. Precise asymptotics with the large deviation scalings

In this section, we follow the argument in S.Lalley[**8**], and M. Babillot and F. Ledrappier[**1**].

To obtain more precise asymptotics of the probability $P_{\xi,n}(A) := P_x(\xi_n - n\xi \in A)$ for $\xi \in \mathcal{D}$ and $A \subset \Gamma \otimes \mathbb{R}$, we extend the notion of twisted transition operator for complex 1-forms w, i.e. L_w with $w \in \mathrm{Hom}(\Gamma, \mathbb{C})$, and use the complex Perron-Frobenius theorem (cf. [**10**]). As in §3, we identify $w \in \mathrm{Hom}(\Gamma, \mathbb{C})$ as a $\mathbb{C}$-valued one-form of X_0 by

$$w(e) := \langle w, d\Phi(e)\rangle.$$

The *twisted transition operator* $L_w : C(X_0, \mathbb{C}) \to C(X_0, \mathbb{C})$ for $w = u + iv$ is defined by

$$(L_w f)(x) = \sum_{e \in E_x} p(e) \exp(w(e)) f(t(e)).$$

Since $C(X_0, \mathbb{C})$ is a finite dimensional vector space ($C(X_0, \mathbb{C}) \cong \mathbb{C}^N$ with $N = \#X_0$), L_ω is nothing but a complex Perron-Frobenius type matrix of finite size. Note

$$(L_w^n f)(x) = \exp(\langle w, \Phi(x)\rangle) \sum_{c \in \Omega_{x,n}} p(c) \exp(\langle w, \Phi(t(c))\rangle) f(t(c)).$$

We remark, due to the rationality of Φ, that L_w is periodic in v, i.e. there is an integer $M \in \mathbb{N}$ such that if $v_2 - v_1 \in M\Gamma'$, then $L_{u+iv_1} = L_{u+iv_2}$.

For simplicity, from now on, we assume X_0 is non-bipartite (discussion is easily generalized to a random walk with period K.) As seen in §2, for a real valued 1-form u, L_u has the maximum positive eigenvalue $\alpha(u)$ with the right eigenfunction ϕ_u and the left eigenfunction ψ_u which are positive valued and $\langle \phi_u, \psi_u\rangle = 1$.

Put $V := \{\varphi \mid \langle \varphi, \psi_u\rangle = 0\}$, an $N-1$ dimensional subspace of $C(X_0, \mathbb{C}) \cong \mathbb{C}^N$. It is preserved by L_u, as

$$\langle L_u \varphi, \psi\rangle = \langle \varphi, L_u^* \psi\rangle = \alpha(u)\langle \varphi, \psi\rangle = 0.$$

Hence, according to the decomposition $C(X_0, \mathbb{C}) = < \phi_u > \oplus V$, we have

$$L_u = \begin{pmatrix} \alpha(u) & 0 \\ 0 & K_u \end{pmatrix} : < \phi_u > \oplus V \to < \phi_u > \oplus V,$$

with $\|K_u\| < \alpha(u)(1 - \epsilon)$.

Now for the case $w = u + iv$;

LEMMA 4.1. (*see e.g.* [**10**] *for details*) *L_w has the eigenvalue $\alpha(w)$ of the maximal module with $|\alpha(w)| \le \alpha(u)$ and the equality holds if and only if $v = 0$. For a small v, $\alpha(w)$ is the simple eigenvalue and hence analytic.*

PROOF. For a real 1-form u, the eigenvalue $\alpha(u)$ is characterized as

$$\alpha(u) = \min\Big\{\frac{L_u\varphi(x)}{\varphi(x)} \mid \varphi \ge 0, x \text{ with } \varphi(x) \ne 0\Big\}.$$

For $w = u + iv$, if $L_w \varphi = \lambda \varphi$, then

$$|\lambda||\varphi| = \Big|\sum_{e \in E_x} p(e) e^{u(e)+iv(e)} \varphi(t(e))\Big| \le \sum_{e \in E_x} p(e) e^{u(e)} |\varphi|(t(e)) = L_u|\varphi|(x).$$

Thus $|\lambda| \le \alpha(u)$.

The equality $|\lambda| = \alpha(u)$ holds if and only if there exists a function θ of X_0 such that,

$$e^{iv(e)}\varphi(t(e)) = e^{i\theta(x)}\phi_u(t(e)), \quad (\forall e \in E_x). \tag{4.1}$$

We shall prove that $v = 0$ if (4.1) is satisfied. We can put $\varphi(x) = e^{if(x)}\phi_u(x)$ with a function f of X_0 and $\lambda = \alpha(u)e^{i\theta_0}$ with a constant θ_0. From (4.1), we obtain

$$v(e) = \theta(o(e)) - f(t(e)) \quad (\forall e \in E_0).$$

Putting this into the equation $L_w\varphi = \lambda\varphi$, we get $f(x) = \theta(x) - \theta_0$. That is

$$v(e) = \theta(o(e)) - \theta(t(e)) + \theta_0.$$

Let us consider a path $c = (e_1, e_2, \dots, e_n)$ which represents $\sigma \in \Gamma$. Then

$$\langle v, \sigma\rangle = v(e_1) + \cdots + v(e_n) = \theta(o(c)) - \theta(t(c)) + |c|\theta_0 = |c|\theta_0.$$

Take $\bar{c}$ as a path representing $-\sigma$ and proceed the same argument. Then we have

$$-\langle v, \sigma\rangle = |\bar{c}|\theta_0 = |c|\theta_0 = \langle v, \sigma\rangle,$$

which implies $\langle v, \sigma\rangle = 0$ for all $\sigma \in \Gamma$. Thus we conclude $v = 0 \in \Gamma \otimes \mathbb{R}$. □

As we have noted earlier that L_w is periodic in v, together with the perturbation theory, we have

LEMMA 4.2. *There is a small neighborhood U of the origin in $\Gamma' \otimes \mathbb{R}/M\Gamma'$ such that*

(1) *If $v \notin U$, $\|L_w\| < \alpha(u)(1-\epsilon)$,*
(2) *If $v \in U$, there is a regular matrix P whose entries are functions of X_0 such that*

$$L_w = P\begin{pmatrix} \alpha(w) & 0 \\ 0 & K_w \end{pmatrix} P^{-1},$$

with $\|K_w\| < \alpha(u)(1-\epsilon)$.

Now we are ready to study asymptotics of $P_{\xi,n}$ as n goes to the infinity. Recall for $\xi \in \mathcal{D}$, there is a real valued 1-form u_ξ and the left eigenfunction m_ξ of L_{u_ξ} such that $\xi = \nabla\beta(u_\xi)$. We define a family of measures

$$M_{n,\xi}(h) := m(x)\int_{(\Gamma\otimes\mathbb{R})'} (L^n_{u_\xi+iv}1)(x)e^{-n\langle u_\xi+iv,\xi\rangle}\hat{h}(v)dv, \quad (h \in C_0(\Gamma \otimes \mathbb{R})),$$

where $\hat{h}$ is the Fourier transform of h.

THEOREM 4.3. *For $h \in C_0(\Gamma \otimes \mathbb{R})$,*

$$M_{n,\xi}(h) \sim \left(\frac{2\pi}{n}\right)^{d/2} e^{-nI(\xi)}m(x)\sqrt{\det(\mathrm{Hess}\, I)(\xi)}\int_{\Gamma\otimes\mathbb{R}} h(\zeta)d\lambda(\zeta),$$

where $d\lambda$ is the Lebesque measure of $\Gamma \otimes \mathbb{R}$ with respect to the Albanese metric (see [**6**] *for definition).*

First we see that, once we obtain Theorem 4.3, we have

COROLLARY 4.4. *Let $A \subset \Gamma \otimes \mathbb{R}$ and $\xi \in \mathcal{D}$.*

$$P_x(\xi_n - n\xi \in A) \sim \left(\frac{2\pi}{n}\right)^{d/2} e^{-nI(\xi)}m(x)\sqrt{\det(\mathrm{Hess}\, I)(\xi)}\int_A e^{-\langle u_\xi,\zeta\rangle}d\lambda(\zeta).$$

PROOF. Take $h(\zeta) = e^{-\langle u_\xi, \zeta\rangle} I_A(\zeta)$ and apply Theorem 4.3. Since

$$(L^n_{u_\xi+iv}1)(x) = \sum_{c\in\Omega_{x,n}} p(c)e^{\langle u_\xi+iv, \Phi(t(c))\rangle},$$

we have

$$\begin{aligned}
M_{n,\xi}(e^{-\langle u_\xi,\cdot\rangle} I_A) &= m(x)\int_{(\Gamma\otimes\mathbb{R})'} \sum_{c\in\Omega_{x,n}} p(c)e^{\langle u_\xi+iv,\Phi(t(c))-n\xi\rangle}\hat{h}(v)dv \\
&= m(x)\sum_{c\in\Omega_{x,n}} p(c)e^{\langle u_\xi,\Phi(t(c))-n\xi\rangle}\int_{(\Gamma\otimes\mathbb{R})'} e^{i\langle v,\Phi(t(c))-n\xi\rangle}\hat{h}(v)dv \\
&= m(x)\sum_{c\in\Omega_{x,n}} p(c)e^{\langle u_\xi,\Phi(t(c))-n\xi\rangle}e^{-\langle u_\xi,\Phi(t(c))-n\xi\rangle\rangle} I_A(\Phi(t(c))-n\xi) \\
&= m(x)\sum_{c\in\Omega_{x,n}} p(c)I_A(\Phi(t(c))-n\xi) = P_x(\xi_n - n\xi\in A).
\end{aligned}$$

Due to Theorem 4.3, it is asymptotic to

$$\left(\frac{2\pi}{n}\right)^{d/2} e^{-nI(\xi)} m(x)\sqrt{\det(\operatorname{Hess} I)(\xi)}\int_A e^{-\langle u_\xi,\zeta\rangle}d\lambda(\zeta).$$

□

Put

$$\nu_{\xi,n} = \left(\frac{n}{2\pi}\right)^{d/2} e^{nI(\xi)} M_{\xi,n}.$$

By the standard measure theory, Theorem 4.3 follows from the following proposition 4.5.

PROPOSITION 4.5. *Set*

$$\mathcal{H} := \{h \text{ non-negative} \mid \hat{h}\in C_0(\Gamma'\otimes\mathbb{R})\}.$$

Then for $h\in\mathcal{H}$, *we have*

$$\lim_{n\to\infty}\nu_{\xi,n}(h) = m(x)\sqrt{\det(\operatorname{Hess} I)(\xi)}\lambda(h),$$

uniformly in $\xi\in\mathcal{D}$,

PROOF. Let U be the small open neighborhood of the origin in $\Gamma'\otimes\mathbb{R}/M\Gamma'$ as in Lemma 4.2 and then we identify U with the open set in the fundamental domain. Let us redefine U as the open set in $\Gamma'\otimes\mathbb{R}$ by the orbit of U under the $M\Gamma'$ action so that $\|L_{u+iv}\| < \alpha(u)(1-\epsilon)$ for $v\in U^c$. When $v\in U$, an arbitrary function φ of X_0 can be written as $\varphi = \langle\varphi,\psi\rangle\phi + \varphi'$, where ϕ and ψ are the right and left eigenfunctions of L_{u+iv} and $\varphi'\in V$ respectively. Thus we have

$$L^n_w\varphi = \alpha(w)^n\langle\varphi,\psi\rangle\phi + K^n_w\varphi',$$

with $\|K_w\| < \alpha(u)(1-\epsilon)$.

Now we take $w = u_\xi + iv$, where u_ξ is the solution of $\nabla\beta(u) = \xi$. First we shall see the term $K^n_w\varphi'$ does not contribute to the limit. Actually,

$$\begin{aligned}
&\left|\left(\frac{n}{2\pi}\right)^{d/2} e^{nI(\xi)} m(x)\int_U K^n_w\varphi' e^{-\langle u_\xi+iv,\xi\rangle}\hat{h}(v)dv\right| \\
&\qquad\le \left(\frac{n}{2\pi}\right)^{d/2}\|\hat{h}\|_{L^1}\|\varphi'\|e^{n\beta(u_\xi)}(1-\epsilon)^n e^{n(I(\xi)-\langle u_\xi,\xi\rangle)}.
\end{aligned}$$

Since $I(\xi) = \langle u_\xi, \xi\rangle - \beta(u_\xi)$, the right hand side is equal to

$$\left(\frac{n}{2\pi}\right)^{d/2} \|\hat{h}\|_{L^1}\|\varphi'\|(1-\epsilon)^n$$

which goes to zero as n goes to the infinity. In the same way, we see that the integral over U^c doesn't contribute to the limit.

Thus we find

$$\nu_{\xi,n}(h) \sim \left(\frac{n}{2\pi}\right)^{d/2} e^{nI(\xi)} m(x) \int_U \langle 1, \psi\rangle \phi(x) e^{n(\beta(u_\xi+iv)-\langle u_\xi+iv,\xi\rangle)}\hat{h}(v)dv. \tag{4.2}$$

We, on the other hand, know that, for $v \in U$,

$$\begin{aligned}
&\beta(u_\xi + iv) - \langle u_\xi + iv, \xi\rangle \\
&\quad = \beta(u_\xi) + i\langle v, \nabla\beta\rangle - \frac{1}{2}\operatorname{Hess}\beta(u_\xi)(v,v) - \langle u_\xi + iv, \xi\rangle - \text{higher terms} \\
&\quad = \beta(u_\xi) - \langle u_\xi, \xi\rangle - f(v) = -I(\xi) - f(v),
\end{aligned}$$

with $\Re f(v) = \frac{1}{2}\operatorname{Hess}\beta(u_\xi)(v,v) + O(|v|^4)$ and $\Im f(v) = O(|v|^3)$.

The right hand side of (4.2), then becomes

$$\left(\frac{n}{2\pi}\right)^{d/2} m(x) \int_U e^{-nf(v)}\langle 1, \psi\rangle\phi(x)\hat{h}(v)dv.$$

Note that $|\langle 1, \psi\rangle \phi_k(x)|$ is bounded as ϕ, ψ are the functions on a finite graph.

We change coordinate $v = v'/\sqrt{n}$ of the right hand side to obtain

$$\left(\frac{1}{2\pi}\right)^{d/2} m(x) \int_{\sqrt{n}U} e^{-nf(v'/\sqrt{n})}\langle 1, \psi\rangle\phi(x)\hat{h}(v'/\sqrt{n})dv',$$

where $\psi = \psi_{u_\xi + v'/\sqrt{n}}$ and $\phi = \phi_{u_\xi+v'/\sqrt{n}}$.

Notice that the integrand is bounded by an integrable function independent of n, since

$$|e^{-nf(v'/\sqrt{n})}\langle 1, \psi\rangle\phi(x)\hat{h}(v'/\sqrt{n})| \le c_1 e^{-\operatorname{Hess}\beta(u_\xi)(v',v')} e^{-c_2/n}|\hat{h}(v'/\sqrt{n})|.$$

We apply Lebesgue's convergence theorem and obtain

$$\begin{aligned}
&\lim_{n\to\infty} \nu_{\xi,n}(h) \\
&\quad = \left(\frac{1}{2\pi}\right)^{d/2} m(x)\langle 1, \psi_{u_\xi}\rangle\phi_{u_\xi}(x) \int_{\Gamma'\otimes\mathbb{R}} e^{-\operatorname{Hess}\beta(u_\xi)(v',v')}\hat{h}(0)dv' \\
&\quad = \frac{m(x)}{\sqrt{\det\operatorname{Hess}\beta(u_\xi)}}\langle 1, \psi_{u_\xi}\rangle\phi_{u_\xi}(x)\hat{h}(0) = \frac{m(x)}{\sqrt{\det\operatorname{Hess}\beta(u_\xi)}}\hat{h}(0).
\end{aligned}$$

Here, at the last equality, we use $\langle 1, \psi_{u_\xi}\rangle = \sum_{X_0} m_\xi(x) = 1$ and $\phi_{u_\xi} = 1$. Since $\hat{h}(0) = \int_{\Gamma\otimes\mathbb{R}} h(\zeta)d\lambda(\zeta)$ and $\operatorname{Hess}\beta(u_\xi) = (\operatorname{Hess} I(\xi))^{-1}$, we obtain the Proposition 4.5. □

References

[1] M. Babillot and F. Ledrappier, *Lalley's theorem on periodic orbits of hyperbolic flows*, Ergod. Th. & Dynam. Sys.**18**(1998), 17– 39.

[2] A. Dembo and O. Zeitouni, *Large Deviations Techniques and Applications*, 2nd Ed. Springer, New Yourk, 1998.

[3] M. Gromov, Metric Structures for Riemannian and Non-Riemannian Spaces, Birkhäuser, 1999.

[4] A. Krámli and D. Szász, *Random walks with internal degrees of freedom I, Local limit theorems*, Z. Wahrscheinlichleitsth. verw. Geb.**63**(1983), 85–95.

[5] M. Kotani, T. Shirai and T. Sunada, *Asymptotic behavior of the transition probability of a random walk on an infinite graph*, J. Funct. Anal.**159**(1998), 669–689.
[6] M. Kotani and T. Sunada, *Albanese maps and off diagonal long time asymptotics for the heat kernels*, *Comm.Math.Phys.***209**(2000), 633–670.
[7] M. Kotani and T. Sunada, *Geometric aspects of large deviations for random walks on a crystal lattice*, preprint
[8] S. Lalley, *Closed geodesics in homology classes on surfaces of variable negative curvature*, Duke Math. J.**58** (1995), 351–360.
[9] F. Otto and C. Villani, *Generalization of an inequality by Talagrand and links with the logarithmic Sobolev inequality*, J. Funct. Anal. **173**(2000), 361-400.
[10] W. Parry and M. Pollicott, *Zeta functions and the periodic orbit structure of hyperbolic dynamics*, Astérisque **187-188**(1990).
[11] K. T. Sturm and M. K. von Renesse, *Transport inequalities, gradient estimates, entropy and Ricci curvature*, preprint.

Mathematical Institute, Graduate School of Sciences, Tohoku University, Aoba, Sendai 980-77, Japan

E-mail address: kotani@math.tohoku.ac.jp

Contemporary Mathematics
Volume **347**, 2004

Heat Kernel Estimates and Law of the Iterated Logarithm for Symmetric Random Walks on Fractal Graphs

Ben M. Hambly and Takashi Kumagai

ABSTRACT. We study two-sided heat kernel estimates on a class of fractal graphs which arise from a subclass of finitely ramified fractals. These fractal graphs do not have spatial symmetry in general, and we find that there is a dependence on direction in the estimates. We will give a new form of expression for the heat kernel estimates using a family of functions which can be thought of as a "distance for each direction". As an application, we give a law of the iterated logarithm which shows that the directional dependence leads to non-uniform behaviour in the typical paths of the random walk.

1. Introduction

There is a long history of work on Gaussian bounds for the heat kernel on spaces with smooth structure. The development of analysis on fractals showed that there are sets where heat kernels naturally have sub-Gaussian bounds. Graphs provide a setting in which to further investigate the generality of these results. One area of recent research activity is in determining analytic conditions which are equivalent to heat kernel estimates of sub-Gaussian type ([**BB1, BCG, GT1, GT2**]). Our interest here will be in exploring the range of behaviour that can be exhibited by the heat kernel in a class of self-similar graphs (cf. [**BB2, HK2, Jo**]).

The class of graphs we consider consists of infinite graphs generated using a set of contraction maps all with the same scale parameter. Instead of iterating forward the set of maps to obtain a fractal set, we take one of the contractions and invert it, then apply it to a finite set (the natural boundary of the fractal) and iterate to obtain an infinite self-similar set of vertices. These can then be connected in a way compatible with the fractal structure to form a fractal graph. The graphs we consider will be finitely ramified in the sense that any subset can be disconnected by removing a finite number of points and hence we will call them uniformly finitely ramified graphs. Some examples are shown in Figure 1 in Section 3. Note that

1991 *Mathematics Subject Classification.* Primary 60J10, 60J35; Secondary 60G17, 31C25.

Key words and phrases. Random walks on graphs, sub-Gaussian heat kernel bounds, law of iterated logarithm, geodesic distance, diffusion on fractals.

The second author was supported in part by Japan Association for Mathematical Sciences and Grant-in-Aid for Scientific Research (C)(2) 14540113.

these graphs are based on triangles but they do not have the full symmetry of the triangle and this leads to a variety of interesting behaviour.

In a previous paper [**HK2**] we considered the class of uniform finitely ramified fractal graphs and obtained an expression for the heat kernel. It was shown that in the case of nested fractal graphs, in which the underlying fractal set is invariant under the full symmetry group of the basic cell, this expression is equivalent to the usual formulation of sub-Gaussian estimates. That is there are constants d_s, d_w and $c_{1.1}, c_{1.2}$ (which differ in the upper and lower bounds), such that

$$p_k(x,y) \asymp c_{1.1} k^{-d_s/2} \exp(-c_{1.2}(\frac{d(x,y)^{d_w}}{k})^{1/(d_w-1)}), \quad \forall x,y, d(x,y) \le k,$$

where $d(x,y)$ is the usual graph distance (see (2.8) for the meaning of $\asymp$ here). The exponent $d_s = 2d_f/d_w$ is called the spectral dimension, d_w is the walk dimension, which describes the scaling in the time for the random walk to exit balls and d_f is the Hausdorff dimension of the underlying fractal. We also showed that, without symmetry, the estimates cannot in general be expressed so simply in terms of the graph metric. Indeed, for fractals of this type, where the random walk on the graph is strongly recurrent, the natural metric is the resistance metric, in the sense that the parabolic Harnack inequality holds with respect to this metric. When we write the estimates in terms of the usual graph distance there is a dependence on direction. In this paper we will give a more complete description of this dependence and show how the estimates can be expressed in a form involving functions, one for each direction in the graph, which are like distance functions.

In particular, for the case where there are two directions (each corresponds to a certain type of edge in the graph, for instance diagonal and horizontal for our Sierpinski gasket based examples in Figure 1 in Section 3), we typically have constants $c_{1.3}, c_{1.4}$ (differing in the upper and lower bounds) such that

$$p_k(x,y) \asymp c_{1.3} k^{-d_s/2} \exp(-c_{1.4} \sum_{i=1}^{2} (\frac{d_i(x,y)^{d_w^i}}{k})^{1/(d_w^i-1)}) \quad \forall x,y, d_i(x,y) \le k,$$

where the d_w^i is the walk dimension in the direction i and d_i is a function which is similar to a metric. Though this is typical, it may also be the case that there is a logarithmic correction as well. The full result for the two type case is given in Theorem 3.5.

Finally, once we have such estimates for the heat kernel, we can determine what information they give concerning the random walk itself. Here we will show that the directional dependence leads to two laws of the iterated logarithm, one for each direction. This means that extreme fluctuations of the path in one direction are larger than those in the other.

We believe that similar expressions for the heat kernel estimates are possible for all the p.c.f. graphs and p.c.f. fractals which have a non-degenerate harmonic structure by extending the arguments in this paper. We discuss the diffusion case in Section 5 and partially answer in the affirmative two conjectures made in [**HK3**].

2. Uniform finitely ramified graphs and their quadratic forms

2.1. Uniform finitely ramified graphs. For $\alpha > 1$ and $I = \{1, 2, \cdots, N\}$, let $\{\Psi_i\}_{i \in I}$ be a family of *α-similitudes* on $\mathbb{R}^D$. An α-similitude is a map $\Psi_i \mathbf{x} = \alpha^{-1} U_i \mathbf{x} + \gamma_i$, $\mathbf{x} \in \mathbb{R}^D$ where U_i is a unitary map and $\gamma_i \in \mathbb{R}^D$. We will impose several assumptions on this family. First, we assume

(H-0) $\{\Psi_i\}_{i\in I}$ satisfies the open set condition,

that is there is a non-empty, bounded open set W such that $\{\Psi_i(W)\}_{i\in I}$ are disjoint and $\cup_{i\in I}\Psi_i(W)\subset W$. As $\{\Psi_i\}_{i\in I}$ is a family of contraction maps, there exists a unique non-void compact set $\hat{K}$ such that $\hat{K}=\cup_{i\in I}\Psi_i(\hat{K})$. We assume

(H-1) $\hat{K}$ is connected.

Let Fix be the set of fixed points of the Ψ_i's, $i\in I$. A point $x\in Fix$ is called an *essential fixed point* if there exist $i,j\in I$, $i\neq j$ and $y\in Fix$ such that $\Psi_i(x)=\Psi_j(y)$. Let I_F be the set of $i\in I$ for which the fixed point of Ψ_i is an essential fixed point. We write $\hat{V}_0$ for the set of essential fixed points. We also let $\Psi_{i_1,\dots,i_n}=\Psi_{i_1}\circ\cdots\circ\Psi_{i_n}$. We make one further important assumption, that the family of contraction maps has a finite ramification property;

(H-2) If $\{i_1,\dots,i_n\},\{j_1,\dots,j_n\}$ are distinct sequences, then

$$\Psi_{i_1,\dots,i_n}(\hat{K})\bigcap\Psi_{j_1,\dots,j_n}(\hat{K})=\Psi_{i_1,\dots,i_n}(\hat{V}_0)\bigcap\Psi_{j_1,\dots,j_n}(\hat{V}_0).$$

DEFINITION 2.1. ([**HK2**]) A (compact) uniform finitely ramified fractal (u.f.r. fractal for short) $\hat{K}$ is a set determined by α-similitudes $\{\Psi_i\}_{i\in I}$ satisfying the assumptions (H-0), (H-1), (H-2) and with $\sharp\hat{V}_0\geq 2$.

If we also assume the following symmetry condition, then $\hat{K}$ is called a (compact) nested fractal, as introduced and discussed in [**Lind, Kus**].

(SYM) If $x,y\in\hat{V}_0$, then the reflection in the hyperplane $H_{xy}=\{z\in\mathbb{R}^D:|z-x|=|z-y|\}$ maps $\hat{V}_n$ to itself, where $\hat{V}_n=\cup_{i_1,\cdots,i_n\in I}\Psi_{i_1,\dots,i_n}(\hat{V}_0)$.

Thus u.f.r. fractals form a class of fractals which is wider than nested fractals, and is included in the class of p.c.f. self-similar sets ([**Kig**]).

Next we define unbounded u.f.r. fractals. We assume without loss of generality that $\Psi_1(\mathbf{x})=\alpha^{-1}\mathbf{x}$ and that $\mathbf{0}$ belongs to $\hat{V}_0$. Let $K=\cup_{n=1}^{\infty}\alpha^n\hat{K}$. Then, clearly $\Psi_1(K)=K$. We call K an unbounded uniform finitely ramified fractal. Let $V=V_0=\cup_{n=0}^{\infty}\alpha^n\hat{V}_n$ and $V_n=\alpha^nV$ for $n\in\mathbb{Z}$. (Note that this labelling is the opposite to the one given in [**HK3**]. As n gets bigger, the graph distance between each vertex of V_n gets bigger and $V_n\subset V_{n-1}$.) Then, $K=Cl(\cup_{n\in\mathbb{Z}}V_n)$. For each $l,n\geq 0$ and $i_1,\cdots,i_l\in I$, we call a set of the form $\alpha^{n+l}\Psi_{i_1,\cdots,i_l}(\hat{V}_0)$ an n-cell and $\alpha^{n+l}\Psi_{i_1,\cdots,i_l}(\hat{K})$ an n-complex.

We now introduce uniform finitely ramified graphs. These will be graphs with vertices V and a collection of edges B. In order to define the edges, we first define e_{xy} for an edge between x and y in V and then set $\hat{B}_0:=\{e_{xy}:x\neq y\in\hat{V}_0\}$. Then inside each 0-cell we place a copy of $\hat{B}_0$ and we denote by B the set of all the edges determined in this way. We call the graph (V,B) a uniform finitely ramified (u.f.r.) graph. If we construct the graph starting from a nested fractal, then it will be called a nested fractal graph.

2.2. Quadratic forms. An *electrical network* (V,C) on an arbitrary infinite connected graph with vertices V is an assignment to each edge e_{xy} of a positive number $C_{xy}=C_{yx}$, the conductance between x and y. For our fractal graphs a *basic electrical network* (V,C) on (V,B) is defined as follows. Firstly we assign a conductance to each edge $e_{xy} \in \hat{B}_0$. The conductance matrix C is constructed on the whole graph by putting on each edge in each 0-cell the same conductance as that of the corresponding edge in $\hat{B}_0$ and setting $C_{xy} = 0$ if $e_{x,y} \notin B$. Note that, for our fractal electrical networks, there exists $c_{2.1}, c_{2.2} > 0$ such that

$$c_{2.1} \le C_{xy} \le c_{2.2} \qquad \text{for all } e_{xy} \in B. \tag{2.1}$$

We next define a quadratic form on (V, B) associated with the electrical network. For $x, y \in V$, we write $x \sim y$ if $e_{xy} \in B$. For each $f, g \in l(V) := \{h :$ h is a function on $V\}$, we define

$$\mathcal{E}_C(f,g) = \frac{1}{2} \sum_{\substack{x,y \in V \\ e_{xy} \in B}} (f(x) - f(y))(g(x) - g(y)) C_{xy}. \tag{2.2}$$

We sometimes abbreviate $\mathcal{E}_C(f,f)$ as $\mathcal{E}_C(f)$. Now, define $\mu_x = \sum_{y \in V} C_{xy}$ for each $x \in V$. Set $\mu(A) = \sum_{x \in A} \mu_x$ for each $A \subset V$; μ is then a measure on V. For each $e_{xy} \in B$, define $P_{xy} = C_{xy}/\mu_x$, the transition probability matrix of the Markov chain corresponding to $\mathcal{E}_C$. To be precise, the process corresponding to $\mathcal{E}_C$ is a continuous time Markov chain in which jumps occur along edge e_{xy} at rate C_{xy}. In this paper, we will consider instead the induced discrete time random walk, the discrete time Markov chain which moves at unit time intervals to any vertex y connected to x with the probabilities for these jumps given by $\{P_{xy}\}$. We denote this induced random walk by $\{X_k\}_{k \ge 0}$. The random walk is reversible with respect to μ, indeed,

$$P_{xy}\mu_x = C_{xy} = C_{yx} = P_{yx}\mu_y.$$

The discrete Laplace operator corresponding to the random walk can be defined as

$$\mathcal{L}f(x) = \sum_y P_{xy} f(y) - f(x) = \frac{1}{\mu_x} \sum_y (\nabla_{xy} f) C_{xy},$$

where $\nabla_{xy} f = f(y) - f(x)$.

Let $\mathcal{Q}_M = \mathcal{Q}_M(\hat{V}_0)$ be the set of all $Q = \{q_{ij} : i, j \in \hat{V}_0$ such that $q_{ii} = 0, q_{ij} = q_{ji} \ge 0$ for any $i, j \in \hat{V}_0\}$. We denote by $\mathrm{Int}(\mathcal{Q}_M)$ the subset of $\mathcal{Q}_M$ such that every Q has $q_{ii} = 0, q_{ij} = q_{ji} > 0$ for all $i \sim j \in \hat{V}_0$. Also, let $\mathcal{Q}_{irr}$ be the set of $Q \in \mathcal{Q}_M$ such that $\mathcal{E}_Q(\xi, \xi) = 0$ if and only if ξ is constant. As our quadratic form on the graph V is constructed from a basic electrical network, the conductances are determined by placing $C \in \mathcal{Q}_M$ within each 0-cell, and we denote the form by $\mathcal{E}_C(\cdot,\cdot)$. Given the graph on V_1 which has the same structure as (V_0, B) and $C' \in \mathrm{Int}(\mathcal{Q}_M(\alpha\hat{V}_0))$, we can define a form on V_1 in the same way:

$$\mathcal{E}^{(1)}_{C'}(f,g) := \frac{1}{2} \sum_{\substack{x,y \in V_1 \\ e_{xy} \in \alpha B}} (f(x) - f(y))(g(x) - g(y)) C'_{xy} \qquad \text{for all } f, g \in l(V_1),$$

where $\{C'_{xy}\}_{x,y}$ is given by placing C' in each 1-cell. As our graph is finitely ramified, we know that for each $C \in \mathcal{Q}_M(\hat{V}_0)$, there exists $C' \in \mathcal{Q}_M(\alpha\hat{V}_0)$ so that

$$\mathcal{E}^{(1)}_{C'}(v) = \inf\{\mathcal{E}_C(f) : f \in l(V_0), f|_{V_1} = v\} \qquad \text{for all } v \in l(V_1),$$

see [**Bar, Kig**] for the proof. We can define a decimation map F from $\mathcal{Q}_M = \mathcal{Q}_M(\hat{V}_0)$ to itself by setting $F(C) = C'$. Note that F is homogeneous, as $F(\theta C) = \theta F(C)$ for all $\theta > 0$ and $C \in \mathcal{Q}_M$, however F is in general a non-linear map. In order to study the asymptotic properties of the form, it is important to observe the dynamics of the iteration of F (see [**Kum1**] and the references therein). By Schauder's fixed point theorem, we know that there exists $Q \in \mathcal{Q}_M$ (with $q_{ij} > 0$ for some $i \neq j$) and $\rho_Q^{-1} > 0$ such that $F(Q) = \rho_Q^{-1} Q$. Throughout this paper, we assume the following.

ASSUMPTION 2.2. (1) *For each* $Q \in \mathcal{Q}_{irr}$, *there exists* $l = l(Q) \in \mathbb{N}$ *such that* $F^n(Q) \in Int(\mathcal{Q}_M)$ *for all* $n \geq l$.
(2) *There exists* $Q_0 \in Int(\mathcal{Q}_M)$ *and* $\rho_{Q_0} > 0$ *such that* $F(Q_0) = \rho_{Q_0}^{-1} Q_0$.

REMARK 2.3. (1) By Corollary 6.20 of [**Bar**], $\rho_{Q_0} > 0$ is uniquely determined, that is if $Q_1, Q_2 \in \mathcal{Q}_{irr}$ satisfies $F(Q_j) = \rho_{Q_j}^{-1} Q_j$ $(j = 1, 2)$ with $\rho_{Q_1}, \rho_{Q_2} > 0$, then $\rho_{Q_1} = \rho_{Q_2} = \rho_{Q_0}$. In the class of fractal graphs we consider, we can prove $\rho_{Q_0} > 1$ (see [**Kig**] etc.). For the rest of the paper we denote ρ_{Q_0} by ρ.
(2) For $x, y \in \hat{K}$, the set of vertices $\{x_0, \cdots, x_m\}$ is called an n-chain from x to y if $x_0 = x, x_m = y$, $x_j \in \hat{V}_n$ for $1 \leq j \leq m-1$ and x_i, x_{i+1} are in the same $(-n)$-complex for $0 \leq i \leq m-1$. A sufficient condition for Assumption 2.2 1) is the following.

> (H-3) There exists $l \in \mathbb{N}$ such that for each $x, y \in \hat{V}_0$, there is an l-chain $\{x_0, \cdots, x_m\}$ from x to y so that for each $1 \leq i \leq m-2$, there is a l-cell C_i with $x_i, x_{i+1} \in C_i$ and $\hat{V}_0 \cap C_i = \emptyset$.

Indeed, if (H-3) holds, it is easy to show $F^n(Q) \in \mathrm{Int}(\mathcal{Q}_M)$ for $n \geq l$, $Q \in \mathcal{Q}_{irr}$ by observing the corresponding Markov chain on $\hat{V}_n$.
(3) Every nested fractal satisfies Assumption 2.2 1) and 2). Indeed, (H-3) can be shown using (SYM) and [**Kus**] Lemma 2.10 (or [**Lind**] Proposition *IV*.11), establishing 1). The existence of the fixed point 2) is proved in [**Kus**] Theorem 3.10 and in [**Lind**] Theorem *V*.5.
(4) Note that any non-degenerate fixed point of ρF is not necessarily unique even for nested fractals. In [**Bar**] Example 6.13, a one parameter family of non-degenerate fixed points on the Vicsek set is given.
(5) Using [**Kum1**] Theorem 3.4, we see that the following, which corresponds to Assumption 2.3 in [**HK2**], always holds under our Assumption 2.2.

For all $Q \in \mathrm{Int}(\mathcal{Q}_M)$, there exist $c_{1,Q}, c_{2,Q} > 0$ such that,

$$c_{1,Q}\rho^{-n}(Q_0)_{ij} \leq (F^n(Q))_{ij} \leq c_{2,Q}\rho^{-n}(Q_0)_{ij} \qquad \text{for all } n \in \mathbb{N},\ i,j \in \hat{V}_0.$$

Note that for the construction of diffusion processes on finitely ramified fractals, the standard approach is to start with quadratic forms (random walks) whose conductances (transition probabilities) are invariant under the decimation map. See [**Bar, Kig**] for details.

The natural metric on the graph obtained by counting the number of steps in the shortest path between points is denoted by $d(x, y)$ for $x, y \in V$.

The effective resistance between $x \neq y \in V$ is defined by

$$R(x,y)^{-1} = \inf\{\mathcal{E}_C(f,f) : f \in l(V), f(x) = 1, f(y) = 0\}. \tag{2.3}$$

We define $R(x, x) = 0$ for each $x \in V$. Note that for any $x \neq y \in V$, $R(x, y)$ is positive and finite. Let $\hat{B}_n = \{e_{xy} : x \neq y \in \alpha^n \hat{V}_0\}$ and define edges B_n by placing a copy of $\hat{B}_n$ in each of the n-cells in V_n. Under Assumption 2.2, there exist constants $c_{2.3}, c_{2.4}$ such that

$$c_{2.3}\rho^n \leq R(x,y) \leq c_{2.4}\rho^n, \quad \text{for all } n \in \mathbb{N},\ e_{xy} \in B_n.$$

Let $B_R(x,r)$ be the ball centred at x and radius r in the resistance metric. Note that $B_R(x,r)$ is not necessarily connected (see Remark 7.19 of [**Bar**]). Let $S = \log N/\log\rho$, the Hausdorff dimension of the fractal in the resistance metric. The mass of balls in the resistance metric is then controlled as there exist $c_{2.5}, c_{2.6}, r_0 > 0$ such that ([**HK2**] Lemma 3.2).

$$c_{2.5}r^S \leq \mu(B_R(x,r)) \leq c_{2.6}r^S \qquad \text{for all } r \geq r_0. \tag{2.4}$$

2.3. Heat kernel estimates. Let $P_k(x,y)$ be the transition function after k steps, $P_k(x,y) = P(X_k = y|X_0 = x)$, where $\{X_k; k \in \mathbb{N}\}$ is the random walk on V associated with the Dirichlet form $\mathcal{E}_C$. The heat kernel $p_k(x,y)$ we will discuss is defined by

$$p_k(x,y) = P_k(x,y)/\mu_y.$$

Note that by the reversibility of $\{X_k; k \in \mathbb{N}\}$, we have $p_k(x,y) = p_k(y,x)$.

In order to give our expression for the heat kernel estimates we will require a shortest path counting function. Let $N_m(x,y)$ denote the number of edges in the shortest path on V_m from x to y. By definition we have $d(x,y) = N_0(x,y)$. If $x, y \notin V_m$, we define the shortest path counting function to be $N_m(x,y) = \max_{x_1\in\partial D_m(x), y_1\in\partial D_m(y)} N_m(x_1,y_1)$, where $D_l(x)$ is an l-complex containing x. Define the time scale factor $\tau = \rho N$ and let

$$l(k,n) = \inf\{j : N_j(x,y)\tau^j \geq k\} \wedge n. \tag{2.5}$$

We can now state our heat kernel estimates.

THEOREM 2.4 ([**HK2**]: Theorem 4.10). *There exist constants $c_{2.7}, \cdots, c_{2.10} > 0$ such that for all $x, y \in V$ and $k \geq d(x,y)$, if $\rho^{n-1} \leq R(x,y) \leq \rho^n$, then*

$$p_k(x,y) \leq c_{2.7}k^{-\frac{S}{S+1}} \exp(-c_{2.8}N_{l(k,n)}(x,y)), \tag{2.6}$$

$$p_k(x,y) + p_{k+1}(x,y) \geq c_{2.9}k^{-\frac{S}{S+1}} \exp(-c_{2.10}N_{l(k,n)}(x,y)). \tag{2.7}$$

From now on we will express the pair of estimates (2.6), (2.7) as

$$p_k(x,y) \asymp c_1 k^{-\frac{S}{S+1}} \exp(-c_2 N_{l(k,n)}(x,y)). \tag{2.8}$$

Note that this theorem contains the diagonal estimate as $0 \leq N_{l(k,n)}(x,x) \leq 1$ for all $n, k \in \mathbb{N}$ and $x \in V$. For $x, y \in V$ such that $c_1\rho^n \leq R(x,y) \leq c_2\rho^n$ for some fixed constants, define a chemical exponent with respect to the resistance metric, for $0 \leq l < n$ as $d_l^c(x,y) = (\log_\rho N_l(x,y))/(n-l)$. For $l = n$ we can choose the exponent arbitrarily and thus define $d_n^c(x,y) = 1$. Using the definition of $l(k,n)$ we can write Theorem 2.4 in terms of the resistance metric and this chemical exponent as follows.

COROLLARY 2.5. *There exist constants $c_{2.11}, c_{2.12} > 0$ such that for $x, y \in V$ and $k \geq d(x,y)$, with n, l as above, then*

$$p_k(x,y) \asymp c_{2.11}k^{-\frac{S}{S+1}} \exp\left(-c_{2.12}\left(\frac{R(x,y)^{S+1}}{k}\right)^{\frac{d^c_{l(k,n)}(x,y)}{S+1-d^c_{l(k,n)}(x,y)}}\right).$$

For nested fractal graphs, where the fractals have spatial symmetry, there are constants $d_c, c_{2.13}, c_{2.14} > 0$ (d_c is called the chemical exponent) such that for $e_{xy} \in B_n$,

$$c_{2.13}\rho^{(n-m)d_c} \leq N_m(x,y) \leq c_{2.14}\rho^{(n-m)d_c}.$$

Hence, writing $d_w = (S+1)/d_c$, we can express our estimates in the following form.

$$(2.9) \qquad p_k(x,y) \asymp c_{2.15} k^{-S/(S+1)} \exp\left(-c_{2.16}\left(\frac{d(x,y)^{d_w}}{k}\right)^{1/(d_w-1)}\right).$$

We note that when there is no spatial symmetry, we do not necessarily have a single exponent d_w (as in (2.9)) for the off-diagonal estimates. See [**HK2**] for details.

We conclude this section with a result about the parabolic Harnack inequality. Let $\beta > 0$. We say $(\mathcal{E}_C, \mathbb{L}^2(V, d\mu))$ satisfies $(PHI(\beta))$, a parabolic Harnack inequality of order β, if whenever $u(n,x) \geq 0$ is defined on $[0,4N] \times \bar{B}(y,2r)$ and satisfies $u(n+1,x) - u(n,x) = \mathcal{L}u(n,x)$ for $(n,x) \in [0,4N] \times B(y,2r)$, then

$$\max_{\substack{N \leq n \leq 2N \\ x \in B(y,r)}} u(n,x) \leq c_{2.17} \min_{\substack{3N \leq n \leq 4N \\ x \in B(y,r)}} (u(n,x) + u(n+1,x)),$$

where $N \geq 2r$ and $c_{2.18} r^{\beta} \leq N \leq c_{2.19} r^{\beta}$ (cf. [**BB1, GT1, GT2, HK2**]). By Corollary 2.5 and a standard argument, we can deduce the following.

PROPOSITION 2.6.
$(\mathcal{E}_C, \mathbb{L}^2(V, d\mu))$ satisfies $(PHI(S+1))$ with respect to the resistance metric.

3. Heat kernel estimates and chemical exponents

In this section we will give a new form of expression for the heat kernel estimates which incorporates the dependence of the estimate on the direction. Our first step is to explain what we mean by a direction for our u.f.r. fractal graphs.

We begin by giving the elements of $\hat{B}_0$ a type. The classification uses the spatial symmetry of the fractal graph, that is, if there is a reflection which maps $\hat{V}_n$ to itself and maps $e \in \hat{B}_0$ to $e' \in \hat{B}_0$, then e and e' are of the same type, otherwise they are different. Let s_0 be the total number of types; clearly $s_0 \leq \sharp V_0(\sharp V_0 - 1)/2$. By labelling each copy of $\hat{B}_0$ using the types, we can classify each element of the set of edges B by type. Also, by self-similarity, we can classify each edge which consists of two elements in an n-cell for any $n \in \mathbb{Z}$.

For $1 \leq i \leq s_0$ and $m \geq 1$, let $\pi_m(i)$ be the set of m-chains from x to y which do not contain multiple points, where $e_{xy} \in \hat{B}_0$ is of type i. Set

$$S^i = \{\mathbf{u}(\pi) \mid \pi \in \pi_1(i), \pi \text{ is minimal}\},$$

where we define $\mathbf{u}(\pi) = (\mathbf{u}(\pi)_1, \cdots, \mathbf{u}(\pi)_{s_0})$ to be an s_0-dimensional vector such that $\mathbf{u}(\pi)_j$ is the number of type j steps (edges) in the path π $(1 \leq j \leq s_0)$. Here, we say $\pi \in \pi_m(i)$ is minimal when there is no $\pi' \in \pi_m(i)$, $\pi' \neq \pi$ such that $\mathbf{u}(\pi')_j \leq \mathbf{u}(\pi)_j$ for all $1 \leq j \leq s_0$. Let

$$M_{mat} = \{A : A \text{ is an } s_0 \times s_0 \text{ matrix such that for each } j,\ (j\text{-th row of } A) \in S^j.\},$$

and define $\lambda_{min} = \min_{A \in M_{mat}}\{\text{maximum eigenvalue of } A\}$. Throughout this section, we call a matrix $A \in M_{mat}$ whose maximum eigenvalue attains λ_{min} the minimal step matrix, and define $M_{min} = \{A \in M_{mat} : A \text{ is a minimal step matrix.}\}$.

In this paper, we will only treat the case where the number of types is 2. This is purely for simplicity, but it does include essentially all the different forms for the heat kernel estimates. Note that the possible structures for the minimal step matrix are the following.

$$\{\begin{pmatrix} a & b \\ c & d \end{pmatrix} : a,b,c,d \in \mathbb{N} \cup \{0\}, a^2+b^2 \geq 1, c^2+d^2 \geq 1\}.$$

FIGURE 1. Examples

We first classify the possible structures.

Case (0) $a = 0$; in this case there are three distinct structures which are possible for the minimal step matrix;

$$\text{(0-1)} \begin{pmatrix} 0 & b \\ c & d \end{pmatrix} \text{(0-2)} \begin{pmatrix} 0 & b \\ c & 0 \end{pmatrix} \text{(0-3)} \begin{pmatrix} 0 & b \\ 0 & d \end{pmatrix},$$

where $b, c, d \in \mathbb{N}$. By simple calculations, we see that the eigenvector corresponding to the maximum eigenvalue is a constant multiple of ${}^t(1, z)$ for some $z > 0$ where ${}^t\mathbf{v}$ denotes the transpose of $\mathbf{v}$. The case (0') $d = 0$ can be treated similarly.

We next consider the case $ad \neq 0$. There are four (essentially three) cases as indicated below;

$$\text{(1)} \begin{pmatrix} a & b \\ c & d \end{pmatrix} \text{(2)} \begin{pmatrix} a & b \\ 0 & d \end{pmatrix} \text{(2')} \begin{pmatrix} a & 0 \\ c & d \end{pmatrix} \text{(3)} \begin{pmatrix} a & 0 \\ 0 & d \end{pmatrix},$$

where $a, b, c, d \in \mathbb{N}$. Cases (2), (2') and (3) can each be decomposed into 3 subcases ($*$-1) $a > d$, ($*$-2) $a = d$ and ($*$-3) $a < d$. Note that cases (2') and (3-3) can be reduced to the cases (2) and (3-1) respectively by switching the role of type 1 and type 2.

We now describe how we choose and fix one element, denoted by A_{min}, from M_{min} (cf. Remark 3.6 (a)):

Case (a); there is a reducible matrix in M_{min}. (A non-negative matrix A is called irreducible if for each i, j, there exists $n = n(i, j)$ such that $(A^n)_{ij} > 0$. A matrix which is not irreducible is called reducible.) By switching the role of type 1 and type 2 if necessary, we can write the matrix as in the case (2).

(a-1); there is a minimal step matrix with $d < \lambda_{min}$ (thus $a = \lambda_{min}$). Then define this matrix as A_{min}. Such a matrix in M_{min}, if it exists, is uniquely determined.

(a-2); (a-1) does not hold (thus $d = \lambda_{min}$ and $a \leq \lambda_{min}$). If there is a matrix in M_{min} with $a < \lambda_{min}$, then take the matrix with $b = 0$ (if it exists) as A_{min}, otherwise take the matrix where $(\lambda_{min} - a)/b$ attains a maximum (if the original matrix is as in the case (2'), take the matrix where $c/(\lambda_{min} - d)$ attains a maximum) within such matrices in M_{min}. If there is no matrix in M_{min} with $a < \lambda_{min}$, then take the matrix with $a = \lambda_{min}$ as A_{min}. Note that in such a case, one can prove that $M_{min} = \{(\begin{smallmatrix} \lambda_{min} & b \\ 0 & \lambda_{min} \end{smallmatrix})\}$ and $\lambda_{min} = \min_{\mathbf{b} \in S^1} \mathbf{b}_1$.

Case (b); all the matrices in M_{min} are irreducible. Then, take an arbitrary element of M_{min} as A_{min}.

In Figure 1, we indicate several concrete examples of the different cases. These are cases where diagonal moves correspond to type 1 paths and horizontal moves correspond to type 2 paths. The left figure is the case (1), the middle figure (cf. Figure 2 in [**HK2**]) is the case (2-1) and the right figure is the case (2-2). If the contraction maps corresponding to the shaded triangles have a rotation of 60 (or -60) degrees (so that the shaded triangles are rotations of the original one), then the right figure is the case (3). We note that it is easy to check that Assumption 2.2 holds for these cases. The corresponding A_{min}'s are as follows.

$$\begin{pmatrix} 9 & 1 \\ 2 & 8 \end{pmatrix}, \begin{pmatrix} 14 & 2 \\ 0 & 12 \end{pmatrix}, \begin{pmatrix} 9 & 1 \\ 0 & 9 \end{pmatrix}, \begin{pmatrix} 10 & 0 \\ 0 & 9 \end{pmatrix}.$$

For the cases (1), (2-3) and (3-2), the eigenvector corresponding to the maximum eigenvalue is again a constant multiple of ${}^t(1, z)$ for some $z > 0$. For the cases (2-1) and (3-1), the maximum eigenvalue is a, the corresponding eigenvector is a constant multiple of ${}^t(1, 0)$ and the Jordan cell for the eigenvalue is 1-dimensional. For the case (2-2), the maximum eigenvalue is a, the corresponding eigenvector is a constant multiple of ${}^t(1, 0)$ and the Jordan cell for the eigenvalue is 2-dimensional.

We now determine the relationship between A_{min} and the asymptotic growth of the minimum number of edges in a path for each direction. For $\mathbf{v} \geq 0$, define

$$(G\ \mathbf{v})_i = \min_{\mathbf{b} \in S^i} \mathbf{b} \cdot \mathbf{v} \qquad i = 1, 2,$$

and hence $G\mathbf{v} = \min_{A \in M_{mat}} A\mathbf{v}$. Let $\lambda_{min} > 0$, $\mathbf{x} \geq \mathbf{0}$ be the maximum eigenvalue of A_{min} and its corresponding eigenvector, so that

$$G\ \mathbf{x} = A_{min}\mathbf{x} = \lambda_{min}\mathbf{x}. \tag{3.1}$$

When $\mathbf{x} \geq \mathbf{0}$ is not strictly positive, we can prove this as follows. Without loss of generality, we may assume that $\mathbf{x}_2 = 0$ and hence the matrix A_{min} is as in the case (2). Thus, $\lambda_{min} = a$. We can further see that $a = \min_{\mathbf{b} \in S^1} \mathbf{b}_1$. Indeed, if this does not hold for $d < \lambda_{min}$, by substituting the element of S^1 whose first element is minimal into the first row of A_{min}, we can obtain a matrix whose maximum eigenvalue is less than $\lambda_{min} = a$, which is a contradiction. For $d = \lambda_{min}$, this fact is guaranteed by our choice of A_{min}. We thus obtain

$$(G \begin{pmatrix} 1 \\ 0 \end{pmatrix})_1 = \min_{\mathbf{b} \in S^1} \mathbf{b} \cdot \begin{pmatrix} 1 \\ 0 \end{pmatrix} = \min_{\mathbf{b} \in S^1} \mathbf{b}_1 = a.$$

Clearly $(G\ {}^t(1, 0))_2 = 0$ and we obtain (3.1). When $\mathbf{x} > \mathbf{0}$, we can prove (3.1) using the properties of non-negative matrices, but omit the somewhat lengthy proof here.

Let $x, y \in V_n$ be in the same n-cell and assume the edge e_{xy} is of type i, $(i = 1, 2)$. Clearly $N_n(x, y) = (G^{\circ n}\mathbf{1})_i$ where $G^{\circ n}$ is the n-th iteration of G and $\mathbf{1} = {}^t(1, 1)$. Thus, when the eigenvector $\mathbf{x}$ of λ_{min} is strictly positive (that is, cases (0), (1), (2-3), (3-2)), using (3.1) we have constants such that

$$c_1\lambda_{min}^n\mathbf{x}_i = c_1(G^{\circ n}\mathbf{x})_i \leq (G^{\circ n}\mathbf{1})_i \leq c_2(G^{\circ n}\mathbf{x})_i = c_2\lambda_{min}^n\mathbf{x}_i.$$

Hence there is no directional dependence for the asymptotic order of the minimum number of edges in the shortest path and the problem is reduced to the case of a single type as discussed in [**Kum2, FHK**]. In particular, the heat kernel estimate has the same form as (2.9).

We thus consider only the cases (2-1), (2-2) and (3-1) in the following. For each case, the n-th power of A_{min} is given by

$$\text{(2-1)} \begin{pmatrix} a^n & \frac{b}{a-d}(a^n - d^n) \\ 0 & d^n \end{pmatrix} \text{(2-2)} \begin{pmatrix} a^n & bna^{n-1} \\ 0 & a^n \end{pmatrix} \text{(3-1)} \begin{pmatrix} a^n & 0 \\ 0 & d^n \end{pmatrix}.$$

Let $\pi(x,y)$ be the shortest path (with respect to the graph distance) from x to y in V. Note that each edge in the path is either of type 1 or type 2. We let $d_1(x,y)$ be the number of type 1 steps in $\pi(x,y)$. We decompose the path $\pi(x,y)$ into connected components C_i that only consist of type 2 steps and define $d_2(x,y) = \max_i |C_i|$. In other words, $d_2(x,y)$ is the maximum length of a connected chain of type 2 steps in the shortest path. Note that $d_i(\cdot,\cdot)$ is not a distance.

In order to define two associated shortest path counting functions, we decompose the shortest path. For $x, y \in V$ such that $\rho^{n-1} \le R(x,y) < \rho^n$, we can decompose the shortest path from x to y as follows. Let x_i be the first element of V_i on the shortest path from x ($x_i = x$ if $x \in V_i$). Define y_i in the same way by looking at the path from y. Then there is an $L = L(x,y) \le n$ such that there in no element in V_{L+1} on the path between x_L and y_L. By the self-similarity of the graph, it is easy to see that either $x_L = y_L$ or the number of elements in V_L on the path between x_L and y_L is uniformly bounded by some positive constant. We define $N^1_m(x,y)$ as the number of type 1 steps in the shortest path on V_m from x_m to y_m if $x_m \neq y_m$. We decompose the shortest path on V_m from x_m to y_m into connected components $C_{m,i}$ that only consist of type 2 steps and define $N^2_m(x,y) = \max_i |C_{m,i}|$ if $x_m \neq y_m$. Finally we set $N^i_m(x,y) = 1$ if $x_m = y_m$ and $m \neq 0$ (this is for consistency with the definition of $N_m(x,y)$). By definition we have $d_i(x,y) = N^i_0(x,y)$ for $i = 1, 2$.

We now change notation and let $\alpha_1 = a, \alpha_2 = d$. Note that $\alpha_1 > \alpha_2$. For $x, y \in V$ such that $\rho^{n-1} \le R(x,y) < \rho^n$, define $n_i = n_i(x,y) \le n$ so that $\alpha_i^{n_i} \le d_i(x,y) < \alpha_i^{n_i+1}$ for $i = 1, 2$. Note that $N^i_{m'}(x,y) = 1$ for $m' > n_i$. Note also that $n_1 \vee n_2 = n$.

Let $\bar{N}^1_m(x,y) = N^1_m(x,y) \log N^1_m(x,y)$ for the case (2-2), $\bar{N}^1_m(x,y) = N^1_m(x,y)$ otherwise. We let $\bar{N}^2_m(x,y) = N^2_m(x,y)$ for all the cases. Then the following holds.

LEMMA 3.1. *There exists $c_{3.1}, c_{3.2} > 0$ such that for all $m \ge 0$ and $x, y \in V$,*

$$c_{3.1} \sum_{i=1}^{2} \bar{N}^i_m(x,y) \le N_m(x,y) \le c_{3.2} \sum_{i=1}^{2} \bar{N}^i_m(x,y).$$

PROOF. Here and in the following, we sometimes abbreviate $N_m(x,y)$ (resp. $\bar{N}^i_m(x,y)$) as N_m (resp. $\bar{N}^i_m$). Firstly, when $x, y \in V_n$ are in the same n-cell and e_{xy} is of type i ($i = 1, 2$), then $d_i(x,y) = \alpha_i^n$ and $N^i_m = \alpha_i^{n-m} = \alpha_i^{-m} d_i(x,y)$ for all $m \le n$. Next, for $x, y \in V$ such that $\rho^{n-1} \le R(x,y) < \rho^n$, we can decompose the shortest path from x to y into $\{x, x_1, \cdots, x_L, y_L, \cdots, y_1, y\}$ as discussed above. Using the first observation and the self similarity of the graph, we see that

$$N^1_m = \left(\sum_{j=m}^{L-1} \alpha_1^{-m} d_1(x_j, x_{j+1}) + \sum_{j=m}^{L-1} \alpha_1^{-m} d_1(y_j, y_{j+1}) + \alpha_1^{-m} d_1(x_L, y_L) \right) \vee 1,$$

$$\max\{\alpha_2^{-m} d_2(x_j, x_{j+1}), \alpha_2^{-m} d_2(y_j, y_{j+1}), \alpha_2^{-m} d_2(x_L, y_L) : m \le j \le L-1\} \le$$

$$N^2_m \le \left(\sum_{j=m}^{L-1} \alpha_2^{-m} d_2(x_j, x_{j+1}) + \sum_{j=m}^{L-1} \alpha_2^{-m} d_2(y_j, y_{j+1}) + \alpha_2^{-m} d_2(x_L, y_L) \right) \vee 1.$$

For $i = 1, 2$, let q_i be the largest $j \ge 0$ such that either $d_i(x_j, x_{j+1})$ or $d_i(y_j, y_{j+1})$ is non-zero. When $d_i(x_L, y_L) \neq 0$, we define $q_i = L$ and when all these values are 0 we define $q_i = m$. Note that there are constants such that $c_1 \alpha_i^j \le d_i(x_j, x_{j+1}) \le c_2 \alpha_i^j$

if it is not zero. The same fact is true for $d_i(y_j, y_{j+1})$ and $d_i(x_L, y_L)$ (in this case $j = L$). Thus, using the estimates of N^i_m above, we have constants such that $c_3\alpha_i^{q_i-m} \le N^i_m \le c_4\alpha_i^{q_i-m}$. Since $\alpha_i^{n_i} \le d_i(x,y) < \alpha_i^{n_i+1}$, we see that $n_i + c_5 \le q_i \le n_i + c_6$ when $n_i \ge m$ and

$$c_7[\alpha_i^{n_i-m} \vee 1] \le N^i_m \le c_8[\alpha_i^{n_i-m} \vee 1] \qquad i = 1,2. \tag{3.2}$$

where $[x]$ is the maximum integer not greater than x.

We next compute N_m. Note that we are in the case $A_{min} = \begin{pmatrix} a & b \\ 0 & d \end{pmatrix}$ with $a \ge d$ ($d = a$ for the case (2-2) and $b = 0$ for the case (3-1)). Firstly, when $x, y \in V_n$ are in the same n-cell and e_{xy} is of type i, ($i = 1,2$), then $N_m = (G^{\circ(n-m)}\ {}^t(1,1))_i$. Clearly,

$$G^{\circ(n-m)}\begin{pmatrix} 1 \\ 1 \end{pmatrix} \le A_{min}^{n-m}\begin{pmatrix} 1 \\ 1 \end{pmatrix} \le \begin{cases} c_9\begin{pmatrix} a^{n-m} \\ d^{n-m} \end{pmatrix} & \text{for (2-1), (3-1),} \\ c_9\begin{pmatrix} (n-m)a^{n-m} \\ a^{n-m} \end{pmatrix} & \text{for (2-2).} \end{cases} \tag{3.3}$$

Now, since $a = \min_{\mathbf{b}\in S^1} \mathbf{b}_1$ and S^1, S^2 are finite sets, by taking $\gamma \ge 1$ large enough, we have $\gamma a + b = \min_{\mathbf{b}\in S^1}(\gamma \mathbf{b}_1 + \mathbf{b}_2)$ and $d = \min_{\mathbf{b}\in S^2}(\gamma\mathbf{b}_1 + \mathbf{b}_2)$. In particular,

$$G\begin{pmatrix} \gamma \\ 1 \end{pmatrix} = \begin{pmatrix} \gamma a + b \\ d \end{pmatrix} \ge d\begin{pmatrix} \gamma \\ 1 \end{pmatrix}. \tag{3.4}$$

On the other hand,

$$G^{\circ(n-m)}\begin{pmatrix} 1 \\ 1 \end{pmatrix} \ge \frac{1}{\gamma+1} G^{\circ(n-m)}\begin{pmatrix} \gamma+1 \\ 1 \end{pmatrix} \ge \frac{a^{n-m}}{\gamma+1}\begin{pmatrix} 1 \\ 0 \end{pmatrix} + \frac{1}{\gamma+1} G^{\circ(n-m)}\begin{pmatrix} \gamma \\ 1 \end{pmatrix},$$

where we apply (3.1) to obtain the last inequality. Applying (3.4), we have the same lower bound (with different constant) as (3.3) for (2-1) and (3-1). For the case of (2-2), we have

$$G\begin{pmatrix} (\gamma a + (l+1)b)a^l \\ a^{l+1} \end{pmatrix} \ge a^{l+1} G\begin{pmatrix} \gamma \\ 1 \end{pmatrix} + (l+1)ba^l G\begin{pmatrix} 1 \\ 0 \end{pmatrix} = \begin{pmatrix} (\gamma a + (l+2)b)a^{l+1} \\ a^{l+2} \end{pmatrix},$$

for each $l \ge 0$. Applying this inductively, we have the same lower bound as (3.3) (with different constant).

Finally we consider the general case $x, y \in V$ with $\rho^{n-1} \le R(x,y) < \rho^n$. By the definition of n_i, $\alpha_i^{n_i} \le d_i(x,y) < \alpha_i^{n_i+1}$ for $i = 1,2$. Using the same decomposition of the shortest path, by similar arguments to the case of N^i_m and by applying the result for $x, y \in V_n$ being in the same n-cell, we have the following (since the arguments are similar, we leave the details for the reader); N_m can be estimated from above and below by some (uniform) constant multiple of the expressions given below (note that $a = \alpha_1, b = \alpha_2$),

$$\text{(2-1) and (3-1)} \sum_{i=1}^{2}[\alpha_i^{n_i-m} \vee 1], \qquad \text{(2-2)}\ (n_1 - m)[\alpha_1^{n_1-m}] + [\alpha_1^{n_2-m} \vee 1].$$

Combining this with (3.2) and by the definition of $\bar{N}^i_m$, we obtain the result. □

For later convenience, we take $c_{3.1} \le 1$ and $c_{3.2} \ge 1$. For each $c > 0$, let

$$l^i_c = l^i_c(k,n) = \inf\{j : \bar{N}^i_j(x,y)\tau^j \ge ck\} \wedge n.$$

We will omit c when $c = 1$.

LEMMA 3.2. (a) *Assume* $l^i < n_i$, *then* $l_{c_{3.1}} \le l^i$ *where* $i = 1, 2$.
(b) *Let* $j_0 = j_0(l, x, y)$ *be such that* $\max_i \bar{N}_l^i = \bar{N}_l^{j_0}$. *Assume* $l < n$, *then* $l^{j_0}_{1/(2c_{3.2})} \le l$.

PROOF. For (a), using Lemma 3.1 and the fact that $l_i < n_i$, we have

$$k \le \bar{N}^i_{l^i} \tau^{l^i} \le \frac{1}{c_{3.1}} N_{l^i} \tau^{l^i},$$

giving the result. For (b), note that $N_l/2 \le c_{3.2} \bar{N}_l^{j_0}$. Since $l < n$, we have

$$\frac{k}{2} \le \frac{N_l}{2} \tau^l \le c_{3.2} \bar{N}_l^{j_0} \tau^l,$$

giving the result. □

LEMMA 3.3. (a) *For* $c \ge 1$, *there exists* $c_{3.3}(c) \ge 1$ *such that*

$$N_{l_c}(x, y) \le N_l(x, y) \le c_{3.3}(c) N_{l_c}(x, y).$$

(b) *For each* $c < 1$, *there exists* $c_{3.4}(c) \le 1$ *such that*

$$c_{3.4}(c) N_{l_c}(x, y) \le N_l(x, y) \le N_{l_c}(x, y).$$

Both (a) *and* (b) *also hold for* $\bar{N}^i$.

PROOF. When $c \ge 1$, $l_c \ge l$ so that $N_{l_c} \le N_l$ and the first inequality of (a) is clear. To prove the second inequality of (a), we first consider the case (2-1) and (3-1). As discussed in the proof of Lemma 3.1, we have constants c_1, c_2 such that

$$c_1 \sum_{i=1}^{2} [\alpha_i^{n_i - m} \vee 1] \le N_m \le c_2 \sum_{i=1}^{2} [\alpha_i^{n_i - m} \vee 1].$$

If $l = n$, then $l_c = n$ and the result is clear. We thus consider the case $l < n$. In this case, $N_l \tau^l \ge k$ so that

$$k \le c_2 \sum_{i=1}^{2} [\alpha_i^{n_i - l} \vee 1] \tau^l \le 2c_2 [\alpha_{j_0}^{n_{j_0} - l} \vee 1] \tau^l,$$

where $[\alpha_{j_0}^{n_{j_0} - l} \vee 1] := \max_j [\alpha_j^{n_j - l} \vee 1]$. Now, note that $N_{n-m} \le c_3 \tau^{m/2}$ for $e_{xy} \in B_n$ (see the paragraph just after Corollary 4.11 in [**HK2**]). From this, we see $\alpha_j^2 \le \tau$ for $j = 1, 2$, especially $\tau \alpha_j^{-1} > 1$. Take $L \in \mathbb{N}$ as the minimum number that satisfies $(\tau \alpha_j^{-1})^L c_1 > 2c_2 c$. If $n_{j_0} - (l + L) > 0$, then

$$ck \le \frac{(\tau \alpha_j^{-1})^L c_1}{2c_2} k \le c_1 \alpha_{j_0}^{n_{j_0} - (l+L)} \tau^{l+L} \le N_{l+L} \tau^{l+L},$$

so that $l_c \le l + L$ and the result holds. On the other hand, if $n_{j_0} - (l + L) \le 0$, then there exists $c_4 > 0$ such that $n \le l + c_4$. Then, $N_l \le c_5$ for some $c_5 > 0$ and the result holds. The case (2-2) can be proved similarly.

(b) can be proved in the same way as (a). □

The following comparison is important.

PROPOSITION 3.4. *There exists* $c_{3.5}, c_{3.6} > 0$ *such that the following holds.*

$$c_{3.5} \sum_{i=1}^{2} \bar{N}^i_{l^i}(x, y) \le N_l(x, y) \le c_{3.6} \sum_{i=1}^{2} \bar{N}^i_{l^i}(x, y).$$

PROOF. We first consider the case $l < n$ and $l_i < n_i$ for $i = 1, 2$. We note that $\bar{N}^i_{m_1} \le \bar{N}^i_{m_2}$ $(i = 1, 2)$ and $N_{m_1} \le N_{m_2}$ for $m_2 \le m_1$. By Lemma 3.1, Lemma 3.2 (a) and Lemma 3.3 (b), we have

$$c_{3.1} \sum_{i=1}^{2} \bar{N}^i_{l^i} \le c_{3.1} \sum_{i=1}^{2} \bar{N}^i_{l_{c_{3.1}}} \le N_{l_{c_{3.1}}} \le (c_{3.4}(c_{3.1}))^{-1} N_l.$$

On the other hand, by Lemma 3.1, Lemma 3.2 (b) and Lemma 3.3 (b), we have

$$N_l \le 2c_{3.2} \bar{N}^{j_0}_l \le 2c_{3.2} \bar{N}^{j_0}_{l^{j_0}_{(2c_{3.2})^{-1}}} \le 2c_{3.2} \sum_{i=1}^{2} \bar{N}^i_{l^i_{(2c_{3.2})^{-1}}} \le c_1 \sum_{i=1}^{2} \bar{N}^i_{l^i}.$$

Combining them, we obtain the desired result.

We next consider the case $l = n$ where $N_l = N_n = 1$. Using Lemma 3.1, we have

$$c_{3.1} \bar{N}^i_j \tau^j \le N_j \tau^j < k, \qquad \forall j \le n, \ i = 1, 2.$$

Thus, $l^i_{1/c_{3.1}} = n$ for $i = 1, 2$. Using Lemma 3.3 (a), we obtain the desired result.

We now consider the case $l^1 = n_1$, where $\bar{N}^1_{l^1} = \bar{N}^1_{n_1} = 1$ and

$$\bar{N}^1_j \tau^j < k, \qquad \forall j \le n \tag{3.5}$$

holds. Assume first that $l^2 = n_2$. Then, using Lemma 3.1, we have $N_j \tau^j / c_{3.2} < 2k$ for all $j \le n$ so that $l_{2c_{3.2}} = n$. Thus, with the help of Lemma 3.3, the result can be proved in the same way as in the case $l = n$. Now assume that $l^2 < n_2$. By the definition of l^2, there exists $c_2 \ge 1$ such that

$$c_2 k > \bar{N}^2_{l^2} \tau^{l^2} \ge k.$$

Combining this with (3.5) and Lemma 3.1, we obtain

$$c_{3.2}(c_2 + 1)k \ge c_{3.2}(\bar{N}^1_{l^2} + \bar{N}^2_{l^2})\tau^{l^2} \ge N_{l^2} \tau^{l^2} \ge c_{3.1}(\bar{N}^1_{l^2} + \bar{N}^2_{l^2})\tau^{l^2} \ge c_{3.1} k.$$

Thus we have $l_{c_{3.1}} \le l^2 \le l_{c_{3.2}(c_2+1)}$. Using Lemma 3.1 and Lemma 3.3, we obtain the result.

The case where $l^2 = n_2$ can be treated similarly and the lemma is proved. □

By (2.8) and Proposition 3.4, we have the following.

$$p_k(x, y) \asymp c_1 k^{-\frac{S}{S+1}} \exp(-c_2 \sum_i \bar{N}^i_{l^i(k,n)}(x, y)). \tag{3.6}$$

For $i = 1, 2$, let $d^i_c := \log_\rho \alpha_i$ and define $d^i_w = (S + 1)/d^i_c$. For the case (0), (1), (2-2), (2-3) and (3-2), define $d_c := \log_\rho \lambda_{min}, d_w = (S + 1)/d_c$ where λ_{min} is the maximum eigenvalue of A_{min}. By Theorem 2.4 and Proposition 3.4, we have the following expression for the heat kernel estimates.

THEOREM 3.5. *There exists constants $c_{3.7}, c_{3.8} > 0$ such that for $x, y \in V$ and $k \ge d(x, y)$, the following holds.*
Case A: In the cases (2-1), (3-1) *and* (3-3):

$$p_k(x, y) \asymp c_{3.7} k^{-S/(S+1)} \exp\left\{ -c_{3.8} \left(\left(\frac{d_1(x, y)^{d^1_w}}{k} \right)^{\frac{1}{d^1_w - 1}} + \left(\frac{d_2(x, y)^{d^2_w}}{k} \right)^{\frac{1}{d^2_w - 1}} \right) \right\}.$$

Case B: In the case (2-2):

$$p_k(x,y) \asymp c_{3.7}k^{-S/(S+1)}$$
$$\times \quad \exp\left\{-c_{3.8}\left(\left(\frac{(d_1(x,y)\log(\frac{d_1(x,y)}{k}))^{d_w}}{k}\right)^{\frac{1}{d_w-1}} + \left(\frac{d_2(x,y)^{d_w}}{k}\right)^{\frac{1}{d_w-1}}\right)\right\}.$$

Case C: In the remaining cases (*i.e.*, (0), (1), (2-3) *and* (3-2)):

$$p_k(x,y) \asymp c_{3.7}k^{-S/(S+1)} \exp\left(-c_{3.8}\left(\frac{d(x,y)^{d_w}}{k}\right)^{\frac{1}{d_w-1}}\right).$$

PROOF. In this proof, we write $a_n \asymp b_n$ if there exists $c_1, c_2 > 0$ such that $c_1 a_n \le b_n \le c_2 a_n$. We first prove Case A. Using Proposition 3.4, it is enough to prove the following,

$$N^i_{l^i} \asymp (\frac{d_i(x,y)^{d^i_w}}{k})^{\frac{1}{d^i_w-1}} \qquad \forall i = 1,2. \tag{3.7}$$

Assume $l^i < n_i$. Then, by the definition of l_i, we have $N^i_{l^i-1}\tau^{l^i-1} < k \le N^i_{l^i}\tau^{l^i}$ so that $k \le N^i_{l^i}\tau^{l^i} < c_3 k$ for some $c_3 > 1$. Combining this with the fact $N^i_m \asymp [\alpha_i^{n_i-m} \vee 1]$, we have

$$\frac{\log(k\alpha_i^{-n_i})}{\log(\tau/\alpha_i)} + c_4 \le l^i \le \frac{\log(k\alpha_i^{-n_i})}{\log(\tau/\alpha_i)} + c_5.$$

Thus, noting that $d_i(x,y) \asymp \alpha_i^{n_i}$, we have

$$\begin{aligned} N^i_{l^i} &\asymp \alpha_i^{n_i-l^i} \asymp \alpha_i^{n_i}\alpha_i^{-\frac{\log(k\alpha_i^{-n_i})}{\log(\tau/\alpha_i)}} = \alpha_i^{n_i}\left(\frac{\alpha_i^{n_i}}{k}\right)^{\frac{\log\alpha_i}{\log(\tau/\alpha_i)}} \\ &\asymp \alpha_i^{n_i\frac{\log\tau}{\log(\tau/\alpha_i)}}k^{-\frac{\log\alpha_i}{\log(\tau/\alpha_i)}} \asymp (\frac{d_i(x,y)^{d^i_w}}{k})^{\frac{1}{d^i_w-1}}. \end{aligned} \tag{3.8}$$

Here we have used the fact that $d^i_w = (S+1)/\log_\rho \alpha_i$ and $\rho^{S+1} = \alpha_i^{d^i_w} = \tau$, so that $d^i_w = \log\tau/\log\alpha_i$, $\log\alpha_i/\log(\tau/\alpha_i) = 1/(d^i_w - 1)$ and $\log\tau/\log(\tau/\alpha_i) = d^i_w/(d^i_w - 1)$. Thus (3.7) is proved when $l^i < n_i$.

Next, assume $l^i \ge n_i$ (thus $l^i = n$). Then $N^i_{l^i} \le c_6$ for some $c_6 > 0$. On the other hand, using the last asymptotic equivalence in (3.8), we see that $(d_i(x,y)^{d^i_w}/k)^{\frac{1}{d^i_w-1}}$ is bounded from above by some constant. Thus (3.7) is proved in this case.

We next prove Case B. Using Proposition 3.4, it is enough to prove the following,

$$\bar{N}^1_{l^1} \asymp (\frac{(d_1(x,y)\log(\frac{d_1(x,y)}{k}))^{d^1_w}}{k})^{\frac{1}{d^1_w-1}}, \qquad N^2_{l^2} \asymp (\frac{d_2(x,y)^{d^2_w}}{k})^{\frac{1}{d^2_w-1}}. \tag{3.9}$$

The second asymptotic equivalence is proved in the same way as Case A, so we consider the first asymptotic equivalence. Assume $l^1 < n_1$. Then, by the definition of l_1 and the fact $\bar{N}^1_m \asymp (n_1 - m)[\alpha_1^{n_1-m}]$, we have $\bar{N}^1_{l^1} \asymp (n_1 - l^1)\alpha_1^{n_1-l^1}$ and $(n_1 - l^1)\alpha_1^{n_1-l^1}\tau^{l^1} \asymp k$. Thus, $(n_1 - l^1)(\alpha_1/\tau)^{n_1-l^1} \asymp k/\tau^{n_1}$. Note that by simple calculations, we see that if for fixed $0 < \lambda < 1$, $x\lambda^x \asymp y$ for all $1 \le x$ and $0 < y < 1$, then there exists $c_7, c_8 > 0$ such that

$$\frac{\log(1/y)}{\log(1/\lambda)} + \frac{\log\log(1/y)}{\log(1/\lambda)} - c_7 \le x \le \frac{\log(1/y)}{\log(1/\lambda)} + \frac{\log\log(1/y)}{\log(1/\lambda)} + c_8.$$

Thus, $x(\lambda\tau)^x \asymp y^{-\log(\lambda\tau)/\log(1/\lambda)}(\log(1/y))^{\log\tau/\log(1/\lambda)}$. Applying this with $x = n_1 - l^1, y = k/\tau^{n_1}$ and $\lambda = \alpha_1/\tau$, we have

$$\bar{N}^1_{l^1} \asymp (n_1 - l^1)\alpha_1^{n_1 - l^1} \asymp (\alpha_1^{n_1 d^1_w}/k)^{\frac{1}{d^1_w - 1}}(\log(\alpha_1^{n_1 d^1_w}/k))^{\frac{d^1_w}{d^1_w - 1}}.$$

Noting that $d_1(x,y) \asymp \alpha_1^{n_1}$, we obtain (3.9). The case $l^1 \geq n_1$ can be shown easily.

We omit the proof of Case C as it is similar and simpler (see [**HK3**] etc.). □

REMARK 3.6. (a) The choice of A_{min} given in this section is not necessarily unique. Even so, if there is more than one choice of A_{min}, then the matrices chosen have the same eigenspace for the maximum eigenvalue. This can be proved by using the properties of non-negative matrices.
(b) In this section we discussed the case where there are two types of steps for simplicity. We note that even if there are more than two types, if all the corresponding minimal step matrices can be decomposed into one or two irreducible components, and if (3.1) holds, then the arguments given in this section can be applied and we have the same type of heat kernel estimates.
(c) We believe that the general case of n edge types could be treated similarly, but there are more possibilities for the form of the heat kernel estimates. For instance, when the Jordan cell for the maximum eigenvalue of some irreducible component of the minimal step matrix is p-dimensional, then the $(1,p)$-th element of the n-th power of the matrix is of order $n^p a^n$. Thus we should define $\bar{N}^1_m := N^1_m(\log N^1_m)^p$ and the corresponding part of the heat kernel estimate (in the exponential) should be as follows

$$-c_{3.9}\left(\frac{\{d_1(x,y)(\log(\frac{d_1(x,y)}{k}))^p\}^{d^1_w}}{k}\right)^{\frac{1}{d^1_w - 1}}.$$

4. Law of the iterated logarithm

In this section, as an application of the heat kernel estimates, we discuss the law of the iterated logarithm for the sample paths of the random walk. We will only treat the case (3) of the last section, as this case already includes interesting aspects of the process we treat.

THEOREM 4.1. *For case* (3), *there exists* $c_{4.1}(i) > 0$ *such that*

$$\limsup_{k\to\infty} \frac{d_i(X_0, X_k)}{k^{1/d^i_w}(\log\log k)^{1-1/d^i_w}} = c_{4.1}(i), \qquad \mathbb{P}^0 - a.s.\ , \ i = 1, 2.$$

The proof will follow from some preliminary results. We begin with a little notation. An n-graph complex is the subgraph of V contained in an n-cell. We write $D^1_n(y)$ for the n-graph complex containing y and the neighbouring n-graph complexes. We will write $\partial_i D^1_n(y)$ for the boundary points $z \in D^1_n(y)$ which have $d_i(y,z) \geq \alpha_i^n$. Let $T_A = \inf\{k : X_k \notin A\}$ be the exit time from a set A and define a sequence of crossing times by $T^m_0 = \inf\{k : X_k \in V_m\}$ and for $i \geq 1$, let $T^m_i = \inf\{k > T^m_{i-1} : X_k \in V_m \backslash \{X_{T^m_{i-1}}\}\}$ and $W^m_i = T^m_i - T^m_{i-1}$.

LEMMA 4.2. *For all* $x \in V$ *there exist constants* $c_{4.2}, c_{4.3}$ *such that for* $i = 1, 2$,

$$P^x(d_i(x, X_k) > \lambda) \leq c_{4.2}\exp\left(-c_{4.3}\left(\frac{\lambda^{d^i_w}}{k}\right)^{1/(d^i_w - 1)}\right), \quad k > \lambda.$$

PROOF. The proof of this result uses hitting time estimates. Let n be such that $\alpha_i^{n-1} \le \lambda \le \alpha_i^n$. Observe that, by the roughly uniform exit law of $D_n(y)$, there exists a c_1 such that

$$\begin{aligned} P^x(d_i(x, X_k) > \lambda) &\le P^x(T_{D_n(x)} < k, X_{T_{D_n(x)}} \in \partial_i D_n^1(y)) \\ &\le c_1 \sup_{y \in V_{n-1}} P^y(T_{D_n(y)} < k | X_{T_{D_n(y)}} \in \partial_i D_n^1(y)). \end{aligned}$$

By the crossing time estimates of [**HK2**] Lemma 4.4 and Proposition 4.5, as used in the proof of [**HK2**] Theorem 4.6, we have the following for all $0 \le m \le n$,

$$\begin{aligned} P^y(T_{D_n(y)} < k | X_{T_{D_n(y)}} \in \partial_i D_n^1(y)) &\le P^y(\sum_{i=1}^{\alpha_i^{n-m}} W_i^m < k) \\ &\le \exp(-c_2 \alpha_i^{n-m} + c_3(\alpha_i^{n-m} \tau^{-m} k)^{1/2}) \end{aligned}$$

If we choose $l := \inf\{j : \alpha_i^{n-j} \tau^j \ge k\} \wedge n$, then

$$P^y(T_{D_n(y)} < k | X_{T_{D_n(y)}} \in \partial_i D_n^1(y)) \le \exp(-c_4 \alpha_i^{n-l}).$$

Rewriting this, using the definitions of l and n, we have the estimate. □

As we are in the case (3) we can find vertices $x_i^\lambda \in V$ such that $d_i(0, x_i^\lambda) = [\lambda]$ and $d_j(0, x_i^\lambda) = 0$ for $j \ne i$. We will write

$$B_i(x, r) = \{y : d_i(0, y) \ge d_i(0, x), R(x, y) \le r^{1/(S+1)}\},$$

and

$$B_i'(x, r) = \{y : d_i(0, y) < d_i(0, x), R(x, y) \le r^{1/(S+1)}\}.$$

LEMMA 4.3. *There exist constants $c_{4.4}, c_{4.5}$ such that for $i = 1, 2$,*

$$P^0(X_k \in B_i(x_i^\lambda, k^{1/(S+1)})) \ge c_{4.4} \exp\left(-c_{4.5}\left(\frac{\lambda^{d_w^i}}{k}\right)^{1/(d_w^i - 1)}\right), \quad k > \lambda.$$

PROOF. For this lower estimate we will use our heat kernel lower bound. Writing $B_i = B_i(x_i^\lambda, k^{1/(S+1)})$, we have

$$\begin{aligned} P^0(X_k \in B_i) &= \sum_{y \in B_i} p_k(0, y) \mu(y) \\ &\ge \mu(B_i) \inf_{y \in B_i} k^{-S/(S+1)} \exp\left(-c \sum_j \left(\frac{d_j(0, y)^{d_w^j}}{k}\right)^{1/(d_w^j - 1)}\right) \\ &\ge c_1 \exp\left(-\sup_{y \in B_i} \sum_j \left(\frac{d_j(0, y)^{d_w^j}}{k}\right)^{1/(d_w^j - 1)}\right). \end{aligned} \tag{4.1}$$

If we now observe that within the set B_i we have constants such that

$$\sup_{y \in B_i} d_j(0, y) \le \begin{cases} c_1 \lambda + c_2 k^{1/d_w^i} & j = i \\ c_3 k^{1/d_w^j} & j \ne i. \end{cases}$$

Thus, placing this in (4.1), we have

$$\sup_{y\in B_i}\left(\frac{d_j(0,y)^{d^j_w}}{k}\right)^{1/(d^j_w-1)} \le \begin{cases} c_5\left(\frac{c_1\lambda^{d^i_w}}{k}+c_4\right)^{1/(d^i_w-1)} & j=i \\ c_3^{1/(d^j_w-1)} & j\neq i \end{cases}$$

and hence the result. □

Let $h^c_i(k) = ck^{1/d^i_w}(\log\log k)^{1-1/d^i_w}$ and write $h_i(k)$ for $h^1_i(k)$.

COROLLARY 4.4. *For each $i = 1, 2$, there exists $N > 1$ such that, if $\tilde{x}^n_i$ satisfies $d_i(0, \tilde{x}^n_i) = [h^c_i(N^n)]$, then there are constants $c_{4.6}, c_{4.7}, c_{4.8}$ such that*

$$P^0(X_{N^n} \in B_i(\tilde{x}^n_i, N^{n/(S+1)})) \ge c_{4.6}n^{-c_{4.5}c^{1/(d^i_w-1)}}, \tag{4.2}$$

and for all $z \in B'_i(\tilde{x}^{n-1}_i, N^{(n-1)/(S+1)})$,

$$P^z(X_{N^n} \in B_i(\tilde{x}^n_i, N^{n/(S+1)})) \ge c_{4.8}n^{-c_{4.7}c^{1/(d^i_w-1)}}. \tag{4.3}$$

PROOF. For (4.2) we substitute $\lambda = h^c_i(N^n)$ in Lemma 4.3.

For (4.3), for any vertex pair $z \in B'_i(\tilde{x}^{n-1}_i, N^{(n-1)/(S+1)}), y \in B_i(\tilde{x}^n_i, N^{n/(S+1)})$, there is a constant c_1 such that $d_i(z, y) \le h^c_i(N^{n-1}) + c_1N^{(n-1)/(S+1)}$. Hence applying the same approach as the proof of Lemma 4.3 will give the result. □

Finally we state a triviality result for tail events associated with our random walk. This is a simple extension to the random walk case of Theorem 8.4 of [**BB3**].

LEMMA 4.5. *Let $A \in \cap_k\sigma(X_n : n \ge k)$ be a tail event, then $P^x(A) = 0$ for all x or $P^x(A) = 1$ for all x.*

PROOF OF THEOREM 4.1. We need to modify the usual proof of the law of the iterated logarithm to allow for the existence of different types. Firstly, by Lemma 4.5, it is enough to show that there are constants c_1, c_2 such that

$$c_1 \le \limsup_{k\to\infty}\frac{d_i(0, X_k)}{h_i(k)} \le c_2, \quad P^0 - a.s.$$

The upper bound is standard as we apply the upper estimate of Lemma 4.2 and the first Borel-Cantelli lemma.

For the lower bound we use a slight variant of the second Borel-Cantelli Lemma. In the case of a Markov process we can use a simple modification of the proof to show that if A_n is a sequence of events, each in $\sigma(X_k : k \le N^n)$, then, if $\sum_n P(A_n\cap A^c_{n-1}) = \infty$, then $P(A_n \;\; i.o.) = 1$, for a discussion see [**HHH**] Lemma 4.5. Let $A_n = \{X_{N^n} \in B(\tilde{x}^n_i, N^{n/(S+1)})\}$, so that $A^c_{n-1} = \{X_{N^{n-1}} \notin B(\tilde{x}^{n-1}_i, N^{(n-1)/(S-1)})\}$. Now $B_{n-1} := \{X_{N^{n-1}} \in B'_i(\tilde{x}^{n-1}_i, N^{(n-1)/(S+1)})\} \subset A^c_{n-1}$ and hence

$$\begin{aligned} P^0(A_n \cap A^c_{n-1}) &\ge P^0(A_n \cap B_{n-1}) \\ &= P^0(A_n|B_{n-1})P^0(B_{n-1}) \end{aligned}$$

We can now apply both estimates of Corollary 4.4, to get

$$P^0(A_n \cap A^c_{n-1}) \ge c_{4.6}c_{4.8}n^{-(c_{4.5}+c_{4.7})c^{1/(d^i_w-1)}}.$$

We can now choose $c = (c_{4.5} + c_{4.7})^{-(d^i_w-1)}$ to obtain

$$P^0(A_n \cap A^c_{n-1}) \ge c_{4.6}c_{4.8}n^{-1}.$$

Thus the sum diverges and we have that the event A_n occurs infinitely often. Thus

$$X_{N^n} \in B(\tilde{x}_i^n, N^{n/(S+1)}) \ \ i.o. \ \ P^0 - a.s.$$

By the construction of the set $B(\tilde{x}_i^n, N^{n/(S+1)})$ we have that there is a constant c' such that $d_i(0, X_{N^n}) \geq c' h_i^c(N^n)$ infinitely often. Hence

$$\limsup_{k\to\infty} \frac{d_i(0,X_k)}{h_i(k)} \geq \limsup_{n\to\infty} \frac{d_i(0,X_{N^n})}{h_i^1(N^n)} \geq c'c, \ \ P^0 - a.s.$$

This completes the proof of our LIL. □

We note that this shows that the size of fluctuations in the different directions is different. The construction of $d_i(.,.)$ ensures that if X_k has moved in the direction of type i, then $d_i(0, X_k) \asymp R(0, X_k)^{d_c^i}$ and hence

$$\frac{d_i(0,X_k)^{d_w^i}}{k} \asymp \frac{R(0,X_k)^{S+1}}{k}.$$

Thus the bulk spread of the random walk is determined by the resistance. However if we look at the large fluctuations we see that the direction has an effect as

$$\frac{R(0,X_k)^{S+1}}{k} \asymp (\log\log k)^{d_w^i - 1} \ \text{ infinitely often.}$$

Thus the size of fluctuations in the different directions is different and we would see that the process makes larger extreme movements in the direction with the largest d_w^i (that is, the smallest α_i).

5. Diffusion case

In this section, we will mention diffusion processes on u.f.r. fractals and their heat kernel estimates. For $u, v \in l(\hat{V}_n)$, define

$$\hat{\mathcal{E}}_{Q_0}^n(u,v) = \rho^n \sum_{i_1,\cdots,i_n \in I} S_{Q_0}(u \circ \Psi_{i_1,\ldots,i_n}, v \circ \Psi_{i_1,\ldots,i_n}), \tag{5.1}$$

where $S_{Q_0}(\xi,\xi) = \sum_{i,j\in I_F} (Q_0)_{ij}(\xi_i - \xi_j)^2/2$ for each $\xi \in l(I_F)$. Let $\hat{\nu}$ be the normalized Hausdorff measure on $\hat{K}$. Then, the following is known (see for example, [**HK3, Kig**]).

THEOREM 5.1. *Let $Q_0 \in Int(\mathcal{Q}_M)$ be as in Assumption 2.2 2), i.e. $F(Q_0) = \rho_{Q_0}^{-1} Q_0$. Then, there is a local regular Dirichlet form $(\hat{\mathcal{E}}, \hat{\mathcal{F}})$ in $\mathbb{L}^2(\hat{K}, d\hat{\nu})$ satisfying the following,*

$$\begin{aligned} \hat{\mathcal{F}} &= \{u \in C(\hat{K},\mathbb{R}) : \sup_n \hat{\mathcal{E}}_{Q_0}^n(u,u) < \infty\}, \\ \hat{\mathcal{E}}(u,v) &= \lim_{n\to\infty} \hat{\mathcal{E}}_{Q_0}^n(u,v) \qquad \text{for } u,v \in \hat{\mathcal{F}}. \end{aligned}$$

For each $m \in \mathbb{N}$, let $K_m = \alpha^m \hat{K}$ and define $\sigma_m : C(K_m, \mathbb{R}) \to C(\hat{K}, \mathbb{R})$ by $\sigma_m u(x) = u(\alpha^m x)$ for $x \in \hat{K}$. Set $\mathcal{F}_{<m>} = \sigma_{-m}\hat{\mathcal{F}}$, $\mathcal{E}_{<m>}(u,v) = \rho^{-m}\hat{\mathcal{E}}(\sigma_m u, \sigma_m v)$ for $u, v \in \mathcal{F}_{<m>}$. Let ν be a Hausdorff measure on K such that $\nu|_{\hat{K}} = \hat{\nu}$ and $N\nu = \nu \circ \Psi_1^{-1}$. Now let

$$\begin{aligned} \mathcal{F}_K &= \{u \in l(K) : u|_{K_m} \in \mathcal{F}_{<m>} \text{ for all } m \in \mathbb{N}, \\ &\qquad \lim_{m\to\infty} \mathcal{E}_{<m>}(u|_{K_m}, u|_{K_m}) < \infty\} \cap \mathbb{L}^2(K, d\nu), \\ \mathcal{E}_K(u,v) &= \lim_{m\to\infty} \mathcal{E}_{<m>}(u|_{K_m}, v|_{K_m}) \qquad \text{for } u,v \in \mathcal{F}. \end{aligned}$$

Then, $(\mathcal{E}_K, \mathcal{F}_K)$ is a local regular Dirichlet form on $\mathbb{L}^2(K, d\nu)$, $\mathcal{F}_K \subset C(K, \mathbb{R})$ and the following scaling property holds,

$$\mathcal{E}_K(u, v) = \rho \mathcal{E}_K(u \circ \Psi_1, v \circ \Psi_1) \qquad \text{for } \ u, v \in \mathcal{F}_K.$$

Let $\{X_t\}_{t\geq 0}$ (resp. $\{\hat{X}_t\}_{t\geq 0}$) be the diffusion process corresponding to $(\mathcal{E}_K, \mathcal{F}_K)$ (resp. $(\hat{\mathcal{E}}, \hat{\mathcal{F}})$). It can be shown that there is a jointly continuous version of the heat kernel $p_t^K(x, y)$ (resp. $p_t^{\hat{K}}(x, y)$) such that $E^x[f(X_t)] = \int_K p_t^K(x, y) f(y) \nu(dy)$ for all $f : K \to \mathbb{R}$ (resp. $E^x[f(\hat{X}_t)] = \int_{\hat{K}} p_t^{\hat{K}}(x, y) f(y) \hat{\nu}(dy)$ for all $f : \hat{K} \to \mathbb{R}$). In the following, we write $p_t(x, y)$ for both $p_t^K(x, y)$ and $p_t^{\hat{K}}(x, y)$ and state the results for K and $\hat{K}$ simultaneously. Let

$$k(n, t) := \inf\{j : N_{n+j}(x, y)\tau^{n+j} \geq t\}.$$

Then the following holds (see [**HK3**] for the case of K; recall that our labelling of V_n is the opposite to the one given in [**HK3**]).

THEOREM 5.2. *There exist constants $c_{5.1}, c_{5.2} > 0$ such that for all $x, y \in K$ (resp. $\hat{K}$) and $t > 0$ (resp. $0 < t < 1$), if $\rho^{n-1} \leq R(x, y) \leq \rho^n$, then*

$$p_t(x, y) \asymp c_{5.1} t^{-\frac{S}{S+1}} \exp(-c_{5.2} N_{k(n,t)}(x, y)).$$

Now, we consider the case of two step types. For each $x, y \in K$ (or $\hat{K}$), define

$$D_i(x, y) := \limsup_{m\to\infty} \frac{N_{-m}^i(x, y)}{\alpha_i^m}, \ i = 1, 2, \qquad D(x, y) := \limsup_{m\to\infty} \frac{N_{-m}(x, y)}{\lambda_{min}^m}.$$

Note that if we take liminf instead of limsup, there is only a (uniform) constant time difference (due to the results in Section 3). By similar arguments to those in Section 3, we can translate Theorem 3.5 to this setting as follows.

THEOREM 5.3. *There exists constants $c_{5.3}, c_{5.4} > 0$ such that for all $x, y \in K$ (resp. $\hat{K}$) and $t > 0$ (resp. $0 < t < 1$), the following holds.*
Case A: In the cases (2-1), (3-1) *and* (3-3):

$$p_t(x, y) \asymp c_{5.3} t^{-S/(S+1)} \exp\left\{-c_{5.4}\left(\left(\frac{D_1(x, y)^{d_w^1}}{t}\right)^{\frac{1}{d_w^1-1}} + \left(\frac{D_2(x, y)^{d_w^2}}{t}\right)^{\frac{1}{d_w^2-1}}\right)\right\}.$$

Case B: In the case (2-2):

$$p_t(x, y) \asymp c_{5.3} t^{-S/(S+1)}$$
$$\times \quad \exp\left\{-c_{5.4}\left(\left(\frac{(D_1(x, y)\log(\frac{D_1(x,y)}{t}))^{d_w}}{t}\right)^{\frac{1}{d_w-1}} + \left(\frac{D_2(x, y)^{d_w}}{t}\right)^{\frac{1}{d_w-1}}\right)\right\}.$$

Case C: In the remaining cases (i.e., (0), (1), (2-3) *and* (3-2)*):*

$$p_t(x, y) \asymp c_{5.3} t^{-S/(S+1)} \exp\left(-c_{5.4}\left(\frac{D(x, y)^{d_w}}{t}\right)^{\frac{1}{d_w-1}}\right).$$

By definition, $d_w^i = (S + 1)/d_c^i$ is the box dimension of the type i path with respect to the resistance metric, where $d_c^i = \log_\rho \alpha_i$ is the chemical exponent of the type i path (with respect to the resistance metric). Thus Theorem 5.3 gives an affirmative answer to Conjecture 6.9 and 6.10 in [**HK3**] for the case of two step types.

We can also obtain a law of the iterated logarithm for the long and short time behaviour of the diffusion process on $\hat{K}$ which corresponds to that of Theorem 4.1.

References

[Bar] M.T. Barlow, *Diffusions on fractals*, Lectures in Probability Theory and Statistics: Ecole d'été de probabilités de Saint-Flour XXV, Springer, New York, 1998.

[BB1] M.T. Barlow and R.F. Bass, *Stability of parabolic Harnack inequalities,* Trans. A.M.S., to appear.

[BB2] M.T. Barlow and R.F. Bass, *Random walks on graphical Sierpinski carpets,* Random walks and discrete potential theory (Cortona, 1997), 26–55, Sympos. Math., XXXIX, Cambridge Univ. Press, Cambridge, 1999.

[BB3] M.T. Barlow and R.F. Bass, *Brownian motion and harmonic analysis on Sierpinski carpets,* Canad. J. Math., **51** (1999), 673–744.

[BCG] M.T. Barlow, T. Coulhon and A. Grigor'yan, *Manifolds and graphs with slow heat kernel decay,* Invent. Math., **144** (2001), 609–649.

[FHK] P.J. Fitzsimmons, B.M. Hambly and T. Kumagai, *Transition density estimates for Brownian motion on affine nested fractals,* Commun. Math. Phys., **165** (1994), 595–620.

[GT1] A. Grigor'yan and A. Telcs, *Harnack inequalities and sub-Gaussian estimates for random walks,* Math. Annalen, **324** (2002), 521–556.

[GT2] A. Grigor'yan and A. Telcs, *Sub-Gaussian estimates of heat kernels on infinite graphs,* Duke Math. J., **109** (2001), 451–510.

[HHH] B.M. Hambly, K. Hattori and T. Hattori, *Self-repelling walk on the Sierpinski gasket,* Probab. Theory Related Fields, **124** (2002), 1–25.

[HK1] B.M. Hambly and T. Kumagai, *Diffusion processes on fractal fields: heat kernel estimates and large deviations,* Probab. Theory Related Fields, **127** (2003), 305–352.

[HK2] B.M. Hambly and T. Kumagai, *Heat kernel estimates for symmetric random walks on a class of fractal graphs and stability under rough isometries,* Proc. of Symposia in Pure Math. (AMS), to appear.

[HK3] B.M. Hambly and T. Kumagai, *Transition density estimates for diffusions on p.c.f. fractals,* Proc. London Math. Soc., **78** (1999), 431–458.

[Jo] O.D. Jones, *Transition probabilities for the simple random walk on the Sierpiński graph,* Stochastic Process. Appl., **61** (1996), 45–69.

[Kig] J. Kigami, *Analysis on Fractals,* Cambridge Univ. Press, Cambridge, 2001.

[Kum1] T. Kumagai, *Homogenization on finitely ramified fractals,* Proc. of the RIMS Project '02 "Stochastic Analysis and Related Topics", Advanced Studies in Pure Math. (MSJ), to appear.

[Kum2] T. Kumagai, *Estimates of transition densities for Brownian motion on nested fractals,* Probab. Theory Relat. Fields, **96** (1993), 205–224.

[Kus] S. Kusuoka, *Diffusion processes on nested fractals,* Statistical Mechanics and Fractals (Lect. Notes Math. **1567**), Springer, 1993.

[Lind] T. Lindstrøm, *Brownian motion on nested fractals,* Mem. Amer. Math. Soc. 420 **83**, 1990.

MATHEMATICAL INSTITUTE, UNIVERSITY OF OXFORD, 24-29 ST GILES, OXFORD OX1 3LB, UK

E-mail address: `hambly@maths.ox.ac.uk`

RESEARCH INSTITUTE FOR MATHEMATICAL SCIENCES, KYOTO UNIVERSITY, KYOTO 606-8502, JAPAN

E-mail address: `kumagai@kurims.kyoto-u.ac.jp`

Contemporary Mathematics
Volume **347**, 2004

Finite representations in the unitary dual and Ramanujan groups

Alexander Lubotzky and Yehuda Shalom

ABSTRACT. We define and study two "opposite" representation theoretic group properties pertaining to the place of the finite representations in the unitary dual. The first describes an abstract group theoretic counterpart to the (asymptotically optimal) Ramanujan graph property. The second, in which the finite representations are dense in the Fell topology, is shown to be shared by some lattices in $SL_2(\mathbb{C})$, thereby answering, in particular, two questions raised by Bekka and Louvet. This, together with various other examples of arithmetic groups satisfying one property or the other, suggests yet another feature of the well known "rank one" versus "higher rank" dichotomy.

1. Introduction

Let Γ be a residually finite discrete group, and $\tilde{\Gamma}$ be its unitary dual, i.e., the set of equivalence classes of **all** unitary Γ-representations, equipped with the Fell topology. The set of finite Γ-representations, namely, those factoring through a finite quotient of Γ, is the simplest collection of Γ-representations, and will be denoted by $\tilde{\Gamma}_F$. The main purpose of this paper is to study the "location" of $\tilde{\Gamma}_F$ in $\tilde{\Gamma}$, as reflected in $F(\Gamma)$ – the closure of $\tilde{\Gamma}_F$ in $\tilde{\Gamma}$ in the Fell topology, i.e., the set of all unitary Γ-representations which are weakly contained (denoted $\prec$) in the direct sum of the representations in $\tilde{\Gamma}_F$.

One may heuristically think of two "opposite sides" of the spectrum $\tilde{\Gamma}$: the first being the regular representation $l^2(\Gamma)$, and the second being the trivial representation 1_Γ. When Γ is amenable, the picture blurs, and the former becomes all of $\tilde{\Gamma}$. But as soon as Γ is non-amenable the two sides are separated, and the situation becomes interesting. In order to make this phenomenon more precise, one is naturally led to discuss norms of the so called "averaging" (or "Hecke") operators, i.e. norms of elements in the group algebra $\mathbb{C}\Gamma$, which act in any Γ-representation. Their intrinsic relation to the Fell topology becomes transparent by Eymard's result [Ey] stating that $\pi \prec \sigma$ iff for any such operator T one has $||T||_\pi \leq ||T||_\sigma$. Thus, any $T \in \mathbb{C}\Gamma$ and any value of α, may be used to isolate a "corner" in $\tilde{\Gamma}$, defined by the (closed) set of Γ-representations π with $||T||_\pi \leq \alpha$.

2000 *Mathematics Subject Classification.* 05C25 20E18 20E05 22E40 22E46.

Key words and phrases. Ramanujan graphs, Ramanujan conjecture, arithmetic groups surface groups.

We can now make our goal more concrete by distinguishing between two extremal types of behavior: the first is that $F(\Gamma) = \tilde{\Gamma}$, i.e., the finite representations are dense in the unitary dual, in which case we shall say that Γ has **Property FD**. The second is that $F(\Gamma)$ lies in the same "corner(s)" as does $l^2(\Gamma)$. Notice however, that we cannot require this to hold for all "corners", nor can we simply ask that $F(\Gamma)$, or parts of it, be contained in (the closure of) $l^2(\Gamma)$, as by Eymard's aforementioned result the two amount to the same, and for non-amenable Γ **no** finite dimensional representation π is weakly contained in $l^2(\Gamma)$. Rather, it appears natural to restrict to positive elements only, i.e, to elements $T \in \mathbb{C}\Gamma$ with non-negative real coefficients. For such operators, on the one hand $||T||$ always achieves its maximal value in 1_Γ, and on the other hand, over a large class of representations, it always achieves its minimal value on $l^2(\Gamma)$ (cf. [Sh2] Lemma 2.3 and the Remark thereafter), so our heuristic "opposite sides" picture becomes sharp and quantitative.

Thus, a first natural attempt to define the opposite type of behavior to property FD would be the property that $||T||_{\pi \in \tilde{\Gamma}_F} \leq ||T||_{l^2(\Gamma)}$ for **all positive** T, of course, with the restriction that π has no non-zero invariant vectors. Unfortunately, as appealing as it may seem, we are currently very far away from constructing any example of a (residually finite, non-amenable) group Γ satisfying this property, and it is not clear whether such an example exists at all. Instead, we shall require that the above inequality holds for at least **one** fixed positive (non-trivial) operator, and in fact, often also restrict to π's coming from some prescribed (non-trivial) family of finite quotients. This leads to the following notion, which as is well known by now, already involves some deep mathematics:

DEFINITION. Let Γ be a finitely generated residually finite group, $T \in \mathbb{C}\Gamma$ a positive operator whose support generates Γ, and $\mathcal{L} = \{\Gamma_i\}_{i \in I}$ a family of finite index normal subgroups which is closed under finite intersections, and has trivial intersection. We say that the group Γ is $(T, \mathcal{L})-$**Ramanujan** if for every unitary Γ-representation π without invariant vectors and factoring through a quotient Γ/Γ_i, one has: $||T||_\pi \leq ||T||_{l^2(\Gamma)}$ $(*)$

We shall say for brevity that Γ is **Ramanujan** if there exists some T and $\mathcal{L}$ for which the above is satisfied.

The outstanding examples of Ramanujan groups are free groups, as follows from the construction of Ramanujan graphs by Lubotzky-Phillips-Sarnak [LPS]. Thus, from the point of view of combinatorics it may be natural to consider the more special operators T arising as an average over a finite (symmetric) generating subset. However, in representation theoretic perspective the above definition is both more natural, and has one additional important advantage: Recall that in the construction of [LPS], one should exclude from $(*)$ a certain 1-dimensional (± 1-valued) character π_0, which gives rise to the maximal possible value in the left hand side of $(*)$. However, the proof in [LPS] can be modified to show that in fact any positive operator $T \in \mathbb{C}\Gamma$ which is spherically symmetric in the associated Cayley graph of the free group Γ, satisfies $(*)$ for any $\pi \neq \pi_0$ (see Cor. 4.2 below). Thus, as any element of the sphere of radius 1 (resp. 2) in Γ acts by the scalar -1 (resp. 1) in π_0, one can easily define a positive linear combination of the two spheres which will be annihilated in π_0, so that now $(*)$ is satisfied for π_0 as well. While in the present case this seems like a mere technicality, in other examples of

Ramanujan groups constructed below there are (finitely many) more exceptional 1-dimensional representations, which can then be handled similarly (although we shall not elaborate here further on this issue). Finally, to put matters in perspective, we remark that for any family $\mathcal{L}$ as in the Definition, the set of representations π appearing in the left hand side of (*) **contains weakly** $l^2(\Gamma)$ (see Lemma 2.1 below), hence for **any** operator $T \in \mathbb{C}\Gamma$ the supermum over the representations π in the left hand side of (*) is always $\geq$ the right hand side of (*). This is what makes the Ramanujuan property so highly non-trivial.

In this paper we shall construct various groups Γ with, and without, one of the above two properties, all coming from the distinguished class of ($S-$)arithmetic groups. Property FD, discussed in Sections 2 and 3, deserves some additional comments. In [Ch] Choi proved that for any free group Γ, the set of **Finite Dimensional** Γ-representations is dense in the Fell topology of $\tilde{\Gamma}$. Modulo natural modifications, it remained the only group known to satisfy this property, which is obviously weaker (at least a priori) than property FD. In [BL] Bekka and Louvet asked whether that property is satisfied also by surface groups and the Picard group $SL_2(\mathbb{Z}[i])$. We shall answer affirmatively their questions in Section 2, by showing directly that these groups (along with some other arithmetic groups $\Gamma < SL_2(\mathbb{C})$) have the stronger property FD. The proof involves a mixture of algebraic, geometric and representation theoretic tools. The Ramanujan property is discussed in Sections 4 and 5. We shall see that a product of non-amenable groups is **never** Ramanujan, and observe that the deep number theoretic tools related to the so-called Ramanujan conjecture (including Drinfeld's and Lafforgue's work) can easily be translated into this property for the relevant groups. Combined with the previous result, this provides an example of two groups Γ, Λ generated by subsets $S = S_\Gamma, S_\Lambda$ resp., such that the two associated Cayley graphs are **isometric**; still, with respect to $T = T_S$ one is Ramanujan and the other is not (relative to **any** family $\mathcal{L}$).

2. Groups with property FD

As mentioned in the introduction, when Γ is amenable $l^2(\Gamma)$ is dense in $\tilde{\Gamma}$ and this property in fact characterizes amenability. Thus for Γ which is amenable and residually finite, we have $F(\Gamma) = \tilde{\Gamma}$, since for **any** residually finite group Γ, it is easy to see that $F(\Gamma)$ contains $l^2(\Gamma)$ (see a direct argument at the end of the proof of Lemma 2.4 below). In fact, the latter can be generalized:

LEMMA 2.1. *Assume that the subgroup $H < \Gamma$ is of infinite index, and is closed in the profinite topology defined by $\mathcal{L}$. Then $l^2(\Gamma/H) \prec \oplus_{\Gamma_i \in \mathcal{L}} l_0^2(\Gamma/\Gamma_i)$. In particular, for every $T \in \mathbb{C}\Gamma$ one has: $||T||_{l^2(\Gamma/H)} \leq sup\, ||T||_{l_0^2(\Gamma/\Gamma_i)}$.*

The point of this observation of course is that in the right hand side we have l_0^2- the zero mean functions, rather than l^2. The Lemma also generalizes Alon-Boppana's well known result on the liminf of λ_1 of regular graphs. For a proof see [Sh1] (Part 1 of Theorem 4.1 and Theorem 2.4).

We now establish the first non-trivial class of groups with property FD:

THEOREM 2.2. *If $\Gamma = F_r$ is a free group on r generators, then $F(\Gamma) = \tilde{\Gamma}$.*

Because of the "local nature" of the Fell topology, this automatically extends also to the case $r = \aleph_0$. The heart of the proof lies in the following result, for which we thank Tim Steger:

PROPOSITION 2.3. *Assume that* Γ *as above acts continuously on a compact metric space* X*, preserving a probability measure* μ*. Then the induced unitary* Γ*-representation on* $L^2(X,\mu)$ *belongs to* $F(\Gamma)$.

Proof of the Proposition: The idea of the proof is to show that there are "arbitrarily small perturbations" of any given action, which factor through a finite quotient. First however, observe that we may assume μ has no atoms since for any $\alpha > 0$ there can be only finitely many atoms of weight α (μ has total mass 1), hence they form Γ-invariant subset on which the Γ-action factors through a finite quotient. By removing all the atoms of μ we are reduced to the non-atomic case. Let $S \subseteq \Gamma$ be a free generating subset. Abusing notation, we shall denote both the Γ-action on X and on $L^2(X)$ by ρ, i.e., for $f \in L^2(X)$ we have: $[\rho(g)f](x) = f(\rho(g^{-1})x)$. Our strategy now goes as follows: Because the subspace of continuous functions on X is dense in $L^2(X)$, it is enough, given any finite number of continuous functions $f_1, \cdots, f_n$ on X, and any $\epsilon > 0$, to construct a **new** (μ-)measure preserving Γ-action ρ' on X, which factors through a **finite quotient** of Γ, such that for every $g \in S \cup S^{-1}$ (as well as $g = e$) one has for the L^2 norm: $||\rho(g)f_i - \rho'(g)f_i|| < \epsilon$ for all $1 \leq i \leq n$.

Let then ϵ and $f_1, \cdots, f_n$ be given. Choose $\delta > 0$ so that $d(x,y) < 2\delta$ implies $|f_i(x) - f_i(y)| < \epsilon$ for all i and $x, y \in X$. Now, again using continuity, choose some $0 < \theta < \delta$ such that $d(x,y) < \theta$ implies $d(\rho(g)x, \rho(g)y) < \delta$ for $g \in S \cup S^{-1}$. Next observe that X can be divided into a disjoint union of subsets $B_1, \cdots, B_k$ which have equal measure (of size $1/k$) and diameter less than θ. We shall leave the straightforward, yet rather tedious verification of this fact to the reader, hinting only that it is easier to find first such a subdivision where the measure of every B_i is rational (and then subdivide arbitrarily all B_i's to get pieces of equal measure), which of course uses the fact that μ is non-atomic (it seems most convenient here to homemorphically embed X in $[0,1]^N$ and work with the latter).

We can now define the new Γ-action ρ' on X. By freeness, it is enough to define it for any $g \in S$ and we fix such an element. Consider the two disjoint subdivisions of X: $X = \cup B_i = \cup C_i$, where $C_i = g(B_i)$. Let us call a couple B_i and C_j "matched" if $\mu(B_i \cap C_j) > 0$. Because $\mu(B_i) = \mu(C_j) = 1/k$ for all i, j, it is easy to see that for every family of m B_i's, the family of C_j's matched to at least one of its members has least m elements (and vice versa), that is, Hall's marriage theorem applies. We can therefore find a permutation $\sigma_g \in S_k$ with the property that $g(B_i) \cap B_{\sigma_g(i)}$ is non empty for all i. Notice that defining $\sigma_{g^{-1}} = (\sigma_g)^{-1}$, the latter holds also when g is replaced by g^{-1}.

By freeness the permutation action of S on the set of B_i's extends to a permutation action of Γ, factoring through a finite quotient F ($\subseteq S_k$). In order to define a measure preserving Γ-action on X (and not only on the collection of subsets B_i), which factors through an action of F and induces the previous permutation action on the B_i's, we choose as a "model space" any non-atomic standard Lebesgue space Y on which S_k acts by permuting a disjoint subdivision of it to k subsets Y_i. Identifying measure preservingly each B_i and Y_i, induces an F- (hence also $\Gamma-$)action on X, which in turn induces the previously defined permutations $\{\sigma_g\}$ of the sets B_i. This defines the new Γ-action ρ' on X.

Finally, given any $1 \leq i \leq n$ we show that $||\rho(g)f_i - \rho'(g)f_i|| < \epsilon$ for any $g \in S \cup S^{-1}$. In fact, we show that for all $x \in X$: $|\rho(g)f_i(x) - \rho'(g)f_i(x)| < \epsilon$. Let $1 \leq j \leq k$ be such that $x \in B_j$. Then $\rho(g^{-1})x \in \rho(g^{-1})B_j$, and $\rho'(g^{-1}x) \in B_{\sigma_{g^{-1}}(j)}$.

By the construction of σ the two B_i's on the right hand sides intersect, and by the choice of θ and δ they both have diameter $< \delta$. By the triangle inequality it follows that $d(\rho(g^{-1})x, \rho'(g^{-1})x) < 2\delta$, which by the choice of δ implies $|\rho(g)f_i(x) - \rho'(g)f_i(x)| = |f_i(\rho(g^{-1})x) - f_i(\rho'(g^{-1})x)| < \epsilon$. As this holds for every $x \in X$ and u is a probability measure, this establishes the Proposition.

Proof of Theorem 2.2: Let π be a unitary Γ-representation. As is well known, it is a general fact that there exists a probability measure preserving Γ-action on a standard Lebesgue space (X, μ) such that π is a subrepresentation of the unitary Γ-representation on $L^2(X, \mu)$ (cf. [Zi] 5.2.13). It is also a general fact that every such action admits a topological model, i.e., measure theoretically it can always be realized on a compact metric Γ-space on which Γ acts continuously (cf. [Zi] 2.1.19). Our result now follows immediately from the previous Proposition.

It is not difficult to see that property FD passes to subgroups (see Lemma 3.2 below). Our next purpose is to provide situations where the converse phenomenon takes place.

LEMMA 2.4. *Let Γ be a discrete group and $\Lambda < \Gamma$ be a subgroup. Assume that Λ has property FD and that every finite index subgroup of Λ is an intersection of finite index subgroups of Γ. Then for every unitary Γ-representation π, $F(\Gamma)$ contains the representation $\pi \otimes l^2(\Gamma/\Lambda)$.*

PROOF. Consider the restriction $\pi|_\Lambda$, which by assumption belongs to $F(\Lambda)$. Thus, for a sequence of finite index subgroups $M_i < \Lambda$ we have a weak containment $\pi|_\Lambda \prec \oplus l^2(\Lambda/M_i)$. Inducing both sides to Γ (and using "continuity of induction in the Fell topology") yields: $\pi \otimes l^2(\Gamma/\Lambda) \prec \oplus l^2(\Gamma/M_i)$. Therefore it only remains to be shown that for every finite index subgroup $M < \Lambda$, $l^2(\Gamma/M)$ is in $F(\Gamma)$, which is quite standard (compare with the less obvious Lemma 2.1 above). Indeed, by assumption one can write $M = \cap N_i$ for a decreasing sequence of finite index subgroups $N_i < \Gamma$, which means that for every finite subset $S \subseteq \Gamma$ one can find i large enough so that for any $\gamma \in S$: $\langle \gamma 1_e, 1_e \rangle_{l^2(\Gamma/M)} = \langle \gamma 1_e, 1_e \rangle_{l^2(\Gamma/N_i)}$, where 1_e denotes the Dirac function at the identity coset of the corresponding space. Therefore every matrix coefficient associated with functions which are finite linear combination of Γ-translations of 1_e (which form a dense subspace of $l^2(\Gamma/M)$), can be approximated by matrix coefficients in $\oplus l^2(\Gamma/N_i)$, as required.

□

COROLLARY 2.5. *Let Γ be a discrete group and $\Lambda < \Gamma$ be a normal subgroup such that Γ/Λ is amenable. Assume either that Λ has property FD and all its finite index subgroups are closed in the profinite topology of Γ; or, in case Λ is not finitely generated, that every finitely generated subgroup of Λ satisfies the same properties. Then Γ has property FD.*

PROOF. Fix a unitary Γ-representation π; we show that $\pi \in F(\Gamma)$. In the first case this follows immediately from Lemma 2.4, since by assumption we have $1 \prec l^2(\Gamma/\Lambda)$, hence: $\pi \prec \pi \otimes l^2(\Gamma/\Lambda) \in F(\Gamma)$. As for the second, denoting by $\Lambda_i < \Lambda$ an increasing exhausting sequence of finitely generated subgroups, we have by the Lemma for any i: $\pi \otimes l^2(\Gamma/\Lambda_i) \in F(\Gamma)$, hence $\pi \otimes (\oplus l^2(\Gamma/\Lambda_i)) \cong \oplus(\pi \otimes l^2(\Gamma/\Lambda_i)) \in F(\Gamma)$. However since $1_\Lambda \prec \oplus l^2(\Lambda/\Lambda_i)$, inducing to Γ yields $l^2(\Gamma/\Lambda) \prec \oplus l^2(\Gamma/\Lambda_i)$. Since $1_\Gamma \prec l^2(\Gamma/\Lambda)$ we deduce $1_\Gamma \prec \oplus l^2(\Gamma/\Lambda_i)$, so tensoring with π and combining with the previous conclusion completes the proof. □

Remark. As is clear from the proof, the normality of Λ in Γ is not necessary; assuming that $1_\Gamma \prec l^2(\Gamma/\Lambda)$ would suffice for our purposes.

The property that every finite index subgroup of Λ is closed in the profinite topology of Γ is not easily achieved. The next Lemma gives a situation in which it is satisfied:

LEMMA 2.6. *Assume Γ is a discrete group, and Λ is a finitely generated normal subgroup of Γ such that $Z(\hat{\Lambda}) = \{1\}$ (when $Z(\hat{\Lambda})$ denotes the center of the profinite completion $\hat{\Lambda}$ of Λ). Then the profinite topology of Γ induces the profinite topology of Λ. In particular, if in addition Γ/Λ is residually finite, then every finite index subgroup of Λ is the intersection of finite index subgroups of Γ.*

PROOF. The conclusion of the first (main) statement is equivalent to the assertion that the natural map i from $\hat{\Lambda}$ to $\bar{\Lambda}$ (= the closure of Λ in $\hat{\Gamma}$) is an isomorphism (it is always continuous and onto; the point is its injectivity).

As Λ is finitely generated, $Aut(\hat{\Lambda})$ is a profinite group (see below). Hence we have the natural map $\hat{\jmath} : \hat{\Gamma} \to Aut(\hat{\Lambda})$ extending the map $j : \Gamma \to Aut(\Lambda)$ given by conjugation within Γ. Now, $\hat{\jmath}$ restricted to $\bar{\Lambda}$ (in $\hat{\Gamma}$) is just the extension of the map sending each element of Λ to the inner automorphism determined by it. It follows that $\bar{\Lambda}$ is mapped under $\hat{\jmath}$ onto the image of $\hat{\Lambda}$ in $Aut(\hat{\Lambda})$, which by the center freeness assumption, may be identified with $\hat{\Lambda}$. Hence $\psi = \hat{\jmath} \circ i$ is an epimorphism from the finitely generated profinite group $K = \hat{\Lambda}$ onto itself, which must therefore be a topological isomorphism, since every epimorphism from a finitely generated profinite group K onto itself is an isomorphism. Indeed, as K is finitely generated (namely, it has a dense finitely generated subgroup), for any n it has finitely many closed subgroups of index at most n, so their intersection $K_n \triangleleft K$ has finite index. Because ψ is onto, for every finite index subgroup $K_0 < K$ one has $[K : K_0] = [K : \psi^{-1}(K_0)]$, so $\psi(K_n) \subseteq K_n$. Thus ψ induces a homomorphism, $\tilde{\psi} : K/K_n \to K/K_n$, which is onto (as is ψ). Since K/K_n is finite, $\tilde{\psi}$ is injective. Now, given any $e \neq k \in K$, there is some n for which $k \notin K_n$, and by the above $\tilde{\psi}[k] \neq [e]$ in K/K_n, so $\psi(k) \neq e$ thus showing the injectivity of ψ. (Note also that with these notations, $Aut(K)$ is the inverse limit of $Aut(K/K_n)$, hence it is indeed a profinite group, as claimed at the beginning of the proof.) Thus the first statement is established, whereas the second follows easily since the assumption made there implies that Λ itself, being the intersection of finite index subgroups of Γ, is closed in the profinite topology of Γ. □

We remark that the finite generation assumption on Λ is necessary. Indeed, as shown below, all other assumptions hold when Γ is a finitely generated free group and Λ – the kernel of its abelianization – is infinitely generated. However, in this case the profinite topology on Λ does not have a countable basis, hence it cannot be induced by that of Γ.

The following result provides a class of examples where the condition in the previous Lemma can be verified:

PROPOSITION 2.7. *If Λ is a group presented by d generators and e relations with $d - e \geq 2$, then $Z(\hat{\Lambda}) = 1$. In particular, any non abelian free or surface group has this property.*

PROOF. Suppose by contradiction that there exists a non-trivial element z in $Z(\hat{\Lambda})$, and take $\overline{N}$ normal open in $\hat{\Lambda}$ so that $z \notin \overline{N}$. Then the action of $\hat{\Lambda}/\overline{N}$ on $\overline{N}/[\overline{N},\overline{N}]$ is not faithful. But $\hat{\Lambda}/[\overline{N},\overline{N}]$ is the profinite completion of the residually finite group $\Lambda/[N,N]$, when $N = \overline{N} \cap \Lambda$. Hence $\Lambda/N \simeq \hat{\Lambda}/\hat{N}$ acts non-faithfully on $N/[N,N]$. However this is impossible in our case. Indeed, by [JR, Theorem 3], for every finite index normal subgroup N of Λ, with G denoting the quotient Λ/N, there exists $T \lhd \Lambda$ with $T \subseteq N$ and $N/T \simeq \mathbb{Z}^m$ for some m (in particular, $T \supseteq [N,N]$), such that $(\mathbb{Q}G)^{d-e-1}$ appears as a direct summand of the $\mathbb{Q}G$-module $\mathbb{Q} \otimes_{\mathbb{Z}} N/T$. In particular, by our assumption, $G = \Lambda/N$ itself must act faithfully on N/T, hence also on $N/[N,N]$, a contradiction. □

THEOREM 2.8. *The groups Γ defined below all have property FD:*
(1) $\Gamma = \Lambda \rtimes \mathbb{Z}$ *– a cyclic extension of the finitely generated free group Λ. In particular, the Picard group $\Gamma = SL_2(\mathbb{Z}[\sqrt{-1}])$ and the group $SL_2(\mathbb{Z}[\sqrt{-3}])$ have property FD.*
(2) Γ *is a surface group.*
(3) *The same as in (1), where Λ denotes now a surface group. In particular, the fundamental group of any closed hyperbolic 3-manifold which fibers over the circle has property FD.*

Remark. As mentioned in the Introduction, for the Picard and surface groups, the Theorem answers questions raised by Bekka and Louvet in [BL].

PROOF. Part (1) follows immediately from Theorem 2.2 after applying the three preceding results. The fact that (a finite index subgroup of) the Picard group fits in the form treated by (1) is well known, by realizing the fundamental group of the whitehead link as its finite index subgroup (cf. [Wi]), and using the fact that this link is fibered (cf. [Ro] p. 338 Ex. 5). The case of $SL_2(\mathbb{Z}[\sqrt{-3}])$ can be dealt with similarly, this time using the figure 8 knot (see [Sh1] Section 5 and the references therein for an explicit matrix realization of the claimed free and cyclic subgroups). For part (2) we wish to apply the second part of Corollary 2.5 in the case where $\Lambda = F_\infty$ is the kernel of the abelianization of Γ, again using Theorem 2.2 above for finitely generated free groups. The fact that the second condition in the statement of the Corollary is satisfied as well, is not trivial; it is given by the following result of Peter Scott [Sc] which, interestingly, is established by geometric methods: A surface group Γ is LERF (= locally extended residually finite), i.e., every finitely generated subgroup of it is closed in the profinite topology of Γ. Finally, part (3) follows now exactly as in (1), from (2).

□

We conclude this section by remarking that since property FD passes to subgroups, parts (1) and (3) yield many other new Kleinian groups having property FD (compare also with the discussion in Section 6 below, and in particular Conjecture 6.4).

3. Groups without property FD

Unlike the case of amenable and free groups, property FD turns out not to be typical in the class arithmetic groups. The following aims to place the Theorem (and method) of Bekka [Be] within its largest natural framework:

THEOREM 3.1. *Let k be a global field, $\mathcal{O}$ its ring of integers, S a finite set of valuations of k containing all the archimedean ones, $\mathcal{O}_S = \{x \in k \mid v(x) \geq 0$ for* *all $v \notin S\}$ and let $\Gamma = G(\mathcal{O}_S)$ where G is a simple, simply-connected k-algebraic group. Assume that both*

1) *k-rank$(G) \geq 1$, and*
2) *S-rank(G) $\big(= \sum_{v \in S} k_v-$rank$(G)\big) \geq 2$.*

Then Γ does not have property FD.

LEMMA 3.2. *Let Γ be a discrete group and Λ a subgroup of Γ. If F is a dense subset of $\tilde{\Gamma}$ then $F|_\Lambda$ (= the representations in F restricted to Λ) is dense in $\tilde{\Lambda}$.*

PROOF. If $\rho \in \tilde{\Lambda}$ then by assumption $Ind_\Lambda^\Gamma(\rho) \in \tilde{\Gamma}$ is the limit of a subset of representations $\{\pi_i\}$ in F. Since $\rho \subseteq Ind_\Lambda^\Gamma(\rho)\big|_\Lambda = \lim_i \pi_i\big|_\Lambda$, the Lemma follows. □

Notice that this Lemma in particular shows that property FD passes to subgroups. The following provides some kind of "converse" in the special framework of a lattice subgroup:

LEMMA 3.3. *Let Λ be a lattice in a second countable locally compact group G. If F is a dense subset of $\tilde{\Lambda}$ then $F_G = \{Ind_\Gamma^G(\rho)\big|\rho \in F\}$ is dense in $\tilde{G}$.*

PROOF. Let $\rho \in \tilde{G}$. Then $\rho|_\Gamma = \lim_i \pi_i$ where $\{\pi_i\} \subseteq F$. Hence $\rho \otimes L^2(G/\Gamma) \cong Ind_\Gamma^G(\rho|_\Gamma) = \lim_i Ind_\Gamma^G(\pi_i)$. The first contains ρ since Γ is a lattice in G, so ρ is indeed in the closure of F_G. □

LEMMA 3.4. *Let $k, \mathcal{O}$ and S as in Theorem 3.1 and $\Lambda = SL_2(\mathcal{O}_S)$. Let $\mathcal{C}(\Lambda)$ be the set of congruence representations of Λ, i.e., the representations factoring through a congruence quotient of Λ. Then $\mathcal{C}(\Lambda)$ is not dense in $\tilde{\Lambda}$.*

PROOF. The group Λ is a lattice in $\prod_{v \in S} SL_2(k_v)$. The well known Theorem of Selberg ("$\lambda_1 \geq \frac{3}{16}$") and its extensions ("the Selberg property" or "property τ with respect to congruence subgroups" - cf. [Lu1], [Lu2], [Cl]) implies in the characteristic 0 case that some of the complementary series representations of $SL_2(k_v)$ are not in the closure of $C_0 = \{Ind_\Gamma^G(\rho)\big|\rho \in \mathcal{C}(\Lambda)\}$. The same applies in characteristic $p > 0$ due to the Theorem of Drinfeld who proved the Ramanujan conjecture in this case. Lemma 3.3 now finishes the Proof. □

Proof of Theorem 3.1 . Let G and Γ be as in the Theorem. As k-rank$(G) \geq 1$, G has a k-algebraic subgroup H of type A_1. This implies that Γ has a subgroup $\Lambda = H(\mathcal{O}_S)$ which is commensurable to $SL_2(\mathcal{O}_S)$ and the congruence topology of Γ induces the congruence topology of Λ. Now, Γ has the congruence subgroup property (cf. [PR]), thus all its finite representations factor through congruence subgroups, and $\tilde{\Gamma}_F$ restricted to Λ gives $C(\Lambda)$ in the notations of Lemma 3.4. As $C(\Lambda)$ is not dense in $\tilde{\Gamma}$, Lemma 3.2 implies that $F(\Gamma) \neq \tilde{\Gamma}$. □

Remark: Note that $\Gamma = G(\mathcal{O}_S)$ is a lattice in $\prod_{v \in S} G(k_v)$, which is a semisimple group of rank ≥ 2 by our assumptions. Γ is a non-uniform lattice since we assume k-rank $(G) \geq 1$, and the theorem covers all the non-uniform lattices by Margulis' arithmeticity theorem. The Theorem and the proof are still valid for **some** uniform arithmetic lattices in higher rank semi-simple groups but not for all. The same proof will apply for G of k-rank zero (i.e. anisotropic) if (a) Γ satisfies the congruence subgroup property and (b) G has a k-subgroup of type A_1. Note that Condition (a) is currently known for all groups whose type is different from A_n. See Section 6 below for more in this direction.

4. Ramanujan groups

In this section we place all the examples of Ramanujan groups under one unified construction, which also simplifies (conceptually and technically) the work of Lubotzky-Phillips-Sarnak [LPS].

Let G be a locally compact second countable group, $\Gamma < G$ a discrete co-compact torsion free subgroup, and $K < G$ a compact subgroup. Let $X \subseteq G$ be a bounded K-invariant fundamental domain for Γ, i.e. $G = X\Gamma$, and assume that the Haar measure m of G is normalized on X. Associated naturally to this X is a cocycle $\alpha_X : G \times X \to \Gamma$, defined by $\alpha_X(g, x) = \gamma$ iff $gx\gamma \in X$, and a measure preserving G-action on it, denoted $(g, x) \to g \cdot x$ (distinguished from the usual multiplication in G), which is obtained by identifying X with G/Γ. Finally, to any compactly supported probability measure μ on G, we may define the probability measure $\nu = \alpha_X(\mu)$ on Γ, by pushing the measure $\mu \times m$ to Γ via the map $\alpha_X : G \times X \to \Gamma$. Notice that because μ is compactly supported and X is bounded, $\alpha_X(\mu)$ is finitely supported. To simplify notation, given a unitary G-representation π we shall denote by $\pi(\mu)$ the μ-averaging (or "convolution") operator acting in the representation π, and by $||\pi(\mu)||$ its norm (similarly for measures on Γ).

In the rest of the paper we denote the regular representation of a group H by ρ_H. Our main tool for constructing Ramanujan groups is the following general result:

PROPOSITION 4.1. *Retain the above notations for $G, \Gamma, \alpha_X, \mu, \nu$. Let σ be a unitary Γ-representation and $\pi = Ind_\Gamma^G \sigma$.*
(1) *We have $||\sigma(\nu)|| \leq ||\pi(\mu)||$, and if μ is bi-K-invariant then $||\pi(\mu)||$ may be replaced by $sup||\pi_x^K(\mu)||$, where π_x^K is the set of those unitary G-representations π_x having a non-zero K-fixed vector in some direct integral decomposition $\pi = \int \pi_x dx$.*
(2) *If μ is bi-K-invariant and $X = K$, (i.e. $G = K\Gamma$), then equality holds: $||\sigma(\nu)|| = ||\pi(\mu)||$ (and $||\pi(\mu)||$ may be replaced by $sup||\pi_x^K(\mu)||$ as in (1)). In particular, (taking $\sigma = \rho_\Gamma$) we have: $||\rho_\Gamma(\nu)|| = ||\rho_G(\mu)||$.*

The point of the Theorem is that it enables one to deduce that certain groups are Ramanujan without making any explicit numerical calculations of norms, as the following shows:

COROLLARY 4.2. *Retain the above notations and assume that μ is bi-K-invariant and $X = K$, as in (2).*

1) *If σ is a Γ-representation for which $\pi = Ind_\Gamma^G \sigma \prec \rho_G$ then $||\sigma(\nu)|| \leq ||\rho_\Gamma(\nu)||$. In fact, the same conclusion holds for any σ for which we assume the weak containment only for the π_x^K above in place of π.*
2) *Consequently, if $\mathcal{L} = \{\Gamma_i\}$ is a family of finite index subgroups of Γ such that for any i one has: $L_0^2(G/\Gamma_i) \prec \rho_G$, then the group Γ is $(T_\nu, \mathcal{L})$-Ramanujan, where the operator $T_\nu \in \mathbb{C}\Gamma$ is the one corresponding to ν. In fact, the same conclusion holds if we assume the above weak containment in ρ_G only for the "K-spherical part" of the representations $L_0^2(G/\Gamma_i)$ (as in (1) of 4.1).*

3) *In particular, if $\mathcal{L}$ is as in (2) above, then for any $g \in G$ the Γ-averaging operator over the set $KgK \cap \Gamma$, or more generally, the averaging operator over $F \cap \Gamma$ for* **any** *bi-K-invariant compact subset $F \subseteq G$, satisfies the conclusion of (2) above.*

Indeed, (1) follows immediately since by (1) and (2) of the Proposition we have: $||\sigma(\nu)|| \leq ||\pi(\mu)|| \leq ||\rho_G(\nu)|| = ||\rho_\Gamma(\nu)||$.

Parts (2) and (3) of the Corollary make abstract a large part of the proof of the existence of Ramanujan graphs (or groups). All that remains to be done is to plug in the (deep) number theoretic tools which ensure that its condition is satisfied, and in the list of examples below we recall various cases where it can indeed be done.

Finally, before proving the Proposition, we remark that when applied to appropriate finite dimensional but infinite Γ-representations, it immediately gives a unified proof also for the part of work of Lubotzky-Phillips-Sarnak pertaining to uniform points distribution on the sphere (cf. [Lu1] Ch. 9). Moreover, using Margulis superrigidity and conditional on the conjectural congruence subgroup property, one could deduce that some of the examples below satisfy the Ramanujan property with respect to **all** finite dimensional representations, not only the finite ones.

Proof of Proposition 4.1: Retain all the notations preceding the statement of the Proposition. Consider part (1). Recall that the representation space for $Ind_\Gamma^G \sigma$ may be identified with $V_\pi = L^2(X, V_\sigma)$, where G operates on $f \in L^2(X, V_\sigma)$ by $[\pi(g)f](x) = \sigma \circ \alpha(g,x)[f(g^{-1} \cdot x)]$. For unit vectors $v, w \in V_\pi$, consider the constant functions $f_v, f_w \in L^2(X, V_\sigma)$, taking the constant value v,w resp. Note that because X is K-invariant, so are f_v, f_w. For any $g \in G$ we have by definition of the scalar product in V_π: $\langle \pi(g) f_v, f_w \rangle_{V_\pi} = \int_X \langle \sigma \circ \alpha(g,x)v, w\rangle_{V_\sigma} dm(x)$, hence integrating both sides over g with respect to μ and recalling the definition of the measure ν on Γ, gives:

$$(*) \qquad \langle \pi(\mu) f_v, f_w \rangle_{V_\pi} = \int_G \int_X \langle \sigma \circ \alpha(g,x)v, w \rangle_{V_\sigma} dm(x) d\mu(g) = \langle \sigma(\nu)v, w\rangle_{V_\sigma}$$

Since v and w are arbitrary, we can choose them so that the right hand side of $(*)$ is arbitrarily close to $||\sigma(\nu)||$, so the first part of (1) follows. The second part would follow from the assertion that in $(*)$ one can replace in the very left hand side π by π^K, where $\pi^K \subseteq \pi$ is the subrepresentation supporting the π_x^K's. Indeed, by K-invariance of μ, $\pi(\mu)$ vanishes on the orthogonal complement to π^K, and f_v, f_w are K-invariant, so the whole computation goes through with π^K in place of π.

For part (2) we wish to establish an inequality in the opposite direction. If we knew that in $(*)$, **any** functions $f_1, f_2 \in V_\pi$ can be obtained in the form f_v, f_w for some $v, w \in V_\sigma$, then the very same argument as in the previous proof could be reversed to show, by appropriately choosing f_1, f_2, that the opposite inequality holds. Notice that the assumption that μ is bi-K-invariant shows, as before, that it is actually enough to consider only f_1, f_2 which are K-invariant. However, since we now assume $X = K$, a K-invariant function is (essentially) constant on X, so it is obvious that the K-invariant functions are exactly those functions of the form f_v, thereby completing the proof.

Examples:

(1) Let H be the standard quaternion algebra, let p be a prime with $p \equiv 1(4)$ and $H(\mathbb{Z}[\frac{1}{p}])$ the $\mathbb{Z}[\frac{1}{p}]$-points of H , i.e.

$$\alpha = x_0 + x_1 i + x_2 j + x_3 k \text{ with } x_i \in \mathbb{Z}\left[\frac{1}{p}\right].$$

For such α denote $\|\alpha\|^2 = \alpha \cdot \bar{\alpha} = \sum_{i=0}^{3} x_i^2$. Let $\Gamma = H(\mathbb{Z}[\frac{1}{p}])^*/Z$, i.e., the invertible elements of $H(\mathbb{Z}[\frac{1}{p}])$ (i.e., those α with $\|\alpha\|^2 \in p^{\mathbb{Z}}$) modulo the center. Γ is a cocompact lattice in $G = PGL_2(\mathbb{Q}_p)$. It is shown in [Lu1] that $\Gamma(2) = \{\alpha \in \Gamma | \alpha \equiv 1(mod 2)\}$ acts simply transitively on the Bruhat-Tits building of $PGL_2(\mathbb{Q}_p)$, which is a $(p+1)$-regular tree. Thus $G = K\Gamma(2)$ with $K = PGL_2(\mathbb{Z}_p)$. The Ramanujan-Petterson conjecture (proved by Deligne) together with the Jacquet-Langlands correspondence, implies that for every congruence subgroup $\Gamma(2m)$ in $\Gamma(2)$, the induced representation $Ind_{\Gamma(2)}^{G} l_0^2(\Gamma(2)/\Gamma(2m))$ is tempered, and is thus weakly contained in $L^2(G)$ (see [Lu1]). Therefore the assumptions of Prop. 4.1 and Cor. 4.2 are satisfied, showing that $\Gamma(2)$ is a Ramanujan group with respect to every spherically symmetric operator T and $\mathcal{L}$ being the family of congruence subgroups.

(2) In [Mor] Morgenstern showed that for every positive characteristic local field F, $G = PGL_2(F)$ has a suitable arithmetic group Γ such that Γ acts simply transitive on the Bruhat-Tits building of G which is a $(q+1)$-regular tree, where $q = |\mathcal{O}/\pi\mathcal{O}|$. Here $\mathcal{O}$ is the valuation ring of F and π a uniformizer in $\mathcal{O}$. Moreover, the analogue Ramanujan-Petterson conjecture was proved in this case by Drinfeld. Similarly to the previous example, this now shows in particular that a free group F_r with $r = \frac{1}{2}(p^n + 1)$ and p an odd prime, is a Ramanujan group.

(3) Let $\mathcal{S} = \{p_1, \ldots, p_\ell\}$ be a set of ℓ-primes with $p_i \equiv 1(mod 4)$ for every $i = 1, \ldots, \ell$. Let $\mathbb{Z}_{\mathcal{S}} = \mathbb{Z}\left[\frac{1}{p_1}, \ldots, \frac{1}{p_\ell}\right]$, $H(\mathbb{Z}_{\mathcal{S}})$ - the $\mathbb{Z}_{\mathcal{S}}$-points of the standard quaternion algebra H (see example 1). Let $\Gamma = H(\mathbb{Z}_{\mathcal{S}})^*/Z$ the invertible elements of $H(\mathbb{Z}_{\mathcal{S}})$ modulo its center. This is a lattice in the product $G = \prod_{i=1}^{\ell} G_i$ with $G_i = PGL_2(\mathbb{Q}_{p_i})$. The congruence subgroup $\Gamma(2)$ acts simply-transitive on the vertices of $X = \prod_{i=1}^{\ell} T_i$ where T_i is the $(p_i + 1)$-regular tree which is the Bruhat-Tits building of G_i. Thus $G = K\Gamma$, where $K = \prod_{i=1}^{\ell} K_i$, $K_i = PGL_2(\mathbb{Z}_{p_i})$. As shown by Jordan and Livne [JL], Deligne's solution of the Ramanujan-Petterson conjecture implies that the assumptions of (Prop 4.1 and) Cor 4.2 are satisfied for $\Gamma(2)$, and $\mathcal{L}$ being its congruence subgroups. The reader is referred to [JL] for more on examples of this kind.

(4) Let F be a local field of positive characteristic, $\mathcal{O}$ and q as in example 2. In [CS1], Cartwright and Steger constructed, for every $d \geq 2$, a cocompact lattice in $G = PGL_d(F)$, which acts simply-transitively on the vertices of the Bruhat-Tits building $\mathcal{B}_d(F)$, which is a building of type $\tilde{A}_{d-1}$. So $G = K\Gamma$ when $K = PGL_d(\mathcal{O})$. Let Γ be the Cartwright-Steger lattice. In [LSV1], Lubotzky, Samuels and Vishne used Lafforgue solution of the Ramanujan-Peterson conjecture for GL_d

in positive characteristic to show: if $\Gamma(m)$ is a congruence subgroup of Γ, then the representation $\tilde{\pi}$, which is the K-spherical part of $\pi = Ind_\Gamma^G(l_0^2(\Gamma/\Gamma(m))$, is weakly contained in the regular representation of G provided at least one of the following two conditions is satisfied: either (a) d is a prime number or (b) m is prime to at least one of the ramification primes of the division algebra defining Γ (in the notations of [CS2], this means that m is a polynomial in $\mathbb{F}_q[y]$ which is prime to $z = y + 1$; see [LSV1] and [LSV2] for a detailed discussion).

We mention in passing that unlike examples 2, 3 and 4, in the case d is not a prime, for some (in fact, for infinitely many) congruence subgroups, the "Ramanujan Conjecture" is not satisfied. Still for infinitely many – it does. In any event, whenever condition (a) or (b) is satisfied, we get (by Proposition 4.1 and Corollary 4.2) quotients making Γ Ramanujan. Since these constructions are very recent, here are some further details for the convenience of the reader. First we describe sets S, for which according to part (3) of the Corollary, $T = T_S$ satisfies the condition of the Ramanujan property:

For $1 \leq k \leq d - 1$, let $g_k = diag(\pi, \pi, \ldots, \pi, 1, \ldots, 1)$ when π appears k times. Let $S_k = Kg_kK \cap \Gamma$ and $A_k = \sum_{s \in S_k} s$ be the corresponding element of the group algebra $\mathbb{C}\Gamma$. In general, A_k is not a self adjoint operator (when acting in unitary Γ-representations), but (i) it is a normal operator, and (ii) $A_k + A_{d-k}$ is self-adjoint. The spectrum of A_k on $l^2(\Gamma)$ (which can be identified with $l^2(B_n^0(F))$, $B_n^0(F)$ being the 0-skeleton of the building $B_n(F)$), is a subset of $\mathbb{C}$ which is described in [CS2] as:

$$C_{d,k} = \{q^{k(d-k)/2}\sigma_k(z_1, \ldots, z_k)\Big| For\ i = 1, \ldots, d, \quad z_i \in \mathbb{C}, |z_i| = 1 \text{ and } \prod_{i=1}^{d} z_i = 1\}$$

Here σ_k is the k-symmetric function i.e.,

$$\sigma_k(z_1, \ldots, z_d) = \sum_{i_1 < i_2 < \cdots < i_k} z_{i_1} z_{i_2} \cdot \cdots \cdot z_{i_k}.$$

In [LSV1] it is shown that for the congruence subgroups satisfying either (a) or (b) the "non-trivial" eigenvalues of A_k on $l^2(\Gamma/\Gamma(m))$ are in $C_{d,k}$. Moreover, for $A_k + A_{d-k}$, it means that its norm on $l_0^2(\Gamma/\Gamma(m))$ is bounded by $2\binom{d}{k}q^{k(d-k)/2}$. This last part follows also directly from Corollary 4.2 above. The A_k's commute with each other and we may take $S = \bigcup_{k=1}^{d-1} S_k$ and $A = \sum_{k=1}^{d-1} A_k$. This is the usual adjacency operator with respect to S. Thus, for the family $\mathcal{L} = \Gamma(m)$, the group Γ is $(T_S, \mathcal{L})$-Ramanujan whenever (a) or (b) is satisfied.

5. Non-Ramanujan groups

Lemma 2.1 above is often useful in showing that a group Γ is **not** Ramanujan with respect to a given family of subgroups $\mathcal{L}$. This is reminiscent of the Burger-Li-Sarnak method [BLS] to show the non-validity of the naive Ramanujan conjecture, of which the Lemma may be viewed as a "discrete" analogue. Here is a useful illustration:

PROPOSITION 5.1. *Let $\Gamma = H_1 \times H_2$ be a direct product of two non-amenable groups H_1 and H_2. Then Γ is not Ramanujan (i.e.,* **no** *$T, \mathcal{L}$ as in the Definition make it Ramanujan).*

PROOF. Assume to the contrary that Γ is Ramanujan with respect to some $\mathcal{L} = \{\Gamma_i\}$ and some positive $T \in \mathbb{C}\Gamma$. We claim that for $j = 1, 2$, $\overline{H_j}$ is of finite index in Γ, where $\overline{H_j}$ is the closure of H_j in the profinite topology on Γ defined by $\mathcal{L}$. Indeed, otherwise by Lemma 2.1 we have: $||T||_{l^2(\Gamma/\overline{H_j})} \leq sup\, ||T||_{l^2_0(\Gamma/\Gamma_i)} \leq ||T||_{\rho_\Gamma}$. However, since $\overline{H_j}$ are non-amenable and still normal in Γ, by a well known result of Kesten [Ke]: $||T||_{\rho_\Gamma} < ||T||_{l^2(\Gamma/\overline{H_j})}$, which is a contradiction. Thus $\overline{H_j}$ must be of finite index in Γ for both $j = 1, 2$, and the two must commute, as do the H_j's. This proves that Γ is virtually abelian, in contradiction to the assumption that the H_j's are non-amenable. □

Let $\Gamma_2 = F_{\frac{p_1+1}{2}} \times F_{\frac{p_2+1}{2}}$ be the product of two free groups on $\frac{p_1+1}{2}$ and $\frac{p_2+1}{2}$ generators, with set of free generators S' and S'', respectively. Here p_1 and p_2 are two primes with $p_1 \equiv p_2 \equiv 1(mod 4)$. Let $S_2 = S' \cup S''$. Then $Cay(\Gamma_2; S_2)$ is the direct product of the $(p_1 + 1)$-regular tree and the $(p_2 + 1)$-regular tree. By Proposition 5.1, Γ_2 is not Ramanujan. Compare this to Γ_1- the congruence subgroup mod 2 of $H^*(\mathbb{Z}[\frac{1}{p_1}, \frac{1}{p_2}])/Z$ discussed in Example 3 cf §4. This group acts simply transitive on the product of the trees associated to $PGL_2(\mathbb{Q}_{p_1})$ and $PGL_2(\mathbb{Q}_{p_2})$. Hence, its Cayley graph with respect to the generators which take any fixed "origin" to its neighbours (of distance 1) is also the product of (p_1+1)-regular tree and the $(p_2 + 1)$-regular tree (for more on these groups see [Moz]). We thus obtained two groups Γ_1 and Γ_2 with systems of generators S_1 and S_2, respectively, such that $Cay(\Gamma_1; S_1)$ is isometric to $Cay(\Gamma_2; S_2)$. Now, Γ_1 is Ramanujan with respect to the averaging operator T_{S_1} (this certainly holds with respect to the family of congruence subgroups, although notice that by the yet conjectural congruence subgroup property of this group, there should be no others), while $(\Gamma_2; S_2)$ is not Ramanujan.

Finally, by direct considerations we can get some information about the outstanding example of an arithmetic group, namely, the group $SL_n(\mathbb{Z})$:

PROPOSITION 5.2. *Let $\Gamma_n = SL_n(\mathbb{Z})$, and S_n be the set of elementary matrices and their inverses, i.e. $S_n = \{I \pm E_{ij} | 1 \leq i \neq j \leq n\}$. Then for n large enough, Γ_n is not $(T_n, \mathcal{L})$ Ramanujan with respect to any $\mathcal{L}$, where T_n is the averaging operator over S_n.*

PROOF. Denote by ρ_n the regular representation of Γ_n. Our strategy is to show that there exists some constant $C < 1$ such that $||T_n||_{\rho_n} < C$ for all n, whereas for any n we can find a finite representation π_n of Γ_n with no invariant vectors, such that $||T_n||_{\pi_n} \to 1$ as $n \to \infty$.

To establish the first claim, we divide the set S_n into $(n^2 - n)/2$ subsets of 4 elements, each consisting of the elements $I \pm E_{ij}, I \pm E_{ji}$ for all different choices of $1 \leq i < j \leq n$. Notice that every such subset generates a copy of a subgroup isomorphic to $SL_2(\mathbb{Z})$, in a way that maps the 4 elements to the $(\pm)$ upper and lower unit elementary matrices of $SL_2(\mathbb{Z})$. Let C denote the norm of the averaging operator on the latter 4-element subset of $SL_2(\mathbb{Z})$, acting in its $(SL_2(\mathbb{Z})$-) regular representation. By non-amenability of $SL_2(\mathbb{Z})$ we have $C < 1$. We claim that this C is the one required in our first claim. Indeed, grouping the elements of S_n in $(n^2 - n)/2$ 4-element subsets as above, and then iterating the triangle inequality $(n^2 - n)/2$ times on the ρ_n-norm, gives the bound $||T_n||_{\rho_n} \leq \frac{1}{2}(n^2 - n) \cdot (C/\frac{1}{2}(n^2 - n)) = C$, as soon as we show that the norm of the (normalized by 4) averaging operator on each 4-element subset acting in ρ_n is C. This, however, is obvious from the definition of C, as the restriction of ρ_n to the subgroup ($\cong SL_2(\mathbb{Z})$) generated by each 4-element subset is a multiple of its regular representation.

We now establish the second ingredient of our strategy. Fix n, and consider a "maximal parabolic" subgroup $\Lambda_n < \Gamma_n$, i.e., the stabilizer in Γ_n of the vector $(1, 0, \cdots, 0)$ for the standard linear action of Γ_n on $\mathbb{R}^n$. Consider the natural unitary Γ_n-representation τ_n on $l^2(\Gamma_n/\Lambda_n)$, and the Dirac function $f = 1_e$ at the identity coset. Since all but $2(n-1)$ elements of S_n are contained in Λ_n, they fix f, while the other $2(n-1)$ elements take f to an orthogonal vector. Therefore we have $\langle T_n f, f\rangle_{\tau_n} = \frac{1}{|S_n|}(2(n^2-n) - 2(n-1)) = \frac{2}{|S_n|}(n^2 - 2n + 1) = \frac{n-1}{n} = 1 - \frac{1}{n}$. Consequently, $||T_n||_{\tau_n} \geq 1 - \frac{1}{n} \to 1$. Thus it only remains to show that for every n, the representation τ_n can be approximated in the Fell topology by finite representations of the form $l_0^2(\Gamma_n/\Gamma_n^i)$ for a sequence of finite index subgroups $\Gamma_n^i < \Gamma_n$. However, since it is easy to see that Λ_n is closed in the congruence topology of Γ_n, i.e., it is the intersection of finite index (congruence) subgroups of Γ_n, this follows now from Lemma 2.1.

We have therefore proved the result when $\mathcal{L}$ is the family of all congruence subgroups. It is not difficult to see that the groups Λ_n used in the proof are in fact closed in any profinite topology on $SL_n(\mathbb{Z})$ which is weaker than the congruence one. Together with the congruence subgroup property, this establishes the result in general. We leave the verification of some missing details here to the reader. $\square$

Question: Is $SL_n(\mathbb{Z})$ Ramanujan for some $n \geq 3$?

6. Concluding remarks and open questions

Since all the groups we studied in this paper are S-arithmetic, it is natural to discuss a relation between their Ramanujan and FD properties, and those of the ambient algebraic groups, while giving a natural meaning to the latter. For simplicity, let us concentrate here only on the arithmetic case, i.e., we let $\mathbf{G}$ be a semisimple algebraic group defined over $\mathbb{Q}$, and consider the group $\Gamma = \mathbf{G}(\mathbb{Z})$. Recall that the **Automorphic Spectrum** Aut $\mathbf{G}$ of $\mathbf{G}$ is defined to be the closure in $\widetilde{\mathbf{G}(\mathbb{R})}$, of the set of all representations of the form $L^2(\mathbf{G}(\mathbb{R})/\Gamma)$, where Γ varies over all the congruence subgroups of $\mathbf{G}(\mathbb{Z})$. Denote also by $\text{Aut}_0\mathbf{G}$ the same space, but with the trivial representation deleted. Much efforts have been devoted to identifying Aut $\mathbf{G}$, which is one of the most interesting objects of number theory. Analogues to the Ramanujan and FD properties for the arithmetic groups, one can now define them for the ambient $\mathbb{Q}$-group by saying that $\mathbf{G}$ has the Ramanujan property if $\text{Aut}_0\mathbf{G}$ is contained in the tempered spectrum (i.e. it is weakly contained in the regular representation), and has property FD if it is the whole dual. These definitions make sharpest the marked difference between the two properties. Indeed, $\mathbf{G}$ is Ramanujan iff it satisfies the naive Ramanujan conjecture (e.g., Selberg's 1/4 conjecture in the case of $\mathbf{G} = SL_2$), that is to say, the automorphic spectrum is as restricted as it can possibly be, and it has property FD if nothing can be said about the "location" of automorphic representations. Although the naive Ramanujan conjecture is of course known to be false in general (cf. [BLS] and the references therein), one certainly expects that the automorphic representations should not be spread out over the whole unitary dual, namely, the following seems plausible:

6.1 Conjecture. Let $\mathbf{G}$ be a semisimple algebraic group defined over $\mathbb{Q}$ and $\mathbb{R}$-isotropic. Then $\mathbf{G}$ does not have property FD.

For example, this is known for $\mathbf{G} = SL_2$. Of course, one can expect property FD of the algebraic group to be analogous to that of the arithmetic group only where the family of (finite) **congruence** representations are involved. Indeed, we have:

Observation: If the family of finite representations of $\Gamma = \mathbf{G}(\mathbb{Z})$ which factor through some congruence quotient is dense in $\tilde{\Gamma}$, then $\mathbf{G}$ has property FD. In particular, if Γ has the congruence subgroup property and property FD, then $\mathbf{G}$ has property FD.

This follows from Lemma 3.3. Of course, it is a natural question whether the converse (of the first statement) holds as well, although as we expect that no $\mathbf{G}$ has property FD, this may be of limited interest. However, combined with Conjecture 6.1, the observation gives:

6.2 Conjecture. Let Γ be an arithmetic group. If Γ satisfies the congruence subgroup property then it does not have property FD.

In fact, by our reasoning above, the Conjecture becomes a Theorem whenever the ambient group $\mathbf{G}$ is known not to have property FD. Since there are numerous groups for which the latter can be shown (using known results on the automorphic spectrum), the observation can be made useful in showing that certain arithmetic groups do not have property FD (in a way independent of Section 3 above and Bekka's method [Be]).

Unfortunately, the relation between the Ramanujan property for the algebraic and arithmetic group seems less understood at the present. Even though it appears that the Ramanujan property for the algebraic groups is stronger, we were able to deduce from it the Ramanujan property of the lattice subgroup only under a strong restriction on the fundamental domain. Thus the following still remains:

6.3 Question: Does $\mathbf{G}$ being Ramanujan imply that (some finite index subgroup of) $\Gamma = \mathbf{G}(\mathbb{Z})$ is as well, at least in the uniform case, i.e., when $\mathbb{Q}$-rank$\mathbf{G} = 0$?

Of particular interest seems the case of a surface group.

To conclude this direction, we remark that among free groups, we currently have the Ramanujan property only for some of them, even though we do not require the relevant operator to be an average over a free generating subset. However, notice that the list of Ramanujan groups consists not only of those obtained by Morgenstern [Mor], but also of **every index two subgroup** of these. This is obtained by using Proposition 4.1 and Corollary 4.2 for spherical operators which are supported on spheres of **even** radius, whose elements generate (though not freely !) a subgroup of index two. At any rate, one would like to know for which values of r, the free group F_r is Ramanujan, which may be viewed also as a motivation for the general problem of understanding to what extent the Ramanujan property passes to finite index subgroups and overgroups.

Turning to a different point of view, the results in §3 show that "most" lattices in higher rank simple Lie groups, do not have property FD. On the other hand from Theorems 2.2 and 2.8 it follows that every lattice Γ in $SL_2(\mathbb{R})$ does, and moreover, the same conclusion holds for some lattices in $SL_2(\mathbb{C})$. A well known conjecture of Thurston asserts that every irreducible 3-dimensional hyperbolic manifold is fibered over a circle. If true, it would imply that every lattice in $SL_2(\mathbb{C})$ has a finite index subgroup of the form $\Lambda \rtimes \mathbb{Z}$, when Λ is a surface or a free group. In either case, Theorem 2.8 shows that such a group has property FD. In fact, more might be expected:

6.4 Conjecture: Every finitely generated discrete subgroup of $SO(n, 1)$ has property FD.

Observe that from the definition it follows that any group Γ which has both property τ and FD, has Kazhdan's property (indeed, the family of finite representations is then both dense and isolated from 1_Γ). Consequently, if an arithmetic lattice in $SO(n, 1)$ or $SU(n, 1)$ has property FD, it cannot have τ, and hence it has a negative solution to the congruence subgroup problem. Indeed, it is known that Γ has the Selberg property (i.e., property (τ) with respect to the congruence subgroups – see [Lu1], [Lu2] and [Cl]). This means that among the non-trivial finite representations of Γ, the **congruence** yet **not all** ones are bounded away from 1_Γ. Thus, Conjecture 6.4 implies Serre's conjecture [Se] for the arithmetic groups of $SO(n, 1)$, while making an abstract group theoretical statement. For the current status of Serre's conjecture for lattices in $SO(n, 1)$ see [Lu2].

It may be quite reasonable to include $SU(n, 1)$ as well in Conjecture 6.4. On the other hand we conjecture that the opposite holds for lattices in higher rank simple Lie groups (a conjecture which is strongly supported by the results in §3). One would hesitate regarding lattices in $Sp(n, 1)$ and $F_4^{(-20)}$. The lattices in these rank one Lie groups sometimes behave as the other rank one groups $SO(n, 1)$ and $SU(n, 1)$, and sometimes as higher rank lattices. In particular, the following seems very intriguing:

6.5 Question: Does there exist an infinite discrete Kazhdan group with property FD ?

Recall that by a result of Wang [Wa], for Kazhdan groups not only the trivial, but in fact **any** finite dimensional unitary representation is isolated in the unitary dual. This may suggest a negative answer to the question. Such an answer would also account for many of the foregoing results (and questions).

References

[Be] Bekka, M. B., *On the full C^*-algebras of arithmetic groups and the congruence subgroup problem*, Forum Math. **11**(1999), no. 6, 705–715.

[BL] Bekka, M. B., and Louvet, N., *Some properties of C^*-algebras associated to discrete linear groups, in: C^*-algebras*, (Munster, 1999), 1–22, Springer, Berlin, (2000).

[BLS] Burger, M., Li, J.-S., and Sarnak, P., *Ramanujan duals and automorphic spectrum*, Bull. Amer. Math. Soc. (N.S.) **26** (1992), no. 2, 253–257.

[Ch] Choi, M. D., *The full C^*-algebra of the free group on two generators*, Pacific J. Math. **87**(1980), no. 1, 41–48.

[Cl] Clozel, L., *Demonstration de la conjecture τ*, Invent. Math. **151** (2003), no. 2 , 297–328.

[CS1] Cartwright, D., and Steger, T., *A family of $\tilde{A}_n$-groups*, Israel J. Math. **103**(1998), 125–140.

[CS2] Cartwright, D., and Steger, T., *Elementary symmetric polynomials in numbers of modulus 1*, Canad. J. Math. **54** (2002), no. 2, 239–262.

[Ey] Eymard, P., *L'algebra de Fourier d'un groupes localement compact*, Bull. Soc. Math. France **92** (1964), 181–236.

[JR] Jarden, M., and Ritter, J., *Normal automorphisms of absolute Galois groups of $\mathfrak{p}$-adic fields*, Duke Math. J. **47** (1980), no. 1, 47–56.

[JL] Jordan, B. W., and Livne, R., *The Ramanujan property for regular cubical complexes*, Duke Math. J. **105** (2000), no. 1, 85–103.

[Ke] Kesten, H., *Symmetric random walks on groups*, Trans. AMS **92** (1959), 336–354.

[Lu1] Lubotzky, A., *Discrete Groups, Expanding Graphs and Invariant Measures*, Progress in Mathematics, 125, Birkhaüser Verlag, Basel, 1994.

[Lu2] Lubotzky, A., *Eigenvalues of the Laplacian, the first Betti number and the congruence subgroup problem*, Annals of Math. **144** (1996), no. 2, 441–452.

[LPS] Lubotzky, A., Phillips, R., and Sarnak, P., *Ramanujan graphs*, Combinatorica **8** (1988), no. 3, 261–277.

[LSV1] Lubotzky, A., Samuels, B., and Vishne, U., *Ramanujan complexes of type* $\tilde{A}_d$, preprint.

[LSV2] Lubotzky, A., Samuels, B., and Vishne, U., *Explicit constructions of Ramanujan complexes of type* $\tilde{A}_d$, preprint.

[Mor] Morgenstern, M., *Existence and explicit constructions of* $q+1$ *regular Ramanujan graphs for every prime power* q, J. Combin. Theory Ser. B **62** (1994), no. 1, 44–62.

[Moz] Mozes, S., *On closures of orbits and arithmetic of quaternions*, Israel J. Math., **86** (1994), no. 1–3, 195–209.

[PR] Platonov, V., and Rapinchuk, A., *Algebraic Groups and Number Theory*, Pure and Applied Mathematics **139**, Academic Press, (1984).

[Ro] Rolfsen, D., *Knots and Links*, Mathematics Lecture Series no. 7, Publish or Perish, Berkeley (1976).

[Se] Serre, J.P., *Le probleme des groupes congruence pour SL2*, Ann. of Math., **92** (1970), 487–529.

[Sc] Scott, P., *Subgroups of surface groups are almost geometric*, J. London Math. Soc. (2) **17** (1978), no. 3, 555–565.

[Sh1] Shalom, Y., *Expander graphs and invariant means*, Combinatorica **17** (1997), no. 4 555–575.

[Sh2] Shalom, Y., *Rigidity, unitary representations of semisimple groups, and fundamental groups of manifolds with rank one transformation group*. Annals of Math. **152** (2000), no.1, 113–182.

[Wa] Wang, S.P., *On isolated points in the dual spaces of locally compact groups*, Math. Ann. **218** (1975), 19–34.

[Wi] Wielenberg, N., *The structure of certain subgroups of the Picard group*, Math. Proc. Camb. Phil. Soc., **84** (1978), no.3 427–436.

[Zi] Zimmer, R., *Ergodic Theory and Semisimple Groups*, Monographs in Mathematics **81**, Birkhauser (1984).

Institute of Mathematics, The Hebrew University of Jerusalem, 91904 Israel

School of Mathematical Sciences, Tel-Aviv University, Tel-Aviv 69978 Israel

Contemporary Mathematics
Volume **347**, 2004

Stabilization for SL_n in bounded cohomology

Nicolas Monod

ABSTRACT. We prove that for all local fields SL_n is stable over n in terms of continuous bounded cohomology. We complement this by various computations in low degree, showing notably $\mathrm{H}^3_\mathrm{b}(\mathrm{SL}_n(\mathbf{R})) = 0$ for all $n \in \mathbf{N}$. We link the corresponding vanishing for p-adic fields to a question on prime numbers.

1. Introduction

The continuous Eilenberg-MacLane cohomology of classical simple Lie groups is well known. By results going back to É. Cartan and W.T. van Est, it can notably be realized as relative Lie algebra cohomology or as the cohomology of the compact dual symmetric space [**4**]. For instance, the continuous cohomology $\mathrm{H}^\bullet_\mathrm{c}(\mathrm{SL}_n(\mathbf{C}))$ is an exterior algebra over the Borel classes. In this example, there is moreover a *stabilization*: the standard inclusions $\mathrm{SL}_n(\mathbf{C}) \hookrightarrow \mathrm{SL}_{n+1}(\mathbf{C})$ induce an isomorphism on H^q_c for n large enough compared to q; in particular, one obtains a description of the cohomology of the direct limit $\mathrm{SL}_\infty(\mathbf{C})$ which is notably of relevance for Bott periodicity [**9**].

The continuous cohomology of semi-simple Lie groups is also essential for the understanding of the more complicated cohomology of their discrete subgroups [**4, 3**]. On the other hand, the theory of *bounded cohomology*, which plays an important role for discrete groups in a number of contexts [**14, 12, 6, 7, 16, 20, 21, 22**], is reputedly much less accessible to computation – and mostly unknown as yet.

However, we find that for the continuous bounded cohomology $\mathrm{H}^\bullet_\mathrm{cb}$ of Lie groups, more tools are available, at least in low degrees. Our first result is a stabilization statement:

THEOREM 1.1. *Let k be a local field and $0 \leq m \leq n$. Then the standard embeddings $\mathrm{GL}_m(k) \hookrightarrow \mathrm{GL}_n(k)$ induce isomorphisms*

$$\mathrm{H}^q_\mathrm{cb}(\mathrm{GL}_n(k)) \cong \mathrm{H}^q_\mathrm{cb}(\mathrm{GL}_m(k))$$

for any $0 \leq q \leq m$. Similarly, one has

$$\mathrm{H}^q_\mathrm{cb}(\mathrm{SL}_n(k)) \cong \mathrm{H}^q_\mathrm{cb}(\mathrm{GL}_m(k))$$

if $0 \leq q \leq m-1$.

2000 *Mathematics Subject Classification.* Primary 22E41, 46M20; Secondary 46H25, 57R20.
Key words and phrases. Characteristic classes, bounded cohomology.

Theorem 1.1 makes it all the more desirable to gain an understanding of the low degree bounded cohomology for the low dimensional linear groups. Let us first review what is known. In degree one, $\mathrm{H}^\bullet_{\mathrm{cb}}$ always vanishes. In degree two, the situation is fully understood:

THEOREM (see [**6, 7**]). *Let G be a connected simple Lie group or more generally the group of k-points of a connected isotropic almost simple algebraic group over a local field k of any characteristic. Then $\mathrm{H}^2_{\mathrm{cb}}(G) \cong \mathrm{H}^2_{\mathrm{c}}(G)$.*

(In particular, this space is one dimensional if k is Archimedean and the associated symmetric space is of Hermitian type; it vanishes in all other cases).

Beyond degree two, no complete results are known. For the case of $\mathrm{SL}_2(\mathbf{C})$, we have shown with M. Burger [**8**, Theorem 1.2] that a result of S. Bloch [**2**] implies

THEOREM (see [**8**]). *The space $\mathrm{H}^3_{\mathrm{cb}}(\mathrm{SL}_2(\mathbf{C}))$ is naturally isomorphic to $\mathrm{H}^3_{\mathrm{c}}(\mathrm{SL}_2(\mathbf{C}))$ (and thus has dimension one).*

We have also proved the following [**8**, Theorem 1.5]

THEOREM (see [**8**]). $\mathrm{H}^3_{\mathrm{cb}}(\mathrm{SL}_2(\mathbf{R})) = 0$.

Further, A. Goncharov's study of functional equations for the trilogarithm [**13**] shows that the Borel class generating $\mathrm{H}^5_{\mathrm{c}}(\mathrm{SL}_3(\mathbf{C}))$ can be represented by a bounded cocycle and hence $\mathrm{H}^5_{\mathrm{cb}} \to \mathrm{H}^5_{\mathrm{c}}$ is onto for this group. Goncharov gives also formulae expressing cocycles for higher Borel classes in terms of higher polylogarithms ; we could however not check whether theses cocycles are bounded (this would settle for $\mathrm{SL}_n(\mathbf{C})$ a conjecture of J.L. Dupont [**10**], see [**20**, 9.3.9]). The only other surjectivity results in higher rank that we are aware of are R.P. Savage's proof [**23**] that the volume form associated to $\mathrm{SL}_n(\mathbf{R})$ is in the image of $\mathrm{H}^d_{\mathrm{cb}}$ (for $d = n(n+1)/2 - 1$) and the fact that the Euler class for $\mathrm{GL}^+_n(\mathbf{R})$ is in the image of $\mathrm{H}^n_{\mathrm{cb}}$ as can be deduced *e.g.* from the Ivanov-Turaev cocycle [**17**].

However, none of this information can be immediately used in Theorem 1.1 as it stands because of the assumption $0 \leq q \leq m-1$ or m. We shall show:

THEOREM 1.2. $\mathrm{H}^3_{\mathrm{cb}}(\mathrm{SL}_n(\mathbf{R})) = 0$ *and* $\mathrm{H}^3_{\mathrm{cb}}(\mathrm{GL}_n(\mathbf{R})) = 0$ *for all* $n \in \mathbf{N}$.

It turns out that a similar investigation for the field $\mathbf{Q}$ (Proposition 4.1 below) leads to an intriguing question:

For a prime p, denote by v_p the valuation normalized by $v_p(p^n) = n$. If q is another prime, define $D_{p,q} : \mathbf{Q} \setminus \{0,1\} \to \mathbf{Z}$ by

$$D_{p,q}(x) = v_p(x)v_q(1-x) - v_q(x)v_p(1-x).$$

This function is obviously unbounded because of the Chinese remainder theorem; on the other hand, one can form arbitrary combinations of such $D_{p,q}$ by varying the primes p, q since the sum is finite at each x. We propose the following

QUESTION 1.3. Is the function $\sum_{p<q} \alpha_{p,q} D_{p,q}$ unbounded on $\mathbf{Q} \setminus \{0,1\}$ for every family of real numbers $\{\alpha_{p,q}\}$ (unless they are all zero)?

Using an adèle argument, the Bloch-Suslin complex and Theorem 1.1, we show:

THEOREM 1.4. *A positive answer to Question 1.3 implies $\mathrm{H}^3_{\mathrm{cb}}(\mathrm{GL}_n(\mathbf{Q}_p)) = 0$ for all $n \in \mathbf{N}$ and all primes p.*

Acknowledgments I am indebted to D. Zagier for suggesting and discussing the reference [**24**] used in the last section. Warm thanks are due to the organisers of the first JAMS symposium. Work partially supported by grants of the FNS and the NSF.

2. Cohomology with Coefficients on Projective Spaces

Let k be a local field, *i.e.* a locally compact non-discrete field, and let $n \geq 1$. We write

$$G_n = \mathrm{GL}_n(k),\ G_n^+ = \mathrm{SL}_n(k),\ \mathbf{P}_{n-1} = \mathbf{P}^{n-1}(k) = (k^n \setminus \{0\})/k^*.$$

We endow $\mathbf{P}_{n-1}$ with the measure class determined by the Haar measures on k. The natural G_n-action preserves this class and therefore we obtain a coefficient module $L^\infty(\mathbf{P}_{n-1})$, and more generally for every $m \geq 0$ the diagonal action on the direct product turns $L^\infty(\mathbf{P}_{n-1}^{m+1})$ into a coefficient module (a *coefficient module* is a dual isometric Banach representation with separable continuous predual; for background see [**20**]). We agree that G_n, G_n^+ are trivial groups for $n \leq 0$.

An important ingredient for Theorem 1.1 is the following.

PROPOSITION 2.1. *Let k be a local field and n, p, q non-negative integers. There are canonical isomorphisms*

(i) $\mathrm{H}^q_{\mathrm{cb}}\big(G_n, L^\infty(\mathbf{P}_{n-1}^{p+1})\big) \cong \mathrm{H}^q_{\mathrm{cb}}(G_{n-p-1})$ *for* $0 \leq p \leq n$.
(ii) $\mathrm{H}^q_{\mathrm{cb}}\big(G_n^+, L^\infty(\mathbf{P}_{n-1}^{p+1})\big) \cong \mathrm{H}^q_{\mathrm{cb}}(G_{n-p-1})$ *for* $0 \leq p \leq n-1$.
(iii) $\mathrm{H}^q_{\mathrm{cb}}\big(G_n, L^\infty(\mathbf{P}_{n-1}^{p+1})\big) = \mathrm{H}^q_{\mathrm{cb}}\big(G_n^+, L^\infty(\mathbf{P}_{n-1}^{p+1})\big) = 0$ *for* $p \geq n-2$ *and* $q \neq 0$.

REMARK 2.2. The case $p \geq n-2, q = 0$ is not contained in the proposition because indeed the very definition of $\mathrm{H}^0_{\mathrm{cb}}$ yields $\mathrm{H}^0_{\mathrm{cb}}\big(G_n^+, L^\infty(\mathbf{P}_{n-1}^{p+1})\big) \cong L^\infty(\mathbf{P}_{n-1}^{p+1})^{G_n^+}$ and $\mathrm{H}^0_{\mathrm{cb}}\big(G_n, L^\infty(\mathbf{P}_{n-1}^{p+1})\big) \cong L^\infty(\mathbf{P}_{n-1}^{p+1})^{G_n}$ for all p. In particular, comparing with (i) and (ii), we recover the basic facts

$$L^\infty(\mathbf{P}_{n-1}^{p+1})^{G_n} \cong \mathbf{R} \qquad (\forall p \leq n), \tag{2.1}$$

$$L^\infty(\mathbf{P}_{n-1}^{p+1})^{G_n^+} \cong \mathbf{R} \qquad (\forall p \leq n-1). \tag{2.2}$$

REMARK 2.3. An elementary consequence of the proposition is that for all $n, p \geq 0$ one has

$$\mathrm{H}^1_{\mathrm{cb}}\big(G_n, L^\infty(\mathbf{P}_{n-1}^{p+1})\big) = 0,$$
$$\mathrm{H}^1_{\mathrm{cb}}\big(G_n^+, L^\infty(\mathbf{P}_{n-1}^{p+1})\big) = 0.$$

Indeed, for bounded cohomology with *trivial* coefficients one has always $\mathrm{H}^1_{\mathrm{cb}}(-) = 0$, so that for $p \leq n-1$ we are done by (i) and (ii); for other p, apply (iii).

We prepare now the proof of Proposition 2.1. For $v \in k^n \setminus \{0\}$ we denote by $\bar{v}$ the corresponding element of $\mathbf{P}_{n-1}$. We denote by $(e_1, \ldots, e_n)$ the canonical base of k^n and consider by abuse of notation the corresponding inclusion $k^n \subset k^{n+1}$. We let

$$Q_n = \mathrm{Stab}_{G_n}(\bar{e}_n), \qquad Q_n^+ = \mathrm{Stab}_{G_n^+}(\bar{e}_n).$$

If we write N_n and N_n^+ for the corresponding kernels of the Q_n- and Q_n^+-actions on k^{n-1}, we have semi-direct product decompositions $Q_n = N_n \rtimes G_{n-1}$ and $Q_n^+ = N_n^+ \rtimes G_{n-1}$.

LEMMA 2.4. *Let $\sigma : G_{n-1} \to Q_n$ be a section, E a coefficient Q_n-module and $\iota : E^{N_n} \to E$ the inclusion map. Then we have for all $q \in \mathbf{N}$ topological isomorphisms*

$$\mathrm{H}^q_{\mathrm{cb}}(G_{n-1}, E^{N_n}) \xleftarrow{\sigma^*} \mathrm{H}^q_{\mathrm{cb}}(Q_n, E^{N_n}) \xrightarrow{\iota_*} \mathrm{H}^q_{\mathrm{cb}}(Q_n, E).$$

Moreover, the restriction map σ^* does not depend on the choice of σ.

PROOF. Let $\pi : Q_n \to G_{n-1}$ be the quotient map. The amenability of N_n implies that ι_* as well as the *inflation* map

$$\pi^* : \mathrm{H}^q_{\mathrm{cb}}(G_{n-1}, E^{N_n}) \longrightarrow \mathrm{H}^q_{\mathrm{cb}}(Q_n, E^{N_n})$$

are topological isomorphisms, see Corollary 8.5.2 in [**20**]. The statements about σ^* now follow from $\sigma^*\pi^* = Id$. □

Similarly, we have

LEMMA (2.4$^+$). *Let $\sigma : G_{n-1} \to Q_n^+$ be a section, E a coefficient Q_n^+-module and $\iota : E^{N_n^+} \to E$ the inclusion map. Then we have for all $q \in \mathbf{N}$ topological isomorphisms*

$$\mathrm{H}^q_{\mathrm{cb}}(G_{n-1}, E^{N_n^+}) \xleftarrow{\sigma^*} \mathrm{H}^q_{\mathrm{cb}}(Q_n^+, E^{N_n^+}) \xrightarrow{\iota_*} \mathrm{H}^q_{\mathrm{cb}}(Q_n^+, E).$$

Moreover, the restriction map σ^* does not depend on the choice of σ. □

For $n \geq 2$, the canonical projection $k^n \to k^{n-1}$ which omits e_n induces a Q_n-equivariant map

$$\pi_n : \mathbf{P}_{n-1} \setminus \{\bar{e}_n\} \longrightarrow \mathbf{P}_{n-2}.$$

Since the N_n-action on $\mathbf{P}_{n-2}$ is trivial, π_n induces for every $p \geq 1$ a surjective Q_n-map

$$\overline{\pi_n} : \left(\mathbf{P}_{n-1} \setminus \{\bar{e}_n\}\right)^p \big/ N_n \longrightarrow \mathbf{P}^p_{n-2}.$$

Since k is non-discrete, $\mathbf{P}_{n-1} \setminus \{\bar{e}_n\}$ has full measure in $\mathbf{P}_{n-1}$. Denoting by a double slash the standard quotients in the category of measure class spaces, we have

LEMMA 2.5. *For all $0 \leq p \leq n$, the map $\overline{\pi_n}$ yields a Q_n-equivariant isomorphism*

$$\mathbf{P}^p_{n-1} /\!/ N_n \longrightarrow \mathbf{P}^p_{n-2}.$$

of measure class spaces.

PROOF OF LEMMA 2.5. The case $p = 0$ is void, so let $p \geq 1$. It is enough to prove that the map $\overline{\pi_n}$ is injective outside a null set, which can be achieved by a dimension argument because k is non-discrete:

Take two families $u^1, \ldots u^p$ and $v^1, \ldots v^p$ of elements of $k^p \setminus ke_n$ and suppose $\pi_n(\overline{u^i}) = \pi_n(\overline{v^i})$ for all $1 \leq i \leq p$. Thus, there are $\lambda_1, \ldots, \lambda_p \in k^*$ such that, working in the canonical base (e_j), we have

$$u^i_j = \lambda_i v^i_j \qquad \forall 1 \leq i \leq p, 1 \leq j \leq n-1.$$

In this base, we have an identification $k^{n-1} \rtimes k^* \cong N_n$ which takes $\nu = (\nu_1, \ldots, \nu_{n-1}; \nu_n)$ to $\begin{pmatrix} & & 0 \\ & Id & \vdots \\ & & 0 \\ \nu_1 & \cdots \ \nu_{n-1} & \nu_n \end{pmatrix}$. Thus the image of u^i under ν is

$$\begin{pmatrix} u_1^i \\ \vdots \\ u_{n-1}^i \\ \langle \nu | u^i \rangle \end{pmatrix} = \begin{pmatrix} \lambda_i v_1^i \\ \vdots \\ \lambda_i v_{n-1}^i \\ \langle \nu | u^i \rangle \end{pmatrix}$$

where $\langle \cdot | \cdot \rangle$ is the k-bilinear form associated to (e_j). In order to show that the p-tuples $(\overline{u^1}, \ldots, \overline{u^p})$ and $(\overline{v^1}, \ldots, \overline{v^p})$ are in the same N_n-orbit, one must solve simultaneously the equations $\langle \nu | u^i \rangle = \lambda_i v_n^i$ $(1 \leq i \leq p)$ in ν with $\nu_n \neq 0$. This system can be solved if the (u^i) are linearly independent in k^n; this condition fails only for a set of positive codimension in $\mathbf{P}_{n-1}^p$ since $p \leq n$. The additional condition $\nu_n \neq 0$ is also generic in (u^i). □

We observe that with the above notations there is an identification $N_n^+ \cong k^{n-1}$ under which $\nu \in N_n^+$ is equivalent to $\nu_n = 1$. Repeating the preceding proof with this additional restriction decreasing by one the bound on p, we get:

LEMMA 2.6. *For all $0 \leq p \leq n-1$, the map $\overline{\pi_n}$ yields a Q_n^+-equivariant isomorphism $\mathbf{P}_{n-1}^p /\!\!/ N_n^+ \to \mathbf{P}_{n-2}^p$ of measure class spaces.* □

COROLLARY 2.7. (i) *For all $0 \leq p \leq n$, the map $\overline{\pi_n}$ induces an identification $L^\infty(\mathbf{P}_{n-1}^p)^{N_n} \cong L^\infty(\mathbf{P}_{n-2}^p)$ of coefficient Q_n-modules (or G_{n-1}-modules).*

(ii) *For all $0 \leq p \leq n-1$, the map $\overline{\pi_n}$ induces an identification $L^\infty(\mathbf{P}_{n-1}^p)^{N_n^+} \cong L^\infty(\mathbf{P}_{n-2}^p)$ of coefficient Q_n^+-modules (or G_{n-1}-modules).* □

PROOF OF PROPOSITION 2.1. We may assume $n \geq 2$ since otherwise all groups are Abelian and hence have vanishing bounded cohomology in all positive degrees.

(i) and (ii): The identifications

$$\mathbf{P}_{n-1} \cong G_n/Q_n, \quad \mathbf{P}_{n-1} \cong G_n^+/Q_n^+ \tag{2.3}$$

yield coefficient modules identifications

$$L^\infty(\mathbf{P}_{n-1}^{p+1}) \cong L_{\mathrm{w}*}^\infty(G_n/Q_n, L^\infty(\mathbf{P}_{n-1}^p)),$$

where $L_{\mathrm{w}*}^\infty$ denotes the space of weak-* measurable bounded function classes; for background on this and the following, we refer to [**20**]. The right hand side above is isomorphic to the induced module $\mathrm{Ind}_{Q_n}^{G_n} L^\infty(\mathbf{P}_{n-1}^p)$, see point (v) in Examples 10.1.2 in [**20**]. Therefore there is a natural *induction* isomorphism

$$\mathrm{H}_{\mathrm{cb}}^q(G_n, L^\infty(\mathbf{P}_{n-1}^{p+1})) \cong \mathrm{H}_{\mathrm{cb}}^q(Q_n, L^\infty(\mathbf{P}_{n-1}^p))$$

by Propositions 10.1.3 and 10.1.5 in [**20**]. Similarly,

$$\mathrm{H}_{\mathrm{cb}}^q(G_n^+, L^\infty(\mathbf{P}_{n-1}^{p+1})) \cong \mathrm{H}_{\mathrm{cb}}^q(Q_n^+, L^\infty(\mathbf{P}_{n-1}^p)).$$

By Lemma 2.4 and Lemma 2.4$^+$, we have restriction isomorphisms

$$\mathrm{H}_{\mathrm{cb}}^q(Q_n, L^\infty(\mathbf{P}_{n-1}^p)) \cong \mathrm{H}_{\mathrm{cb}}^q(G_{n-1}, L^\infty(\mathbf{P}_{n-1}^p)^{N_n}),$$

$$\mathrm{H}_{\mathrm{cb}}^q(Q_n^+, L^\infty(\mathbf{P}_{n-1}^p)) \cong \mathrm{H}_{\mathrm{cb}}^q(G_{n-1}, L^\infty(\mathbf{P}_{n-1}^p)^{N_n^+}).$$

Since we have assumed $p \leq n$ resp. $p \leq n-1$, we may apply both cases of Corollary 2.7 and deduce

$$\begin{aligned}\mathrm{H}^q_{\mathrm{cb}}\big(G_n, L^\infty(\mathbf{P}^{p+1}_{n-1})\big) &\cong \mathrm{H}^q_{\mathrm{cb}}\big(G_{n-1}, L^\infty(\mathbf{P}^{p}_{n-2})\big),\\ \mathrm{H}^q_{\mathrm{cb}}\big(G^+_n, L^\infty(\mathbf{P}^{p+1}_{n-1})\big) &\cong \mathrm{H}^q_{\mathrm{cb}}\big(G_{n-1}, L^\infty(\mathbf{P}^{p}_{n-2})\big).\end{aligned}$$

We may now repeat the argument ℓ times on the common right hand side above, thus decreasing simultaneously n and p to get

$$\mathrm{H}^q_{\mathrm{cb}}\big(G_{n-1-\ell}, L^\infty(\mathbf{P}^{p-\ell}_{n-2-\ell})\big).$$

If $p \leq n-2$ we stop at $\ell = p$ and obtain the term

$$\mathrm{H}^q_{\mathrm{cb}}\big(G_{n-p-1}, L^\infty(\mathbf{P}^{0}_{n-2-p})\big) = \mathrm{H}^q_{\mathrm{cb}}(G_{n-p-1})$$

If $p = n-1$ we stop at $\ell = p-1$ and obtain

$$\mathrm{H}^q_{\mathrm{cb}}\big(G_1, L^\infty(\mathbf{P}_0)\big) = \mathrm{H}^q_{\mathrm{cb}}(G_1) = \mathrm{H}^q_{\mathrm{cb}}(G_0)$$

(G_1 is amenable), whilst if $p = n$ we stop at $\ell = p-2$ and obtain likewise

$$\mathrm{H}^q_{\mathrm{cb}}\big(G_1, L^\infty(\mathbf{P}^2_0)\big) = \mathrm{H}^q_{\mathrm{cb}}(G_1) = \mathrm{H}^q_{\mathrm{cb}}(G_{-1}).$$

In all three cases we have the right hand side of (i) and (ii).

(iii): We claim that for $p \geq n-2$ the diagonal G_n-action on $\mathbf{P}^{p+1}_{n-1}$ is amenable (in the sense of R. Zimmer [**25**]). Indeed (see [**1**]), the amenability of the action is equivalent to the conjunction of

- the equivalence relation associated to the action is amenable,
- the stabilizer of almost every point is an amenable group.

The first condition is satisfied because the orbits are locally closed. As for the second, the identification (2.3) shows that the stabilizer of a point in $\mathbf{P}^{p+1}_{n-1}$ is the intersection of $p+1$ conjugates of Q_n. Since $p+1 \geq n-1$, a generic such intersection is contained in a minimal parabolic subgroup, hence is amenable. This proves the claim.

The amenability of the action is equivalent to the relative injectivity of the coefficient G_n-module $L^\infty(\mathbf{P}^{p+1}_{n-1})$, see Theorem 5.7.1 in [**20**]. Similarly, $L^\infty(\mathbf{P}^{p+1}_{n-1})$ is relatively injective as a coefficient G^+_n-module[1]. This implies the vanishing of the bounded cohomology in every positive degree (Proposition 7.4.1 in [**20**]). ☐

3. A Double Complex

Keep the notation of the previous section. We define a fist quadrant double complex $(L^{\bullet,\bullet}, {}^{I}d, {}^{II}d)$ by

$$L^{p,q} = L^\infty(G^{p+1}_n \times \mathbf{P}^{q+1}_{n-1})^{G_n} \qquad (p, q \geq 0)$$

with ${}^{I}d : L^{p,q} \to L^{p+1,q}$ defined by the homogeneous coboundary d^p associated to G^{p+1}_n; we recall for later use the definition $d^p = \sum_{j=0}^{p}(-1)^j d^p_j$, where d^p_j omits the jth variable. As for ${}^{II}d : L^{q,p} \to L^{q,p+1}$ it is defined by the homogeneous coboundary on $\mathbf{P}^{p+1}_{n-1}$ affected with the sign $(-1)^{(p+1)}$. To such a complex are associated two spectral sequences ${}^{I}\mathrm{E}$, ${}^{II}\mathrm{E}$; for spectral sequences, we follow standard notations, see *e.g.* [**11**, III.7], [**5**, III §14] or the Section 12.1 in [**20**]. The following lemma is straightforward except for a technical point:

[1]One can also argue that for coefficient modules, relative injectivity is preserved by passing to closed subgroups of second countable locally compact groups, see Proposition 5.8.1 in [**20**].

LEMMA 3.1. *The first spectral sequence converges to* $\mathrm{H}^\bullet_{\mathrm{cb}}(G_n)$. *More precisely,* ${}^{I}\mathrm{E}_r^{p,q} = 0$ $(\forall p \geq 0, q \geq 1, r \geq 1)$ *and* ${}^{I}\mathrm{E}_r^{p,0} \cong \mathrm{H}^p_{\mathrm{cb}}(G_n)$ $(\forall p \geq 0, r \geq 2)$.

PROOF. Since integration over the first variable provides a homotopy (cf. proof of [**7**, 1.5.6] or [**20**, 7.5.5]), the cohomology of the complex

$$0 \longrightarrow L^\infty(\mathbf{P}_{n-1}) \longrightarrow L^\infty(\mathbf{P}^2_{n-1}) \longrightarrow L^\infty(\mathbf{P}^3_{n-1}) \longrightarrow \cdots$$

is concentrated in degree zero, where it is $\mathbf{R}$. We have shown in Lemma 8.2.5 of [**20**] that $L^\infty_{\mathrm{w}*}(G_n^{p+1}, -)^{G_n}$ is exact for all $p \geq 0$ with respect to *adjoint* short exact sequences of coefficient G_n-modules. Therefore, since ${}^{I}\mathrm{E}_1^{p,q}$ is defined by

$$\begin{aligned} L^\infty_{\mathrm{w}*}(G_n^{p+1}, L^\infty(\mathbf{P}^q_{n-1}))^{G_n} &\longrightarrow L^\infty_{\mathrm{w}*}(G_n^{p+1}, L^\infty(\mathbf{P}^{q+1}_{n-1}))^{G_n} \longrightarrow \\ &\longrightarrow L^\infty_{\mathrm{w}*}(G_n^{p+1}, L^\infty(\mathbf{P}^{q+2}_{n-1}))^{G_n} \end{aligned}$$

for $q > 0$ and

$$0 \longrightarrow L^\infty_{\mathrm{w}*}(G_n^{p+1}, L^\infty(\mathbf{P}_{n-1}))^{G_n} \longrightarrow L^\infty_{\mathrm{w}*}(G_n^{p+1}, L^\infty(\mathbf{P}^2_{n-1}))^{G_n}$$

when $q = 0$, we deduce canonical identifications ${}^{I}\mathrm{E}_1^{p,0} \cong L^\infty(G_n^{p+1})^{G_n}$, ${}^{I}\mathrm{E}_1^{p,q} = 0$ for all $p \geq 0, q \geq 1$. Thus ${}^{I}\mathrm{E}_2^{p,0}$ is defined by the L^∞ homogeneous resolution for $\mathrm{H}^p_{\mathrm{cb}}(G_n)$ and the statement follows. □

LEMMA 3.2. *There is a canonical identification* ${}^{II}\mathrm{E}_1^{p,q} \cong \mathrm{H}^q_{\mathrm{cb}}(G_n, L^\infty(\mathbf{P}^{p+1}_{n-1}))$ *for all* $p, q \geq 0$.

PROOF. In view of the identification $L^{q,p} \cong L^\infty_{\mathrm{w}*}\big(G_n^{q+1}, L^\infty(\mathbf{P}^{p+1}_{n-1})\big)^{G_n}$, the spaces ${}^{II}\mathrm{E}_1^{p,q}$ are indeed defined by the L^∞ homogeneous resolution for the continuous bounded cohomology of G_n with coefficients in $L^\infty(\mathbf{P}^{p+1}_{n-1})$. □

Recall that we have given in Proposition 2.1 an interpretation of the right hand side in Lemma 3.2. However, in order to compute the second tableau ${}^{II}\mathrm{E}_2$ we need to understand the differentials ${}^{II}\mathrm{E}_1^{p,q} \to {}^{II}\mathrm{E}_1^{p+1,q}$, induced (up to the sign) by the homogeneous coboundary $d^{p+1} : L^\infty(\mathbf{P}^{p+1}_{n-1}) \to L^\infty(\mathbf{P}^{p+2}_{n-1})$.

LEMMA 3.3. *Let* n, p, q *be non-negative integers with* $p \leq n-2$. *If* p *is even, then the differential* ${}^{II}\mathrm{E}_1^{p,q} \to {}^{II}\mathrm{E}_1^{p+1,q}$ *vanishes. If* p *is odd, then the isomorphisms of (the proof of) Proposition 2.1* (i) *conjugate the differential to the* restriction *map*

$$\mathrm{res}:\ \mathrm{H}^q_{\mathrm{cb}}(G_{n-p-1}) \longrightarrow \mathrm{H}^q_{\mathrm{cb}}(G_{n-p-2})$$

induced by the upper left inclusion $G_{n-p-2} \to G_{n-p-1}$.

PROOF. We consider for $0 \leq j \leq p+1$ the map $A^{p+1}_{n,j}$ defined by the commutative diagram

$$\begin{array}{ccc} \mathrm{H}^q_{\mathrm{cb}}(G_n, L^\infty(\mathbf{P}^{p+1}_{n-1})) & \xrightarrow{(d_j^{p+1})_*} & \mathrm{H}^q_{\mathrm{cb}}(G_n, L^\infty(\mathbf{P}^{p+2}_{n-1})) \\ \cong \uparrow \mathrm{ind} & & \cong \uparrow \mathrm{ind} \\ \mathrm{H}^q_{\mathrm{cb}}(Q_n, L^\infty(\mathbf{P}^{p}_{n-1})) & \xrightarrow{A^{p+1}_{n,j}} & \mathrm{H}^q_{\mathrm{cb}}(Q_n, L^\infty(\mathbf{P}^{p+1}_{n-1})). \end{array}$$

We recall from [**20**, p. 134] that the left hand side induction can be defined *e.g.* as follows. Represent $[\alpha] \in \mathrm{H}^q_{\mathrm{cb}}(Q_n, L^\infty(\mathbf{P}^p_{n-1}))$ by a bounded measurable Q_n-equivariant cocycle $\alpha : G_n^{q+1} \to L^\infty(\mathbf{P}^p_{n-1})$ and define

$$\mathbf{i}\alpha :\ G_n^{q+1} \longrightarrow L^\infty(\mathbf{P}^{p+1}_{n-1}) \cong L^\infty_{\mathrm{w}*}(G_n/Q_n, L^\infty(\mathbf{P}^p_{n-1}))$$

by setting for $g_i, h \in G_n, x \in \mathbf{P}^p_{n-1}$

$$\mathbf{i}\alpha(g_0, \cdots, g_q)(hQ_n)(x) = \alpha(h^{-1}g_0, \cdots, h^{-1}g_q)(h^{-1}x). \tag{3.1}$$

A similar formula defines the right hand side induction. A computation with (3.1) shows that for $j \neq 0$ the map $A^{p+1}_{n,j}$ is but $(d^p_{j-1})_*$ induced by $L^\infty(\mathbf{P}^p_{n-1}) \to L^\infty(\mathbf{P}^{p+1}_{n-1})$. It follows that, for the isomorphisms of the proof of Proposition 2.1, we have a commutative diagram for $j \neq 0$

$$\begin{array}{ccc}
\mathrm{H}^q_{\mathrm{cb}}(G_n, L^\infty(\mathbf{P}^{p+1}_{n-1})) & \xrightarrow{(d^{p+1}_j)_*} & \mathrm{H}^q_{\mathrm{cb}}(G_n, L^\infty(\mathbf{P}^{p+2}_{n-1})) \\
\cong \downarrow & & \cong \downarrow \\
\mathrm{H}^q_{\mathrm{cb}}(G_{n-1}, L^\infty(\mathbf{P}^{p}_{n-2})) & \xrightarrow{(d^{p}_{j-1})_*} & \mathrm{H}^q_{\mathrm{cb}}(G_{n-1}, L^\infty(\mathbf{P}^{p+1}_{n-2})).
\end{array}$$

In the case $j = 0$, a computation with (3.1) shows that $A^{p+1}_{n,0}$ is the map $\mathrm{res} \circ \mathrm{ind}$ such that

$$\begin{array}{ccc}
\mathrm{H}^q_{\mathrm{cb}}(G_n, L^\infty(\mathbf{P}^{p+1}_{n-1})) & \xrightarrow{(d^{p+1}_0)_*} & \mathrm{H}^q_{\mathrm{cb}}(G_n, L^\infty(\mathbf{P}^{p+2}_{n-1})) \\
\cong \uparrow \mathrm{ind} & \searrow^{\mathrm{res}} & \cong \uparrow \mathrm{ind} \\
\mathrm{H}^q_{\mathrm{cb}}(Q_n, L^\infty(\mathbf{P}^{p}_{n-1})) & \xrightarrow{A^{p+1}_{n,0}} & \mathrm{H}^q_{\mathrm{cb}}(Q_n, L^\infty(\mathbf{P}^{p+1}_{n-1}))
\end{array}$$

commutes. One checks further that the relevant isomorphisms from the proof of Proposition 2.1 intertwine $A^{p+1}_{n,0}$ with $A^p_{n-1,0}$. Since d^{p+1} is an alternating sum of the d^{p+1}_j, the statement follows by induction on p. In fact the differential is induced by d^{p+1} up to $(-1)^{p+1}$ only, but this is irrelevant since we established that it vanishes for $p+1$ odd. □

Proof of Theorem 1.1 for G_n. It is enough to show that the upper left embedding $\mathrm{GL}_{n-1}(k) \hookrightarrow \mathrm{GL}_n(k)$ induces isomorphisms

$$\mathrm{H}^q_{\mathrm{cb}}(\mathrm{GL}_n(k)) \cong \mathrm{H}^q_{\mathrm{cb}}(\mathrm{GL}_{n-1}(k))$$

for any $0 \leq q \leq n-1$. We do this by induction on n. The statement is trivial for $n \leq 2$. Take $n \geq 3$ and assume the statement true for all smaller values.

Lemma 3.2 and Proposition 2.1 (i) show that ${}^{II}\mathrm{E}^{p,q}_2$ is computed by

$$0 \longrightarrow \mathrm{H}^q_{\mathrm{cb}}(G_{n-1}) \longrightarrow \mathrm{H}^q_{\mathrm{cb}}(G_{n-2}) \longrightarrow \cdots \longrightarrow \mathrm{H}^q_{\mathrm{cb}}(G_0)$$

for $0 \leq p \leq n-2$ and $q \geq 1$. If we add the induction assumption to the conclusion of Lemma 3.3, we deduce that ${}^{II}\mathrm{E}^{0,q}_2 = \mathrm{H}^q_{\mathrm{cb}}(G_{n-1})$ and that ${}^{II}\mathrm{E}^{p,q}_2$ vanishes for $1 \leq p \leq n-2$ and $q \geq 1$. Lemma 3.2 and Proposition 2.1 (iii) imply now that ${}^{II}\mathrm{E}^{p,q}_2$ vanishes for all p, q provided $pq \neq 0$. In particular we get an exact sequence

$${}^{II}\mathrm{E}^{0,q-1}_2 \longrightarrow {}^{II}\mathrm{E}^{q,0}_2 \longrightarrow {}^{II}\mathrm{E}^{q}_\infty \longrightarrow {}^{II}\mathrm{E}^{0,q}_2 \longrightarrow {}^{II}\mathrm{E}^{q+1,0}_2. \tag{3.2}$$

On the other hand, ${}^{II}\mathrm{E}^{p,0}_2$ is computed by the complex

$$0 \longrightarrow L^\infty(\mathbf{P}_{n-1})^{G_n} \longrightarrow L^\infty(\mathbf{P}^2_{n-1})^{G_n} \longrightarrow L^\infty(\mathbf{P}^3_{n-1})^{G_n} \longrightarrow \cdots \tag{3.3}$$

By Remark 2.2 and in view of the definition of the coboundary, this complex has the form

$$0 \longrightarrow \mathbf{R} \xrightarrow{0} \mathbf{R} \xrightarrow{=} \mathbf{R} \xrightarrow{0} \mathbf{R} \xrightarrow{=} \cdots$$

for $0 \leq p \leq n$. Therefore, ${}^{II}\mathrm{E}^{p,0}_2$ vanishes for $1 \leq p \leq n$. Putting everything together in (3.2), we deduce ${}^{II}\mathrm{E}^q_\infty \cong \mathrm{H}^q_{\mathrm{cb}}(G_{n-1})$ for all $q \leq n-1$. This, together with Lemma 3.1, finishes the proof upon checking that the isomorphism is indeed induced by the inclusion. □

PROOF OF THEOREM 1.1 FOR G_n^+. We define a similar double complex by

$$L^{p,q} = L^\infty\big((G_n^+)^{p+1} \times \mathbf{P}_{n-1}^{q+1}\big)^{G_n^+}.$$

Then the proof is identical to the case of G_n except that one has to keep track of the restriction $p \leq n-1$ of Proposition 2.1, Lemma 2.6 and Corollary 2.7 throughout the argument. This results in the restriction $q \leq m-1$ in Theorem 1.1. □

As a by-product of the previous arguments, we obtain also:

PROPOSITION 3.4. *The restriction maps*

$$\mathrm{H}^n_{\mathrm{cb}}(G_n) \longrightarrow \mathrm{H}^n_{\mathrm{cb}}(G_{n-1}) \quad \textit{and} \quad \mathrm{H}^{n-1}_{\mathrm{cb}}(G_n^+) \longrightarrow \mathrm{H}^{n-1}_{\mathrm{cb}}(G_{n-1})$$

are injective for all n. If $k^*/(k^*)^n$ is trivial, then the restriction map

$$\mathrm{H}^n_{\mathrm{cb}}(G_n^+) \longrightarrow \mathrm{H}^n_{\mathrm{cb}}(G_{n-1})$$

is also injective.

PROOF. Our discussion showed that (3.2) yields an exact sequence

$${}^{II}\mathrm{E}_2^{n,0} \longrightarrow \mathrm{H}^n_{\mathrm{cb}}(G_n) \longrightarrow \mathrm{H}^n_{\mathrm{cb}}(G_{n-1}).$$

The first term is computed by (3.3) which vanishes as pointed out in Remark 2.2 – case $p = n$ of (2.1). Likewise, the spectral sequence for G_n^+ yields

$${}^{II}\mathrm{E}_2^{n-1,0} \longrightarrow \mathrm{H}^{n-1}_{\mathrm{cb}}(G_n^+) \longrightarrow \mathrm{H}^{n-1}_{\mathrm{cb}}(G_{n-1}^+),$$

wherein this time ${}^{II}\mathrm{E}_2^{\bullet,0}$ is realized by

$$0 \longrightarrow L^\infty(\mathbf{P}_{n-1})^{G_n^+} \longrightarrow L^\infty(\mathbf{P}_{n-1}^2)^{G_n^+} \longrightarrow L^\infty(\mathbf{P}_{n-1}^3)^{G_n^+} \longrightarrow \cdots$$

Here the first term vanishes by the case $p = n-1$ of (2.1).

Suppose now that $k^*/(k^*)^n$ is trivial. All we have to show is that ${}^{II}\mathrm{E}_2^{n,0}$ vanishes for the spectral sequence associated to G_n^+. Thus it suffices to show that $L^\infty(\mathbf{P}_{n-1}^{n+1})^{G_n^+}$ consist only of constant function classes. The G_n- and G_n^+-actions on $\mathbf{P}_{n-1}$ factor through $\mathrm{PGL}_n(k)$ and $\mathrm{PSL}_n(k)$ respectively. The condition on k implies that these groups coincide and thus we are done by (2.1). □

PROOF OF THEOREM 1.2. First, we observe that the restriction map

$$\mathrm{H}^\bullet_{\mathrm{cb}}(\mathrm{GL}_n(\mathbf{R})) \longrightarrow \mathrm{H}^\bullet_{\mathrm{cb}}(\mathrm{SL}_n(\mathbf{R}))$$

is injective for all n. Indeed, we have a commutative diagram

$$\begin{array}{ccc}
\mathrm{H}^\bullet_{\mathrm{cb}}(\mathrm{PGL}_n(\mathbf{R})) & \longrightarrow & \mathrm{H}^\bullet_{\mathrm{cb}}(\mathrm{PSL}_n(\mathbf{R})) \\
\Big\downarrow{\scriptstyle \mathrm{inf}} & & \Big\downarrow{\scriptstyle \mathrm{inf}} \\
\mathrm{H}^\bullet_{\mathrm{cb}}(\mathrm{GL}_n(\mathbf{R})) & \longrightarrow & \mathrm{H}^\bullet_{\mathrm{cb}}(\mathrm{SL}_n(\mathbf{R}))
\end{array}$$

in which both lateral arrows are isomorphisms [**20**, 8.5.2] and the upper arrow in injective since $\mathrm{PSL}_n(\mathbf{R})$ is closed of finite index in $\mathrm{PGL}_n(\mathbf{R})$, see [**20**, 8.6.2].

As mentioned in the Introduction, the case of $\mathrm{SL}_2(\mathbf{R})$ is a joint result with M. Burger [**8**, 1.5]; therefore $\mathrm{H}^3_{\mathrm{cb}}(\mathrm{GL}_2(\mathbf{R}))$ vanishes as well. Now Proposition 3.4

implies that both $H^3_{cb}(GL_3(\mathbf{R}))$ and $H^3_{cb}(SL_3(\mathbf{R}))$ vanish since $(\mathbf{R}^*)^3 = \mathbf{R}^*$. Theorem 1.1 implies the result for $GL_n(\mathbf{R})$. The case of $SL_4(\mathbf{R})$ is handled by the injectivity of the second restriction map in Proposition 3.4. The remaining cases $SL_n(\mathbf{R})$ for $n \geq 5$ follow from Theorem 1.1. □

REMARK 3.5. With the above arguments, we see that $H^3_{cb}(SL_n(\mathbf{C}))$ is at most one dimensional for all n.

4. p-adic fields and $GL_2(\mathbf{Q})$

The first step for Theorem 1.4 is the following "global" statement:

PROPOSITION 4.1. *A positive answer to Question 1.3 implies* $H^3_b(GL_2(\mathbf{Q})) = 0$.

PROOF. Write $G = GL_2(\mathbf{Q})$ and $\mathbf{P} = \mathbf{P}(\mathbf{Q})$. Since $\mathbf{P}$ is the quotient of G by an amenable subgroup, the complex

$$0 \longrightarrow \ell^\infty(\mathbf{P})^G \longrightarrow \ell^\infty(\mathbf{P}^2)^G \longrightarrow \ell^\infty(\mathbf{P}^3)^G \longrightarrow \cdots$$

computes $H^\bullet_b(G)$, see [**20**, 7.5.9]. Thus any class in $H^3_b(GL_2(\mathbf{Q})) = 0$ can be represented by a bounded G-invariant cocycle $\omega : \mathbf{P}^4 \to \mathbf{R}$ (which we may choose alternating). The cross-ratio maps the space of G-orbits of fourtuples of distinct points in $\mathbf{P}$ onto $\mathbf{Q}^{**} = \mathbf{Q} \setminus \{0, 1\}$ and ω descends to a map $f : \mathbf{Q}^{**} \to \mathbf{R}$. The cocycle relation $d\omega = 0$ translates into the Abel-Spence relation

$$f\left(\frac{y-x}{1-x}\right) - f(y) + f(x) - f\left(\frac{x}{y}\right) + f\left(\frac{x(1-y)}{y(1-x)}\right) = 0. \tag{4.1}$$

On the other hand (see [**24**] for details), one has for any infinite field F an exact sequence

$$K_3^{\mathrm{ind}}(F)_{\mathbf{Q}} \longrightarrow \mathbf{Q}[F^{**}]/R \xrightarrow{\ \delta\ } F^*_{\mathbf{Q}} \wedge F^*_{\mathbf{Q}} \longrightarrow K_2(F)_{\mathbf{Q}},$$

where $\delta([x]) = x \wedge (1-x)$, the subscript $\mathbf{Q}$ means tensorization $\otimes_{\mathbf{Z}}\mathbf{Q}$ and R is the subgroup generated by all

$$\left[\frac{y-x}{1-x}\right] - [y] + [x] - \left[\frac{x}{y}\right] + \left[\frac{x(1-y)}{y(1-x)}\right]$$

(with $x, y \in F^{**}$, $y \neq x, x-1$). In the case $F = \mathbf{Q}$, both K_2 and K_3^{ind} are torsion groups (see [**19**, 11.6] for K_2 and [**15**] for K_3; recall that K_3^{ind} is a quotient of K_3). Thus the map δ above is an isomorphism in the case at hand. Extending f linearly to $\mathbf{Q}[\mathbf{Q}^{**}]$, the relation (4.1) means that it descends to a map f_* modulo R. Now the induced map

$$\mathbf{Q}^* \wedge \mathbf{Q}^* \longrightarrow \mathbf{Q}^*_{\mathbf{Q}} \wedge \mathbf{Q}^*_{\mathbf{Q}} \xrightarrow{\ \delta^{-1}\ } \mathbf{Q}[\mathbf{Q}^{**}]/R \xrightarrow{\ f_*\ } \mathbf{R}$$

is of the form $\sum \alpha_{p,q} D_{p,q}$ as in Question 1.3 and thus a positive answer to Question 1.3 implies $f = 0$, whence $\omega = 0$. □

PROOF OF THEOREM 1.4. Let $\mathbf{A}$ be the ring of adèles of $\mathbf{Q}$ and $G_{\mathbf{A}} = PGL_2(\mathbf{A})$. Since $G = PGL_2(\mathbf{Q})$ embeds as a lattice in $G_{\mathbf{A}}$ (see *e.g.* Theorem 3.2.2 in [**18**, chap. I]), the restriction $H^3_{cb}(G_{\mathbf{A}}) \to H^3_b(G)$ is injective (Example 8.6.3 in [**20**]). The above proposition now implies $H^3_{cb}(G_{\mathbf{A}}) = 0$. Since $PGL_2(\mathbf{Q}_p)$ is a direct factor of $G_{\mathbf{A}}$, the inflation $H^3_{cb}(PGL_2(\mathbf{Q}_p)) \to H^3_{cb}(G_{\mathbf{A}})$ is injective by functoriality and thus $H^3_{cb}(PGL_2(\mathbf{Q}_p))$ vanishes. The inflation associated to $GL_2(\mathbf{Q}_p) \to PGL_2(\mathbf{Q}_p)$ is an isomorphism because the kernel is amenable [**20**, 8.5.2]. Now we have the vanishing of $\mathrm{H}^3_{\mathrm{cb}}(\mathrm{GL}_2(\mathbf{Q}_p))$; Proposition 3.4 yields the case of $\mathrm{GL}_3(\mathbf{Q}_p)$ and Theorem 1.1 handles $\mathrm{GL}_n(\mathbf{Q}_p)$ for $n \geq 4$. □

References

[1] Scot Adams, George A. Elliott, and Thierry Giordano, *Amenable actions of groups*, Trans. Amer. Math. Soc. **344** (1994), no. 2, 803–822.

[2] Spencer J. Bloch, *Higher regulators, algebraic K-theory, and zeta functions of elliptic curves* (1978 manuscript), American Mathematical Society, Providence, RI, 2000.

[3] Armand Borel, *Stable real cohomology of arithmetic groups*, Ann. Sci. École Norm. Sup. (4) **7** (1974), 235–272 (1975).

[4] Armand Borel and Nolan R. Wallach, *Continuous cohomology, discrete subgroups, and representations of reductive groups*, Princeton University Press, Princeton, N.J., 1980.

[5] Raoul Bott and Loring W. Tu, *Differential forms in algebraic topology*, Springer-Verlag, New York, 1982.

[6] Marc Burger and Nicolas Monod, *Bounded cohomology of lattices in higher rank Lie groups*, J. Eur. Math. Soc. (JEMS) **1** (1999), no. 2, 199–235.

[7] ———, *Continuous bounded cohomology and applications to rigidity theory (with an appendix by m. burger and a. iozzi)*, Geom. Funct. Anal. **12** (2002), no. 2, 219–280.

[8] ———, *On and around the bounded cohomology of* SL_2, Rigidity in dynamics and geometry (Cambridge, 2000) (Berlin), Springer, Berlin, 2002, pp. 19–37.

[9] Henri Cartan, *Périodicité des groupes d'homotopie stable des groupes classiques, d'après Bott*, Séminaire ÉNS, Paris (1961).

[10] Johan L. Dupont, *Bounds for characteristic numbers of flat bundles*, Algebraic topology, Aarhus 1978 (Proc. Sympos., Univ. Aarhus, Aarhus, 1978) (Berlin), Springer, Berlin, 1979, pp. 109–119.

[11] Sergei I. Gelfand and Yuri I. Manin, *Methods of homological algebra*, Springer-Verlag, Berlin, 1996, Translated from the 1988 Russian original.

[12] Étienne Ghys, *Groupes d'homéomorphismes du cercle et cohomologie bornée*, The Lefschetz centennial conference, Part III (Mexico City, 1984) (Providence, RI), Amer. Math. Soc., Providence, RI, 1987, pp. 81–106.

[13] Alexander B. Goncharov, *Geometry of configurations, polylogarithms, and motivic cohomology*, Adv. Math. **114** (1995), no. 2, 197–318.

[14] Mikhaïl Gromov, *Volume and bounded cohomology*, Inst. Hautes Études Sci. Publ. Math. (1982), no. 56, 5–99 (1983).

[15] Alexander Grothendieck, *Classes de Chern et représentations linéaires des groupes discrets*, Dix Exposés sur la Cohomologie des Schémas (Amsterdam), North-Holland, Amsterdam, 1968, pp. 215–305.

[16] Alessandra Iozzi, *Bounded cohomology, boundary maps, and rigidity of representations into* $\mathrm{Homeo}^+(\mathbf{S^1})$ *and* $\mathrm{SU}(1,n)$, Rigidity in Dynamics and Geometry, Cambridge 2000, Springer, Berlin, 2002.

[17] Nikolai V. Ivanov and Vladimir G. Turaev, *The canonical cocycle for the Euler class of a flat vector bundle*, Dokl. Akad. Nauk SSSR **265** (1982), no. 3, 521–524.

[18] Grigorii A. Margulis, *Discrete subgroups of semisimple Lie groups*, Springer-Verlag, Berlin, 1991.

[19] John Milnor, *Introduction to algebraic K-theory*, Princeton University Press, Princeton, N.J., 1971, Annals of Mathematics Studies, No. 72.

[20] Nicolas Monod, *Continuous bounded cohomology of locally compact groups*, Lecture Notes in Mathematics 1758, Springer, Berlin, 2001.

[21] Nicolas Monod and Yehuda Shalom, *Cocycle superrigidity and bounded cohomology for negatively curved spaces*, preprint.

[22] ———, *Orbit equivalence rigidity and bounded cohomology*, to appear in Annals of Mathematics.

[23] Richard P. Savage, Jr., *The space of positive definite matrices and Gromov's invariant*, Trans. Amer. Math. Soc. **274** (1982), no. 1, 239–263.

[24] Andrei A. Suslin, *K_3 of a field, and the Bloch group*, Trudy Mat. Inst. Steklov. **183** (1990), 180–199, 229, Translated in Proc. Steklov Inst. Math. **1991**, Vol. 183 no. 4, 217–239, Galois theory, rings, algebraic groups and their applications.
[25] Robert J. Zimmer, *Ergodic theory and semisimple groups*, Birkhäuser Verlag, Basel, 1984.

UNIVERSITY OF CHICAGO
E-mail address: monod@math.uchicago.edu

Contemporary Mathematics
Volume **347**, 2004

Spectral Theory of Certain Arithmetic Graphs

Hirofumi Nagoshi

ABSTRACT. We consider the infinite graphs constructed from principal congruence subgroups over function fields. These graphs are known to be Ramanujan. In this paper we review some spectral properties of these graphs, comparing them with the case of regular finite graphs. Our concerns are the Selberg trace formulas, the Ihara-Selberg zeta functions, limit distributions of eigenvalues, and the prime geodesic theorems.

1. Introduction

Ramanujan graphs were first introduced in [**LPS**], and an explicit construction of an infinite family of Ramanujan graphs were given by Lubotzky-Phillips-Sarnak [**LPS**] and Margulis [**Ma**] independently. Their constructions depend on deep results in number theory. They used p-adic number fields $\mathbb{Q}_p$, the arithmetic of certain quaternion algebras and the so-called Ramanujan conjecture for Hecke operators acting on holomorphic cusp forms of weight 2, which has been proved. See also [**Lu**] and [**Sa2**], and compare these books.

As a related topic to the above and the following, there is a conjecture due to Selberg, which has not yet been proved in general. Selberg conjectured that for congruence subgroups Γ of $PSL(2,\mathbb{Z})$, the first positive eigenvalues $\lambda_1(\Gamma)$ of the Laplacians on $\Gamma\backslash\mathbb{H} = \Gamma\backslash PSL(2,\mathbb{R})/PSO(2,\mathbb{R})$, where $\mathbb{H}$ is the upper half plane, satisfy $\lambda_1(\Gamma) \geq 1/4$. We note that in view of representation theory, the Selberg conjecture and the Ramanujan conjecture above can be formulated in the same way and that they are contained in certain more general conjecture (Generalized Ramanujan Conjecture). While those Riemann surfaces $\Gamma\backslash PSL(2,\mathbb{R})/PSO(2,\mathbb{R})$ are not compact, we have also arithmetic compact surfaces $\Gamma'\backslash PSL(2,\mathbb{R})/PSO(2,\mathbb{R})$, which are constructed in terms of quaternion algebras (see [**Ka**]). By virtue of the so-called Jacquet-Langlands correspondence, the Selberg conjecture implies that the first positive eigenvalues $\lambda_1(\Gamma')$ of these arithmetic compact surfaces are not less than $1/4$. We note that Ramanujan graphs constructed in [**LPS**] and [**Ma**] above can be seen as analogues of these arithmetic compact surfaces $\Gamma'\backslash PSL(2,\mathbb{R})/PSO(2,\mathbb{R})$ to $PGL(2,\mathbb{Q}_p)$ and holomorphic case. For this account see Section 4.4–Section 7.5 in [**Lu**].

2000 *Mathematics Subject Classification.* Primary 11F72, 11F03.

Key words and phrases. Ramanujan graph, infinite graph, function field, Selberg trace formula, Ihara zeta function, eigenvalues of graphs.

Generalizing the notion of Ramanujan graphs, which were given in [**LPS**] for regular finite graphs, Morgenstern [**M1**] introduced *Ramanujan diagrams*, which may be infinite but have finite volumes. Moreover, Morgenstern [**M2**] gave an explicit example of a family of infinite Ramanujan diagrams. His graph is a quotient graph $\Gamma(A)\backslash\mathcal{T}$, where $\Gamma(A)$ is a principal congruence subgroups of $PGL(2,\mathbb{F}_q[t])$ in $PGL(2,\mathbb{F}_q((1/t)))$ and $\mathcal{T}$ is a $(q+1)$-regular tree (see Section 2 and Section 3). In the proof that $\Gamma(A)\backslash\mathcal{T}$ is Ramanujan, he used certain conjecture which had been proved by Drinfeld and which corresponds to the Selberg conjecture above in view of the number-theoretic analogy between number fields and function fields over finite fields. See also the construction of Ramanujan graphs in [**M3**].

The purpose of this survey paper is to give a comparison between regular finite graphs and Morgenstern's infinite graphs from the viewpoint of some spectral properties. Our concerns are the Selberg trace formulas, the Ihara-Selberg zeta functions, limit distributions of eigenvalues, and the prime geodesic theorems for those graphs. These results for Morgenstern's infinite graphs $\Gamma(A)\backslash\mathcal{T}$ are obtained in [**N1**], [**N2**] and [**N3**].

Originally, the Selberg trace formula was formulated in [**S1**] and [**S2**] (cf. [**H1**] [**H2**]) for $SL(2,\mathbb{R})$, and many surveys can be found (e.g. [**H3**] [**Ve**]). This formula is an equality of two kinds of sums; one sum is over the conjugacy classes of a discrete group Γ, and the other is over the spectra of the Laplacian for Γ. The Selberg trace formula can be formulated also in terms of automorphic representations (see e.g. [**GGP**] [**GJ**] [**Ge**]).

In Section 5 we discuss the Selberg trace formulas of graphs. In the case of regular finite graphs, trace formulas were studied in [**Ah**] [**Br**] [**T3**] [**VN**], etc. For a comparison between original trace formulas due to Selberg and that of regular finite graphs, see [**T2**]. Generally speaking, trace formulas for infinite graphs have been not much investigated.

Trace formulas are powerful tools in various fields, and for applications it is often important to find explicit expressions of these formulas. In [**N1**] we discussed the trace formulas for $\Gamma(A)\backslash\mathcal{T}$ above and obtained explicit expressions of them. In comparison with the case of regular finite graphs, these formulas have more terms contributed by parabolic elements and continuous spectra. The continuous spectra are furnished by the Eisenstein series in Section 4. We will show three applications of those formulas in later sections.

The Selberg zeta function $Z_\Gamma(s)$ was first introduced by Selberg [**S1**] for a Fuchsian subgroup Γ of $PSL(2,\mathbb{R})$, and he found some properties of $Z_\Gamma(s)$ through his trace formula attached to Γ. We note that the Selberg zeta function can be considered as a special case of the Ruelle dynamical zeta function. For this account and the study of $Z_\Gamma(s)$ via the theory of dynamical systems, see e.g. [**La**] [**Ru**] [**M**]. Ihara [**Ih**] defined, for some discrete subgroups of $PGL(2,\mathbb{Q}_p)$, Selberg-type zeta functions, which now can be interpreted as the zeta functions attached to $(p+1)$-regular finite graphs. For other zeta functions and L-functions attached to graphs, see [**ST1**] [**ST2**].

In Section 6 we study the Ihara-Selberg zeta functions of graphs and Ihara's formula. Our approach is via the Selberg trace formulas. First we recall the case of regular finite graphs, and next we define the zeta function $Z_{\Gamma(A)}(s)$ of Morgenstern's infinite graph $\Gamma(A)\backslash\mathcal{T}$ and give a certain determinant expression of $Z_{\Gamma(A)}(s)$, in which the contribution of continuous spectra are given in terms of poles of the scattering determinant in Section 4. By this expression we find that the zeta function $Z_{\Gamma(A)}(s)$ is a rational function in q^{-s}.

In Section 7 we discuss limit density of eigenvalues of a family of graphs. Recently, motivated by quantum chaos, various statistics of eigenvalues of graphs have been investigated (see e.g. [**T4**] [**T5**]). For $\Gamma(A)\backslash\mathcal{T}$ we consider two certain families and find that the spectral density of the $(q+1)$-tree and the semi-circle distribution appear as the limit densities of eigenvalues.

In Section 8 we discuss the prime geodesic theorems of graphs, which are analogues of the prime number theorem. For $\Gamma(A)$ the expected result is obtained.

The author would like to thank the organizers of this conference for their hospitality. The author would also like to thank the referee for his helpful comments.

2. Preliminaries

The aim of this section is to exhibit certain arithmetic infinite graphs mentioned in Section 1, which we will treat in this paper.

Let $\mathbb{F}_q$ be the finite field with q elements, where we assume that q is an odd prime power, $\mathbb{F}_q[t]$ be the ring of polynomials in t over $\mathbb{F}_q$, and $k = \mathbb{F}_q(t)$ be its quotient field. The metric in k which we consider is given, for $a = f(t)/g(t) \in k$, by $|a|_\infty := q^{-(\deg(g)-\deg(f))}$. Let k_∞ be the completion of k with respect to this metric, and r_∞ its maximal compact subring $\{x \in k_\infty \mid |x|_\infty \leq 1\}$. Then k_∞ is the field $\mathbb{F}_q((1/t))$ of Laurent series in $1/t$ over $\mathbb{F}_q$, and r_∞ is the ring $\mathbb{F}_q[[1/t]]$ of Taylor series in $1/t$ over $\mathbb{F}_q$. If an element x in k_∞ is written as $\sum_{i=n}^{\infty} a_i t^{-i}$ ($a_i \in \mathbb{F}_q, a_n \neq 0, n \in \mathbb{Z}$), then $|x|_\infty = q^{-n}$. Throughout this paper we put

$$G := PGL(2, k_\infty) = GL(2, k_\infty)/k_{\infty,2}^\times,$$
$$K := PGL(2, r_\infty) = GL(2, r_\infty)/r_{\infty,2}^\times,$$

where $k_{\infty,2}^\times := \{aI_2 \mid a \in k_\infty^\times\}$, $k_\infty^\times := k_\infty - \{0\}$, $r_{\infty,2}^\times := \{aI_2 \mid a \in r_\infty^\times\}$, $r_\infty^\times := \{x \in k_\infty \mid x, x^{-1} \in r_\infty\} = \{x \in k_\infty \mid |x|_\infty = 1\}$ and I_2 is the 2×2 identity matrix. Note that K is a maximal compact subgroup of G.

We note that these k, k_∞, G and K are seen as the counterparts of $\mathbb{Q}$, $\mathbb{R}$, $PSL(2,\mathbb{R})$ and $PSO(2,\mathbb{R})$ respectively, from the viewpoint of the number-theoretic analogy between algebraic number fields and function fields over finite fields.

In this paper we will deal with connected undirected graphs. For a graph X we denote by $V(X)$ the set of vertices of X and by $E(X)$ the set of edges of X.

Next we introduce one example of symmetric spaces associated with G. We recall that $PSL(2,\mathbb{R})/PSO(2,\mathbb{R})$ is identified with the upper half plane $\mathbb{H} = \{z \in \mathbb{C} \mid \mathrm{Im}(z) > 0\}$. On the other hand, it is known that G/K can be regarded as the $(q+1)$-regular tree $\mathcal{T}_{q+1} = \mathcal{T}$. This is the so-called Bruhat-Tits building attached to G (see [**Se**, II.1.1] [**S**]). More precisely, the set of vertices $V(\mathcal{T})$ of the tree $\mathcal{T}$ corresponds to the right coset space G/K, which is represented by the set of matrices

$$\left\{ \begin{pmatrix} t^n & t^{n+1}x \\ 0 & 1 \end{pmatrix} \middle| \, n \in \mathbb{Z}, x \in \mathbb{F}_q[t] \right\},$$

and a vertex gK ($g \in G$) are connected to the $q+1$ vertices gs_iK ($i = 1, \cdots, q+1$), where

$$\{s_1, \cdots, s_{q+1}\} = \left\{ \begin{pmatrix} \frac{1}{t} & \alpha \\ 0 & 1 \end{pmatrix} \middle| \, \alpha \in \mathbb{F}_q \right\} \cup \left\{ \begin{pmatrix} t & 0 \\ 0 & 1 \end{pmatrix} \right\}.$$

Hence the group G acts on the tree $\mathcal{T}$ from the left as a group of automorphisms. The boundary $\partial\mathcal{T}$ of $\mathcal{T}$ is defined by the set of equivalence classes of half-lines, with two half-lines being equivalent if their difference is included in some finite graph. Note that $\partial\mathcal{T}$ can be identified with $\mathbb{P}^1(k_\infty)$ by the correspondence between the elements in $\partial\mathcal{T}$ except the half-line $\{ \left(\begin{smallmatrix} t^n & 0 \\ 0 & 1 \end{smallmatrix}\right) | \, n = 1, 2, \cdots \}$ and $\mathbb{P}^1(k_\infty) - \{\infty\}$ via the limit of the $(1,2)$-entry of the 2×2 matrix $g_m \cdots g_2 g_1 I_2 \in V(\mathcal{T})$, where $g_i \in \{s_1, s_2, \cdots, s_{q+1}\}$ $(i = 1, 2, \cdots, m)$, as $m \to \infty$. The action above of G on $\mathcal{T}$ induces an action of G on the boundary $\partial\mathcal{T}$, which is just the usual fractional linear transformation of G on $\mathbb{P}^1(k_\infty)$, so that $g \in G$ has at most two fixed points in $\partial\mathcal{T}$.

From the view of actions on $\mathcal{T}$ and $\partial\mathcal{T}$, we can classify elements in G which act without inversions on $\mathcal{T}$, as follows:

(1) identity
(2) hyperbolic elements: they have no fixed vertices on $\mathcal{T}$. (Then it is known that it has two fixed points on $\partial\mathcal{T}$.)
(3) elliptic elements: they have fixed vertices on $\mathcal{T}$ and no fixed points on $\partial\mathcal{T}$.
(4) parabolic elements: they have fixed vertices on $\mathcal{T}$ and a fixed point on $\partial\mathcal{T}$.
(5) split hyperbolic elements: they have fixed vertices on $\mathcal{T}$ and two fixed points on $\partial\mathcal{T}$.

The discrete subgroup of G which we will consider in this paper is a principal congruence subgroup $\Gamma(A)$, $A \in \mathbb{F}_q[t]$, of $PGL(2, \mathbb{F}_q[t])$, which is defined by

$$\begin{aligned} \Gamma(A) &:= \ker\left(PGL(2, \mathbb{F}_q[t]) \longrightarrow PGL(2, \mathbb{F}_q[t]/A\mathbb{F}_q[t])\right) \\ &= \{\gamma \in PGL(2, \mathbb{F}_q[t]) \mid \text{for some representative } \tilde{\gamma} \text{ of } \gamma, \tilde{\gamma} \equiv \left(\begin{smallmatrix} 1 & 0 \\ 0 & 1 \end{smallmatrix}\right) \bmod A\}. \end{aligned} \tag{2.1}$$

This is a normal subgroup of $PGL(2, \mathbb{F}_q[t]) = \Gamma(1)$.

The discrete group $\Gamma(A)$ acts without inversions on $\mathcal{T}$, so that it naturally gives rise to the quotient graph $\Gamma(A)\backslash\mathcal{T}$. This graph $\Gamma(A)\backslash\mathcal{T}$ has the shape of the union of a finite graph and finitely many half-lines (cusps) (see [**Se**, II.1.6] [**M2**]).

3. Adjacency Operator and Ramanujan Diagrams

In this section, following [**M1**] and [**M2**], we give the definition of a *diagram* and a *Ramanujan diagram*, and find that our graph $\Gamma(A)\backslash\mathcal{T}$ in Section 2 is a $(q+1)$-regular infinite Ramanujan diagram.

DEFINITION 3.1. *A diagram is a triple $D = (V, E, w)$, where (V, E) is a undirected graph and the weight function $w : V \cup E \to (0, 1]$ satisfies*

(1) $\sum_{v \in V} w(v) < \infty$
(2) *For any $e = (u, v) \in E$, $w(e) \geq \max\{w(u), w(v)\}$.*

For an edge $e = (u, v) \in E$ we put $\theta(u, v) := w(e)/w(u)$, and D is called k-regular if for every $u \in V$, $\sum_{(u,v) \in E} \theta(u, v) = k$.

In the following we assume that $D = (V, E, w)$ is a k-regular diagram. Put

$$L^2(D) := \left\{ f : V \to \mathbb{C} \,\middle|\, \sum_{v \in V} |f(v)|^2 \, w(v) < \infty \right\} \tag{3.1}$$

and in $L^2(D)$ we consider the inner product defined by

$$\langle f, g \rangle_D := \sum_{v \in V} f(v)\overline{g(v)} w(v).$$

We define the operator S on $L^2(D)$ by

$$(Sf)(u) := \sum_{(u,v)\in E} \theta(u,v)f(v) \tag{3.2}$$

(see [**M1**] [**M2**]). Then $S : L^2(D) \to L^2(D)$ is bounded, the norm $\|S\|$ of S satisfies $\|S\| \leq k$, and S is Hermitian.

Let

$$L_0^2(D) := \left\{ f \in L^2(D) \,\middle|\, \sum_{v\in V} f(v)w(v) = 0 \right\},$$

which is the orthogonal complement in $L^2(D)$ of the constant function, and put

$$\begin{aligned}\lambda(D) &:= \sup\{\langle Sf, f\rangle_D \mid f \in L_0^2(D), \langle f, f\rangle_D = 1\} \\ &= \sup \operatorname{Spec}(S|_{L_0^2(D)}).\end{aligned}$$

Then Morgenstern [**M1**] showed that if D is infinite, then $\lambda(D) \geq 2\sqrt{k-1}$. This result is an analog of Alon-Boppana's theorem which treats an infinite family of k-regular finite graphs X_i $(i = 1, 2, \cdots)$ such that $\#V(X_i) \to \infty$ as $i \to \infty$.

Generalizing the notion of *Ramanujan graphs*, which is defined for regular finite graphs, Morgenstern [**M1**] introduced the following definition:

DEFINITION 3.2. *If a k-regular infinite diagram D satisfies $\lambda(D) = 2\sqrt{k-1}$, then D is called a Ramanujan diagram.*

Next we go back to our graph $\Gamma\backslash\mathcal{T}$, where $\Gamma = \Gamma(A)$ in (2.1). We note that the quotient graph $\Gamma\backslash\mathcal{T} = \Gamma\backslash G/K$ can be made into a measure space induced from a Haar measure of G. If we normalize the volume of K to be 1, then we have the atomic measure m on $\Gamma\backslash\mathcal{T}$ such that for a vertex $v \in V(\Gamma\backslash\mathcal{T})$

$$m(v) = (\#\Gamma_v)^{-1},$$

where $\Gamma_v \subset \Gamma$ denotes the stabilizer of v (see [**Se**, II.1.5] [**BL**]). For an edge $e \in E(\Gamma\backslash\mathcal{T})$ we put

$$m(e) := (\#\Gamma_e)^{-1},$$

where $\Gamma_e \subset \Gamma$ is the stabilizer of e. For an edge $e = (v,u) \in E(\Gamma\backslash\mathcal{T})$ we have $\Gamma_e \subset \Gamma_v$. Since the full modular group $\Gamma(1) = PGL(2, \mathbb{F}_q[t])$ is a non-uniform lattice in G (see [**Se**, II.1.6]), $\Gamma(A)$, which is of finite index in $\Gamma(1)$, is also a non-uniform lattice in G and so satisfies $\sum_{v\in V(\Gamma(A)\backslash\mathcal{T})} m(v) < \infty$.

Thus we find that our graph $\Gamma\backslash\mathcal{T}$ has the structure of a $(q+1)$-regular infinite diagram $(V(\Gamma\backslash\mathcal{T})), E(\Gamma\backslash\mathcal{T}), m)$ (see [**M2**, Section 3] [**E1**, p. 451]).

We now introduce the space

$$L^2(\Gamma\backslash\mathcal{T}) :=$$

$$\left\{ f : V(\mathcal{T}) \to \mathbb{C} \,\middle|\, f(\gamma v) = f(v) \text{ for } \gamma \in \Gamma, v \in V(\mathcal{T}), \sum_{v\in V(\Gamma\backslash\mathcal{T})} |f(v)|^2 m(v) < \infty \right\}$$

A function f on $V(\mathcal{T})$ satisfying $f(\gamma v) = f(v)$ for all $\gamma \in \Gamma$ and $v \in V(\mathcal{T}) = G/K$, which is called an automorphic function for Γ, is just a function on the set of vertices of the quotient graph $\Gamma\backslash\mathcal{T}$, and hence $L^2(\Gamma\backslash\mathcal{T})$ is just the space in (3.1) when $D = \Gamma\backslash\mathcal{T}$. In the space $L^2(\Gamma\backslash\mathcal{T})$ we consider the inner product defined by

$$\langle f, g\rangle := \sum_{v\in V(\Gamma\backslash\mathcal{T})} f(v)\overline{g(v)}m(v).$$

As usual, the adjacency operator for $\mathcal{T}$ is defined by

$$(Tf)(u) := \sum_{(u,v)\in E(\mathcal{T})} f(v) \qquad \text{for } f\colon V(\mathcal{T}) \to \mathbb{C}. \tag{3.3}$$

It induces an operator on automorphic functions for Γ, and this induced operator can be represented as

$$(Tf)(u) = \sum_{e=(u,v)\in E(\Gamma\backslash\mathcal{T})} \frac{m(e)}{m(u)} f(v) \qquad \text{for } f\colon V(\Gamma\backslash\mathcal{T}) \to \mathbb{C},$$

which is just the operator S in (3.2).

Then using Drinfeld's result, Morgenstern [**M2**, Theorem 2.1] showed the following theorem:

THEOREM 3.3. *The diagram $\Gamma(A)\backslash\mathcal{T}$ ($A \in \mathbb{F}_q[t]$) is a $(q+1)$-regular infinite Ramanujan diagram.*

More precisely, Morgenstern showed that if λ is an eigenvalue of T on $L^2(\Gamma\backslash\mathcal{T})$ and $\lambda \neq \pm(q+1)$ ($\pm(q+1)$ are trivial eigenvalues) then

$$|\lambda| \le 2\sqrt{q}, \tag{3.4}$$

while it is known that the continuous spectra of T is the segment $[-2\sqrt{q}, 2\sqrt{q}]$ and is furnished by the Eisenstein series in Section 4.

4. Eisenstein Series

In this section we recall the Eisenstein series attached to $\Gamma(A)$ and their basic properties. Let Γ be a principal subgroup $\Gamma(A)$ of $PGL(2, \mathbb{F}_q[t])$, $\mu = \mu(A)$ be the number of inequivalent cusps of $\Gamma\backslash\mathcal{T}$, and $\{\kappa_1 = \infty, \kappa_2, \cdots, \kappa_\mu\} \subset \mathbb{P}^1(k_\infty) = \partial\mathcal{T}$ be a complete set of inequivalent cusps. Take an element $\tilde{\kappa_i} \in PGL(2, \mathbb{F}_q[t])$ such that $\tilde{\kappa_i}\infty = \kappa_i$. We denote the stabilizer in Γ of κ_i by Γ_{κ_i}.

Now we define an important function $\psi_s(g)$ ($g \in G, s \in \mathbb{C}$) by

$$\psi_s(g) := |\det g|_\infty^s h\left((0,1)g\right)^{-2s},$$

where $h\left((x,y)\right) := \sup\{|x|_\infty, |y|_\infty\}$ and (x, y) denotes a 1×2-matrix. We find that the function $\psi_s(g)$ is K-right invariant, so $\psi_s(g)$ is a function on $V(\mathcal{T})$, and that it is N-left invariant, where $N = \{\left(\begin{smallmatrix}1 & x\\ 0 & 1\end{smallmatrix}\right) \in G\}$. We can check that $\psi_s(g)$ satisfies

$$(T\psi_s)(g) = \left(q^s + q^{1-s}\right)\psi_s(g), \tag{4.1}$$

where T is the adjacency operator in (3.3).

Then the Eisenstein series $E_i(g, s)$ for a cusp κ_i is defined by

$$E_i(g, s) := \sum_{\gamma\in\Gamma_{\kappa_i}\backslash\Gamma} \psi_s(\tilde{\kappa}_i^{-1}\gamma g),$$

where $g \in V(\mathcal{T}) = G/K$ and $s \in \mathbb{C}$ with $\mathrm{Re}\,(s) > 1$. By (4.1) we have

$$(TE_i)(g, s) = (q^s + q^{1-s})E_i(g, s).$$

The series $E_i(g,s)$ is an automorphic function for Γ, and hence it can be expanded as a Fourier series at each cusp κ_j. In the case of a principal congruence group $\Gamma(A)$, Li [**L1**] computed explicitly the Fourier series expansion of $E_i(g,s)$ for each cusp κ_j, which can be expressed in terms of L-functions associated to Dirichlet's characters χ on $\mathbb{F}_q[t]$ mod A. In particular, the constant term of the Fourier series of $E_i(g,s) = E_i\left(\left(\begin{smallmatrix} t^n & x \\ 0 & 1 \end{smallmatrix}\right), s\right)$ at a cusp κ_j is of the form

$$\delta_{ij} q^{ns} + \varphi_{ij}(s) q^{n(1-s)},$$

where δ_{ij} is Kronecker's δ. We now define the $\mu \times \mu$-matrix $\Phi(s) := (\varphi_{ij}(s))$, which is called the scattering matrix of Γ. Its determinant $\varphi(s) := \det \Phi(s)$ is called the scattering determinant of Γ.

It is understood that the determinant $\varphi(s)$, in particular the set of its poles, plays an important role. The function $\varphi(s)$ controls the Eisenstein series in the sense that if $\varphi(s)$ is analytic at some point, then so are all the Eisenstein series. Moreover, $\varphi(s)$ appears in the Selberg trace formula described in Section 5 as the contribution of the continuous spectra of $\mathcal{T}$.

As for $E_i(g,s), \varphi_{ij}(s)$ and $\Phi(s)$, the next result is known in [**L1**, p. 241, p. 242, p. 249]:

THEOREM 4.1. *Let $\Gamma = \Gamma(A), A \in \mathbb{F}_q[t]$. Then the function $\varphi_{ij}(s)$ is a rational function in q^{-2s}, and for fixed $g \in V(\mathcal{T})$ the function $E_i(g,s)$ is a rational function in q^{-s}. Moreover $\varphi_{ij}(s)$ and $E_i(g,s)$ with $g \in V(\mathcal{T})$ fixed are holomorphic on* $\mathrm{Re}(s) \geq 1/2$ *except for simple poles at $s = 1 + n\pi i/\log q\,(n \in \mathbb{Z})$. The matrix $\Phi(s)$ is symmetric and satisfies the functional equation*

$$\Phi(s)\,\Phi(1-s) = I_\mu,$$

where I_μ is the $\mu \times \mu$ identity matrix.

In [**N2**] we give more detailed information about the poles of $\varphi(s)$ for $\Gamma(A)$ by a different technique from [**L1**]:

THEOREM 4.2. *For $A \in \mathbb{F}_q[t]$ let $\hat{A}^\times$ be the set of characters of $(\mathbb{F}_q[t]/A\mathbb{F}_q[t])^\times / \mathbb{F}_q^\times$. Let $L(s,\chi)$ be the Dirichlet L-function associated with a character $\chi \in \hat{A}^\times$. Then the poles of $\varphi(s)$ of $\Gamma(A)$ are contained within the zeros of the function*

$$(q^{2s} - q^2) \prod_{\chi \in \hat{A}^\times} L(2s,\chi)^h,$$

where $h := \mu/r, r := \phi(A)/(q-1)$ and $\phi(A)$ is the Euler totient function. Here we count zeros and poles with multiplicity.

By Theorem 4.1 the scattering determinant $\varphi(s)$ is a rational function in q^{2s}, so we put

$$\varphi(s) = c\frac{(q^{2s} - qa_1)(q^{2s} - qa_2)\cdots(q^{2s} - qa_m)}{(q^{2s} - qb_1)(q^{2s} - qb_2)\cdots(q^{2s} - qb_n)}, \tag{4.2}$$

where $c, a_i\ (i = 1, \cdots, m), b_j\ (j = 1, \cdots, n)$ are constants and we assume that the right hand side is written to be irreducible. We note that by the functional equation $\varphi(s)\varphi(1-s) = 1$ in Theorem 4.1 the set of poles of $\varphi(s)$ is one-to-one correspond to the set of its zeros.

5. Trace Formulas

The aim of this section is to discuss Selberg's trace formulas of a regular finite graph and an infinite graph $\Gamma(A)\backslash\mathcal{T}$. Selberg's trace formula is the equality of two sums; one sum is over conjugacy classes, which have geometric information, and the other sum is over spectra or representations.

Let $\mathcal{T}$ be a $(q+1)$-regular tree, where $q \in \mathbb{N}$ and $q \geq 2$, and d be the distance function on $\mathcal{T}$. For a point pair invariant function $k(v,v') = k(d(v,v'))$, $v, v' \in V(\mathcal{T})$, we define the integral operator L_k on functions on $V(\mathcal{T})$ by

$$(5.1)\qquad (L_k f)(v) := \sum_{v' \in V(\mathcal{T})} k(v,v')f(v').$$

Assume f is a function on $V(\mathcal{T})$ satisfying $Tf = \lambda f$, where T is the adjacency operator in (3.3) and $\lambda \in \mathbb{C}$. Then it is known that there exists a constant $\Lambda(\lambda)$ which depends only on k and λ (not on the individual function f) and which satisfies $L_k f = \Lambda(\lambda) f$. We call the transformation $k \mapsto \Lambda(\lambda)$ the Selberg transform. The Selberg transform and its inverse transform is explicitly computed as in the following.

Proposition 5.1. *If we write $\lambda = q^s + q^{1-s}$, $s = \frac{1}{2} + ir$ $(s, r \in \mathbb{C})$ and $\Lambda(\lambda) = h(r)$, then the Selberg transform of a point pair invariant function k is given by the following formulas:*

$$\Lambda(\lambda) = h(r) = \sum_{n=-\infty}^{\infty} c(n) q^{inr},$$

$$c(n) = q^{\frac{|n|}{2}} \left(k(|n|) + \sum_{m=1}^{\infty} (q^m - q^{m-1})\, k(|n| + 2m) \right), \qquad n \in \mathbb{Z}.$$

The inverse transform is given by

$$c(n) = \frac{\log q}{2\pi} \int_{-\frac{\pi}{\log q}}^{\frac{\pi}{\log q}} h(r) q^{-inr} dr$$

$$(c(n) = c(-n) \text{ since } h(r) \text{ is even}),$$

$$k(|n|) = q^{-\frac{|n|}{2}} \left(c(|n|) - (q-1) \sum_{m=1}^{\infty} c(|n| + 2m) q^{-m} \right), \qquad n \in \mathbb{Z}.$$

Next we recall Selberg's trace formula of a regular finite graph (see e.g. [**Ah**] [**Br**] [**T1**] [**VN**]). Let X be a $(q+1)$-regular finite graph $(q \in \mathbb{N}, q \geq 2)$, then the $(q+1)$-regular tree $\mathcal{T}$ is the universal covering space of X. We have $X = \Gamma\backslash\mathcal{T}$, where Γ is the fundamental group of X. Note that Γ is *strictly hyperbolic*, that is, if $\gamma \in \Gamma$ is not identity then γ acts on $\mathcal{T}$ without fixed points (i.e., γ is hyperbolic), and that Γ is a free group. A hyperbolic element γ induces a shift by $\deg\gamma$ on $\mathcal{T}$, where

$$(5.2)\qquad \deg\gamma := \min\{d(v, \gamma v) \mid v \in V(\mathcal{T})\}.$$

We call a hyperbolic element $\gamma_0 \in \Gamma$ *primitive* if γ_0 generates the centralizer Γ_{γ_0}, which is a cyclic group. So any hyperbolic element γ of Γ is written in the form $\gamma = \gamma_0^k$ $(k \geq 1)$ with a certain uniquely determined primitive hyperbolic element γ_0. We denote by $\mathfrak{P}_\Gamma$ the set of primitive hyperbolic conjugacy classes of Γ.

Let a point pair invariant function $k(v,v') = k(d(v,v'))$ satisfy $k(d(v,o)) \in L^1(V(\mathcal{T}))$, where o is the fixed origin of $\mathcal{T}$. Note that $k(d(v,o)) \in L^1(V(\mathcal{T}))$ if and only if $\sum_{n\geq 1} k(n)q^n < \infty$. Then when L_k in (5.1) operates on Γ-invariant functions f on $V(\mathcal{T})$, i.e., functions on $V(X) = V(\Gamma\backslash\mathcal{T})$, we have

$$(5.3) \qquad (L_k f)(v) = \sum_{v' \in V(\Gamma\backslash\mathcal{T})} K(v,v')f(v')m(v'),$$

where $K(v,v') := \sum_{\gamma\in\Gamma} k(v,\gamma v')$ and in the present case $m(v') = 1$ for any $v' \in V(\Gamma\backslash\mathcal{T})$. We calculate the trace of this operator L_k in two ways and we have :

THEOREM 5.2. *Let X be a $(q+1)$-regular finite graph $(q\in\mathbb{N}, q\geq 2)$. For an eigenvalue λ_j $(j = 1,2,\cdots,\#V(X))$ of the adjacency operator on $L^2(X)$, we write $\lambda_j = q^{s_j} + q^{1-s_j}$ and $s_j = 1/2 + ir_j$. Assume that a sequence $c(n) \in \mathbb{C}$ $(n\in\mathbb{Z})$ satisfies $c(n) = c(-n)$ and $\sum_{n\geq 0} q^{n/2}|c(n)| < \infty$. Then*

$$\sum_{j=1}^{\#V(X)} h(r_j) = \#V(X)\, k(0) + \sum_{\{P\}\in\mathfrak{P}_\Gamma} \sum_{l=1}^{\infty} \frac{\deg P}{q^{\frac{l \deg P}{2}}} c(l \deg P).$$

Here the functions $h(\cdot)$ and $k(\cdot)$ are determined by $c(\cdot)$ via the Selberg transform in Proposition 5.1.

Next we discuss the Selberg trace formula attached to the infinite graph $\Gamma(A)\backslash\mathcal{T}$. Let q be an odd prime power. We again consider the integral operator in (5.3). However, we find that the kernel function $K(v,v') = \sum_{\gamma\in\Gamma} k(v,\gamma v')$ becomes too large as v and v' go to the same cusp for an ordinary function k (see [**N1**, Lemma 3.3]) and hence that we cannot take the trace of the operator L_k. So we modify the kernel $K(v,v')$ by using the Eisenstein series to obtain the following :

PROPOSITION 5.3. *Let $\Gamma = \Gamma(A)\,(A \in \mathbb{F}_q[t])$ and μ be the number of inequivalent cusps of $\Gamma\backslash\mathcal{T}$. Assume that a sequence $c(n) \in \mathbb{C}\,(n\in\mathbb{Z})$ satisfies $c(n) = c(-n)$ and $\sum_{n\geq m} q^{n/2}|c(n)| = O(q^{-m})$ for $m\in\mathbb{N}$. We define the kernel function $H(v,v')\,(v,v'\in V(\mathcal{T}))$ by*

$$H(v,v') := \sum_{i=1}^{\mu} \frac{qb\log q}{4\pi q^a} \int_{-\frac{\pi}{\log q}}^{\frac{\pi}{\log q}} h(r)E_i(v,s)\overline{E_i(v',s)}dr,$$

where $s = \frac{1}{2} + ir$, $a := \deg(A)$, and b is $q-1$ if $\deg(A) = 0$ and 1 otherwise. Then

$$\hat{K}(v,v') := K(v,v') - H(v,v')$$

is bounded on $V(\Gamma\backslash\mathcal{T}) \times V(\Gamma\backslash\mathcal{T})$. Moreover, if we denote by $\mathcal{E}$ the set of all $L^2(\Gamma\backslash\mathcal{T})$-eigenfunctions of T, then for any $g\in\mathcal{E}$ we have

$$(5.4) \qquad \sum_{v'\in V(\Gamma\backslash\mathcal{T})} H(v,v')g(v')m(v') = 0.$$

Noting that by this proposition and $\sum_{v\in V(\Gamma\backslash\mathcal{T})} m(v) < \infty$ we have

$$\sum_{v\in V(\Gamma\backslash\mathcal{T})} \hat{K}(v,v)m(v) < \infty,$$

we consider, instead of L_k in (5.3), the integral operator

$$(Lf)(v) := \sum_{v'\in V(\Gamma\backslash\mathcal{T})} \hat{K}(v,v')f(v')m(v').$$

Moreover, by (5.4) we find that the operators L and L_k have the same effect on eigenfunctions $g \in \mathcal{E}$. Thus we obtain

$$\sum_{v \in V(\Gamma \backslash \mathcal{T})} \hat{K}(v,v)m(v) = \sum_{j=1}^{M} \Lambda(\lambda_j), \tag{5.5}$$

where $\{\lambda_j \mid j = 1, 2, \cdots, M\}$ is the set of eigenvalues of T for Γ. See the argument in [**H2**, Chapter 6, Section 13]. Now, as usual, we divide the left-hand side of (5.5) into the sum over conjugacy classes of Γ.

The classification of conjugacy classes of $\Gamma(1) = PGL(2, \mathbb{F}_q[t])$ is known in [**Ak**] analogously to the case of $PSL(2, \mathbb{Z})$ (cf. [**H2**, Chapter 7, Section 2] [**Sa1**]). As for $\Gamma(A)$, recall that $\Gamma(A)$ is a normal subgroup of $\Gamma(1)$ and note that $\Gamma(A)$ has no elliptic elements and no split hyperbolic elements if $\deg(A) \geq 1$. For a hyperbolic element γ in $\Gamma(A)$ we define $\deg \gamma$ as in (5.2).

Then we compute each term in the sum above explicitly, and we get the following formula :

THEOREM 5.4. *For an eigenvalue λ_j $(j = 1, 2, \cdots, M)$ of T on $L^2(\Gamma \backslash \mathcal{T})$, we write $\lambda_j = q^{s_j} + q^{1-s_j}$ and $s_j = 1/2 + ir_j$. Assume that a sequence $c(n) \in \mathbb{C}$ $(n \in \mathbb{Z})$ satisfies $c(n) = c(-n)$ and $\sum_{n \geq m} q^{n/2}|c(n)| = O(q^{-m})$ for $m \in \mathbb{N}$. If $\Gamma = \Gamma(1)$, then we have*

$$\begin{aligned} h\left(\frac{i}{2}\right) + h\left(\frac{i}{2} + \frac{\pi}{\log q}\right) &= \mathrm{vol}(\Gamma \backslash \mathcal{T})k(0) + \sum_{\{P\} \in \mathfrak{P}_\Gamma} \sum_{l=1}^{\infty} \frac{\deg P}{q^{\frac{l \deg P}{2}}} c(l \deg P) \\ &+ 2\sum_{m=1}^{\infty} (q^{-m} + 1)c(2m) - \frac{2}{(q-1)^2(q+1)} k(0) + 2c(0). \end{aligned} \tag{5.6}$$

If $\Gamma = \Gamma(A)$ $(a := \deg A \geq 1)$, then we have

$$\begin{aligned} \sum_{j=1}^{M} h(r_j) &= \mathrm{vol}(\Gamma \backslash \mathcal{T})k(0) + \sum_{\{P\} \in \mathfrak{P}_\Gamma} \sum_{l=1}^{\infty} \frac{\deg P}{q^{\frac{l \deg P}{2}}} c(l \deg P) \\ &+ \left(\mu - \mathrm{Tr}\,\Phi\left(\frac{1}{2}\right)\right)\left(\frac{1}{2}c(0) + \sum_{m=1}^{\infty} c(2m)\right) \\ &+ \frac{1}{4\pi} \int_{-\frac{\pi}{\log q}}^{\frac{\pi}{\log q}} h(r) \frac{\varphi'\left(\frac{1}{2} + ir\right)}{\varphi\left(\frac{1}{2} + ir\right)} dr - \mu\left(a + \frac{1}{q-1}\right)c(0). \end{aligned}$$

We remark that in the case of $\Gamma = \Gamma(1)$, Efrat [**E2**] computed the spectral decomposition of $L^2(\Gamma \backslash \mathcal{T}, w)$ explicitly. The set of eigenvalues of T on $L^2(\Gamma(1) \backslash \mathcal{T})$ consists of two trivial eigenvalues $\pm(q+1)$, which correspond to two terms in the left-hand side of (5.6)

To the author's knowledge, Akagawa [**Ak**] made a study of an explicit trace formula of the full modular group $\Gamma = \Gamma(1)$. However, his direct method seems to be applicable only for $\Gamma(1)$. Carrying out different calculations from those of [**Ak**] and using the structure of the Bruhat-Tits tree $\mathcal{T}$, Nagoshi [**N1**] obtained the explicit trace formula of $\Gamma(A)$ above in more natural way. The calculation in [**N1**] seems to be similar to that of the case of $PSL(2, \mathbb{R})$ in [**H1**] and [**H2**].

To the author's knowledge, it is an open problem to find an explicit form of the trace formula attached to a congruence subgroup of $PGL(2, \mathbb{F}_q[t])$ other than a principal congruence subgroup $\Gamma(A)$ as in Theorem 5.4.

6. Ihara-Selberg Zeta Functions

The aim of this section is to give some results concerning the Ihara-Selberg zeta function of a graph, and our interest is finding a determinant expression of this zeta function in terms of the adjacency operator.

First we recall the case of finite graphs, where we concentrate on regular finite graphs. Let X be a $(q+1)$-regular finite graph and Γ be the fundamental group of X as in Section 5. For a hyperbolic element P of Γ we put

$$N(P) := q^{\deg P}, \tag{6.1}$$

where $\deg P$ is as in (5.2).

Ihara [**Ih**] defined the following Selberg-type zeta function, which we call the Ihara-Selberg zeta function for Γ:

$$Z_\Gamma(u) := \prod_{\{P\}\in\mathfrak{P}_\Gamma} \left(1 - N(P)^{-s}\right)^{-1}, \tag{6.2}$$

where we write $u = q^{-s}$. We remark that this is an analogue of the Euler product expression of the Riemann zeta function $\zeta(s)$, i.e., $\zeta(s) = \prod_p (1-p^{-s})^{-1}$. If we set

$$N_m := \sum_{\substack{\{P\}\in\mathfrak{P}_\Gamma \\ \deg P | m}} \deg P, \qquad m \geq 1,$$

then the zeta function $Z_\Gamma(u)$ can also be written in the form

$$Z_\Gamma(u) = \exp\left(\sum_{m=1}^{\infty} \frac{N_m}{m} u^m\right). \tag{6.3}$$

The next result is known and called Ihara's formula :

THEOREM 6.1. *Let X be a $(q+1)$-regular finite graph. Then we have*

$$Z_\Gamma(u)^{-1} = (1-u^2)^{\chi} \det(1 - Tu + qu^2).$$

Here $\chi := \#V(X)\frac{q-1}{2}$ and $\det(1-Tu+qu^2) := \prod_\lambda (1-\lambda u + qu^2)$, where the product is over the eigenvalues λ of the adjacency operator T.

This theorem is now proved by several approaches, and is generalized to the case of non-regular finite graphs (see e.g. [**Ah**] [**FZ**] [**Ha**] [**KS**] [**MS**] [**Ni**] [**No**] [**ST1**] [**Su**] [**VN**]).

Next we give a generalization of Theorem 6.1 to the case of an infinite graph $\Gamma(A)\backslash\mathcal{T}$. Our approach follows [**Ah**] and [**VN**], which use Selberg's trace formula. For the Ihara-Selberg zeta function of an infinite graph, see also [**CM**].

Let $\Gamma = \Gamma(A)$. We defined the Ihara-Selberg zeta function $Z_\Gamma(u)$ for Γ as in (6.2). Here $N(P)$ for $\{P\} \in \mathfrak{P}_\Gamma$ is defined as in (6.1) and we note that $N(P) = \sup\{|\lambda_i|_\infty^2 \mid \lambda_i \text{ is an eigenvalue of the } 2\times 2\text{-matrix } P\}$. Then we apply Theorem 5.4 to the function

$$c(n) = \begin{cases} -(\log q)\, q^{-|n|(s-\frac{1}{2})} & \text{if } n \neq 0, \\ 0 & \text{if } n = 0, \end{cases}$$

where $s \in \mathbb{C}$ is fixed with $\mathrm{Re}(s) \geq 2$. This gives the following determinant expression of the Ihara-Selberg zeta function $Z_\Gamma(u)$ in terms of T ([**N1**]):

THEOREM 6.2. *For $\Gamma = \Gamma(1)$ we have*

$$Z_\Gamma(u)^{-1} = \frac{1-q^2u^2}{1-qu^2}.$$

For $\Gamma = \Gamma(A)$ $(\deg A \geq 1)$ we have the formula

$$Z_\Gamma(u)^{-1} = (1-u^2)^{\chi}(1-qu^2)^{-\rho}\mathrm{det}_D(1-Tu+qu^2)\,\mathrm{det}_C(T,s), \tag{6.4}$$

where $\chi := \mathrm{vol}(\Gamma\backslash\mathcal{T})\frac{q-1}{2}$, $\rho := \frac{1}{2}\left(\mu - \mathrm{Tr}\,\Phi(\frac{1}{2})\right)$,

$$\mathrm{det}_D(1-Tu+qu^2) := \prod_{j=1}^{M}(1-\lambda_j u+qu^2),$$

$$\mathrm{det}_C(T,s) := \prod_{|b_j|<1}(1-qu^2b_j)\prod_{|b_j|>1}(1-qu^2b_j^{-1})^{-1},$$

and b_j's are as in (4.2).

Let $A = A_1^{e_1}A_2^{e_2}\cdots A_l^{e_l}$ be the decomposition of A into distinct irreducible polynomials in $\mathbb{F}_q[t]$, and put $a := \deg A$ and $a_i := \deg A_i$. For $\Gamma(A)$ with $\deg A \geq 1$, χ and ρ are explicitly computed as

$$\chi = \frac{q^{3a}}{(q+1)(q-1)}\prod_{i=1}^{l}\left(1-q^{-2a_i}\right),$$

$$\rho = \frac{1}{2}\left(\frac{q^{2a}}{q-1}\prod_{i=1}^{l}\left(1-q^{-2a_i}\right) + q^a\prod_{i=1}^{l}\left(1+q^{-a_i}\right)\right).$$

By these and recalling that q is an odd prime power, we have the following corollary of Theorem 6.2 :

COROLLARY 6.3. *For $\Gamma = \Gamma(A)$ the Ihara-Selberg zeta function $Z_\Gamma(u)$ is a rational function in u.*

We should mention the related result of Scheja [**Sc1**]. By combinatorial and representation-theoretic argument, he obtained a determinant expression of the Ihara-Selberg zeta function $Z_{\Gamma(A)}(s)$ which is different from Theorem 6.2. This expression is given by certain deformation $\Delta^*(u)$ of the operator $\Delta(u) = 1-Tu+qu^2$ in Theorem 6.2, where we let $\Delta^*(u)$ act on $\Gamma(A)\backslash\mathcal{T} - \{$ cusps $\}$.

7. Distribution of Eigenvalues

The aim of this section is to give some results concerning the limit distribution of eigenvalues of a family of graphs.

We now prepare two measures on $\mathbb{R}$. One is the measure μ_∞, which is given by

$$d\mu_\infty(x) := \begin{cases} \frac{1}{\pi}\sqrt{1-\frac{x^2}{4}}dx, & \text{if } |x| \leq 2, \\ 0, & \text{if } |x| > 2. \end{cases}$$

This is what is called the Sato-Tate measure in number theory or the Wigner semicircle in Random Matrix Theory (see [**Me**]). The other is the measure μ_q defined for a real number $q\,(>1)$ by

$$d\mu_q(x) := \frac{q+1}{(q^{1/2}+q^{-1/2})^2-x^2}\,d\mu_\infty(x).$$

The measure μ_q is known to be the spectral measure of T on the $(q+1)$-regular tree $\mathcal{T}$.

Let X be a $(q+1)$-regular finite graph and T be the adjacency operator on X. If λ is an eigenvalue of $T' := T/\sqrt{q}$, then $\lambda \in [-r, r]$ with $r := q^{1/2} + q^{-1/2}$. For a suitable family of regular finite graphs, the following result of McKay [**Mc**] is known.

THEOREM 7.1. *Let $\{X_i\}_{i=1}^{\infty}$ be a family of $(q+1)$-regular finite graphs with $\#V(X_i) \to \infty$ as $i \to \infty$, and $E_i \subset [-r, r]$ be the set of eigenvalues of T' on X_i. Assume that for each integer $n \geq 1$, $c_n(X_i)/\#V(X_i) \to 0$ as $i \to \infty$, where $c_n(X_i)$ is the number of closed geodesics of length n in X_i. Then E_i are equidistributed with respect to the probability measure μ_q on $[-r, r]$ as $i \to \infty$. That is, let $C([-r, r])$ be the space of continuous functions on $[-r, r]$, then for any $f(x) \in C([-r, r])$ we have*

$$\lim_{i\to\infty} \frac{1}{\#E_i} \sum_{\lambda \in E_i} f(\lambda) = \int_{-r}^{r} f(x) d\mu_q(x).$$

For other results concerning the distribution of eigenvalues of a family of finite graphs, see e.g. [**L3**], [**T1**], [**T2**] and [**T3**].

Next we discuss the case of Morgenstern's infinite graphs. As described in Section 3, any non-trivial eigenvalues λ of T on $L^2(\Gamma(A)\backslash\mathcal{T})$ satisfies $|\lambda| \leq 2\sqrt{q}$ (see (3.4)). Hence if we set the operator $T' := T/\sqrt{q}$, then non-trivial eigenvalues λ' of T' on $L^2(\Gamma(A)\backslash\mathcal{T})$ satisfies $|\lambda'| \leq 2$. Then the following theorem holds (see [**N2**]):

THEOREM 7.2. *(i) Let an odd prime power q be fixed. Assume that a sequence of polynomials $\{A_i\}$ $(i = 1, 2, \cdots; A_i \in \mathbb{F}_q[t])$ satisfies that $\deg A_i \to \infty$ as $i \to \infty$. Then the non-trivial eigenvalues E_i of $T' = T/\sqrt{q}$ for $\Gamma(A_i)\backslash\mathcal{T}$ are equidistributed with respect to the measure μ_q on $\Omega = [-2, 2]$. That is, let $C(\Omega)$ be the space of continuous functions on Ω, then for any $f(x) \in C(\Omega)$ we have*

$$\lim_{i\to\infty} \frac{1}{\#E_i} \sum_{\lambda \in E_i} f(\lambda) = \int_{\Omega} f(x) d\mu_q(x).$$

(ii) Assume that a sequence of couples $\{q_i, A_i\}$ $(i = 1, 2, \cdots; A_i \in \mathbb{F}_{q_i}[t])$ satisfies that $q_i \to \infty$ and $\deg A_i \to \infty$ as $i \to \infty$. Then the non-trivial eigenvalues E_i of $T' = T/\sqrt{q_i}$ for $\Gamma(A_i)\backslash\mathcal{T}_{q_i+1}$ are equidistributed with respect to the measure μ_∞ on $\Omega = [-2, 2]$. That is, for any $f(x) \in C(\Omega)$ we have

$$\lim_{i\to\infty} \frac{1}{\#E_i} \sum_{\lambda \in E_i} f(\lambda) = \int_{\Omega} f(x) d\mu_\infty(x).$$

We give a sketch of the proof of (i). By the same argument we can also obtain (ii). On $\Omega = [-2, 2]$ we define the polynomials $\{X_{m,q}(x)\}$ $(m = 0, 1, 2, \cdots)$ by the relation $X_{m,q}(x) := X_m(x) - q^{-1}X_{m-2}(x)$, where $X_m(x)$'s are the Chebychev polynomials of the second kind. We note that $\{X_{m,q}(x)\}$ is a set of orthogonal polynomials with respect to the measure μ_q on Ω. Generalizing the adjacency operator T, we define the operator T_m $(m = 1, 2, \ldots)$ by

$$(T_m f)(v) := \sum_{d(v,u)=m} f(u), \quad \text{for } f : V(\mathcal{T}) \to \mathbb{C}.$$

Then we find the relation $T_m/q^{m/2} = X_{m,q}(T')$. So $\sum_{\lambda\in E_i} X_{m,q}(\lambda) = q^{-(m/2)}\mathrm{Tr}\, T_m$, where $\mathrm{Tr}\, T_m$ denotes the sum of the non-trivial eigenvalues of T_m on $\Gamma(A_i)\backslash\mathcal{T}$. Now we take the logarithmic derivative of (6.3) and (6.4) in u to find the formula connecting $\mathrm{Tr}\, T_m$ with N_m's. This formula contains the terms of the contribution of the scattering determinant of $\Gamma(A_i)$, which can be estimated by Theorem 4.2. Note that for principal congruence subgroups $\Gamma(A)$ we have that $N_m \to 0$ as $\deg A \to \infty$ for each m. Combining the above, we find $\frac{1}{\#E_i}\sum_{\lambda\in E_i} X_{m,q}(\lambda) \to 0$ as $i \to \infty$ for each $m \neq 0$. This completes the proof, since the space spanned by the set of polynomials $\{X_{m,q}(x)\}$ $(m = 0, 1, 2, \cdots)$ is dense in $C(\Omega)$.

8. Distribution of Primitive Hyperbolic Conjugacy Classes

Our concern in this section is the distribution of the primitive hyperbolic conjugacy classes attached to a graph. We discuss the so-called *prime geodesic theorem*, which is a geometric analogue of the *prime number theorem.*

Let $\pi(x) := \#\{p : \text{prime} \mid p \le x\}$. Then it is well known that the Riemann hypothesis of the Riemann zeta function $\zeta(s)$ is equivalent to the fact

$$\pi(x) = \mathrm{Li}(x) + O(x^{\frac{1}{2}}\log x), \tag{8.1}$$

where $\mathrm{Li}(x) := \int_e^x \frac{dt}{\log t} = \int_1^{\log x} \frac{e^t}{t}dt.$

Next we turn our attention to the prime geodesic theorem of $\Gamma = PSL(2,\mathbb{Z})$. Let $\mathfrak{P}_\Gamma$ be the set of primitive hyperbolic conjugacy classes of Γ, and for $\{P\} \in \mathfrak{P}_\Gamma$ put $N(P) := \sup\{|\lambda_i|^2 \mid \lambda_i$ is an eigenvalue of the 2×2-matrix $P\}$. We note that a primitive hyperbolic conjugacy class $\{P\} \in \mathfrak{P}_\Gamma$ corresponds to an oriented prime closed geodesic on $\Gamma\backslash\mathbb{H}$ ($\mathbb{H}$ is the upper half plane) whose length is $\log N(P)$. Let

$$\pi_\Gamma(x) := \#\{\{P\} \in \mathfrak{P}_\Gamma \mid N(P) \le x\}, \tag{8.2}$$

which is a geometric analogue of $\pi(x)$, and we aim to give a good asymptotic formula of $\pi_\Gamma(x)$ (see [**H2**, Chapter 10, Section 3] and [**H2**, p. 519]). It is conjectured that

$$\pi_\Gamma(x) = \mathrm{Li}(x) + O_\varepsilon(x^{\frac{1}{2}+\varepsilon}) \qquad \text{for any } \varepsilon > 0. \tag{8.3}$$

The Selberg zeta function $Z_\Gamma(s)$ of Γ is defined by

$$Z_\Gamma(s) := \prod_{\{P\}\in\mathfrak{P}_\Gamma} \prod_{n=0}^{\infty} \left(1 - N(P)^{-(s+n)}\right).$$

In the prime geodesic theorem the Selberg zeta function $Z_\Gamma(s)$ plays a similar role to the Riemann zeta function $\zeta(s)$ in the prime number theory. We remark that for $\Gamma = PSL(2,\mathbb{Z})$ Selberg's conjecture holds, i.e., $\lambda_1(\Gamma) \ge \frac{1}{4}$ and hence that the Selberg zeta function $Z_\Gamma(s)$ satisfies the *Riemann hypothesis*, which means $Z_\Gamma(s)$ has no zeros in $\mathrm{Re}\,(s) > \frac{1}{2}, s \neq 1$. However, unlike (8.1), the equation (8.3) is not implied directly by the Riemann hypothesis of $Z_\Gamma(s)$, because $Z_\Gamma(s)$ has too many zeros on the line $\mathrm{Re}\,(s) = \frac{1}{2}$. In order to get (8.3), we have to verify certain cancellation also. For remarkable progress in this topic, see [**LS**].

Next we discuss the case of $\Gamma = \Gamma(A) \subset PGL(2, \mathbb{F}_q[t]), A \in \mathbb{F}_q[t]$. Let $\pi_\Gamma(x)$ be defined as in (8.2). If we put

$$\mathrm{Li}_q(x) := 2 \sum_{\substack{1\le m\le \log_q x \\ m:\text{even}}} \frac{q^m}{m},$$

which is an analog of $\mathrm{Li}(x)$, then we have the following result without any assumption (see [**N3**]).

THEOREM 8.1. *Let Γ be a principal congruence subgroup $\Gamma(A)$ $(A \in \mathbb{F}_q[t])$ of $PGL(2, \mathbb{F}_q[t])$. Then we have*

$$\pi_\Gamma(x) = \mathrm{Li}_q(x) + O\left(\frac{x^{\frac{1}{2}}}{\log x}\right).$$

In the proof we follow the ordinary method in the prime number theory (see [**In**]). Our advantages are the facts that the number of zeros of the Ihara-Selberg zeta function $Z_\Gamma(s)$ is essentially finite and that $\Gamma\backslash\mathcal{T}$ is a Ramanujan diagram, which concludes that $Z_\Gamma(s)$ satisfies the *Riemann hypothesis*, i.e., $Z_\Gamma(s)^{-1}$ has no zeros in $\frac{1}{2} < \mathrm{Re}\,(s) < 1$.

By a number-theoretic interpretation of $\mathfrak{P}_{\Gamma(1)} = \mathfrak{P}_{PGL(2,\mathbb{F}_q[t])}$, Theorem 8.1 deduces the following asymptotic formula of certain distribution of fundamental units and class numbers in real quadratic fields over k. Let $\mathcal{L}$ be the set of monic polynomials in $\mathbb{F}_q[t]$ and $\mathcal{D}$ the set of monic and square-free polynomials of even degree in $\mathbb{F}_q[t]$. For $d \in \mathcal{D}$ and $l \in \mathcal{L}$ we put $\mathcal{O}_{l\sqrt{d}} = \mathbb{F}_q[t] + \mathbb{F}_q[t]l\sqrt{d}$, which is an order in $k(\sqrt{d})$. We denote by $h_{l\sqrt{d}}$ the narrow class number of $\mathcal{O}_{l\sqrt{d}}$ in $k(\sqrt{d})$ and by $e_{l\sqrt{d}}$ a fundamental unit of $\mathcal{O}_{l\sqrt{d}}$ with $|e_{l\sqrt{d}}|_\infty > 1$.

COROLLARY 8.2.

$$\sum_{|e_{l\sqrt{d}}|_\infty^2 \leq x} h_{l\sqrt{d}} = \mathrm{Li}_q(x) + O\left(\frac{x^{\frac{1}{2}}}{\log x}\right),$$

where the sum is taken over all $(l, d) \in \mathcal{L} \times \mathcal{D}$ such that $|e_{l\sqrt{d}}|_\infty^2 \leq x$,

As a related result to Corollary 8.2, Akagawa [**Ak**, p. 60] obtained the following.

THEOREM 8.3. *Let m be an integer and let $R_{l\sqrt{d}}$ be defined by $|e_{l\sqrt{d}}|_\infty = q^{R_{l\sqrt{d}}}$. Then*

$$\sum_{R_{l\sqrt{d}}|m} h_{l\sqrt{d}} R_{l\sqrt{d}} = q^m(q^m - 1),$$

where the sum is taken over all $(l, d) \in \mathcal{L} \times \mathcal{D}$ such that $R_{l\sqrt{d}}|m$.

This result is obtained by taking in (5.6)

$$c(n) = \begin{cases} 1 & \text{if } n = 2m \\ 0 & \text{if } n \neq 2m \end{cases} \qquad \text{for } n \in \mathbb{N}.$$

For the case of a regular finite graph the prime geodesic theorem can be formulated as follows. Let X be a $(q+1)$-regular finite graph with $n = \#V(x)$ vertices, and Γ, $\mathfrak{P}_\Gamma$ and $N(P)$ for $\{P\} \in \mathfrak{P}_\Gamma$ are as in Section 5 and Section 6. Note that $\mathfrak{P}_\Gamma$ corresponds to the set of oriented prime cycles of X. Let $\pi_\Gamma(x)$ be as in (8.2).

Let $\{\lambda_0 = q+1, \lambda_1, \cdots, \lambda_{n-1}\}$ be the set of eigenvalues of the adjacency operator T of X. We assume that $|\lambda_j| > 2\sqrt{q}$ for $j = 0, 1, \cdots, \ell$, and we call these λ_j exceptional eigenvalues. For each eigenvalue λ_j $(j = 0, 1, \cdots, n-1)$, we define s_j in $D := D_0 \cup D_1 \cup D_2$ by $\lambda_j = q^{s_j} + q^{1-s_j}$, where

$$D_0 := \left\{ s \in \mathbb{C} \,\middle|\, \mathrm{Re}\,(s) = \frac{1}{2},\, 0 \leq \mathrm{Im}\,(s) \leq \frac{i\pi}{\log q} \right\},$$

$$D_1 := \left\{ s \in \mathbb{C} \;\middle|\; \frac{1}{2} < \mathrm{Re}\,(s) \leq 1,\ \mathrm{Im}\,(s) = 0 \right\},$$

$$D_2 := \left\{ s \in \mathbb{C} \;\middle|\; \frac{1}{2} < \mathrm{Re}\,(s) \leq 1,\ \mathrm{Im}\,(s) = \frac{i\pi}{\log q} \right\}.$$

We find that s_j corresponding to an exceptional eigenvalue $\lambda_j\,(j = 0, 1, \cdots, \ell)$ satisfies $s_j \in D_1 \cup D_2$, and $s_0 = 1$.

Then we obtain the following theorem (see [**N3**]), which should be compared with [**H1**, Theorem 6.19].

THEOREM 8.4. *Let X be a $(q+1)$-regular finite graph. Then we have*

$$\pi_\Gamma(x) = \sum_{1 \leq m \leq \log_q x} \frac{q^m}{m} + \sum_{j=1}^{\ell} \left(\sum_{1 \leq m \leq \log_q x} \frac{q^{ms_j}}{m} \right) + O\left(\frac{x^{\frac{1}{2}}}{\log x} \right).$$

For a related topic, see also [**Zh**].

References

[Ah] G. Ahumada, *Fonctions periodiques et formule des traces de Selberg sur les arbres*, C. R. Acad. Sci. Paris **305** (1987), 709–712.

[Ak] S. Akagawa, *On the Selberg zeta functions for modular groups over function fields*, Master thesis, Tokyo University (1978).

[Ba] H. Bass, *The Ihara-Selberg zeta function of a tree lattice*, Internat. J. of Math. **3** (1992), 717–797.

[BL] H. Bass, A. Lubotzky, *Tree Lattices*, Progress in Math. **176**, Birkhäuser, 2001.

[Br] R. Brooks, *The spectral geometry of k-regular graphs*, J. d'Analyse Math. **57** (1991), 120–151.

[CM] B. Clair, S. Mokhtari-Sharghi, *Zeta funtions of discrete groups acting on trees*, J. Algebra **237** (2001), 591–620.

[D] A. Deitmar, *Combinatorial L^2-determinats*, Proc, Edinburgh Math. Soc. **43** (2000), 185–194.

[E1] I. Efrat, *Spectral deformations over graphs of groups*, Invent. Math. **102** (1990), 447–462.

[E2] I. Efrat, *Automorphic spectra on the tree of PGL_2*, Enseign. Math. **37** (1991), 31–43.

[FZ] D. Foata, D. Zeilberger, *A combinatorial proof of Bass's evaluations of the Ihara-Selberg zeta function for graphs*, Trans. Amer. Math. Soc. **351** (1999), 2257–2274.

[Ge] S. Gelbart, *Lectures on the Arthur-Selberg trace formula*, University lecture series 9, AMS, 1996

[GJ] S. Gelbart, H. Jacquet, *Forms of $GL(2)$ from the analytic point of view*, In: Proceedings of Symposia in Pure Mathematics, Volume 33, Part 1, 213–251. AMS, 1979

[GGP] I. M. Gel'fand, M. Graev, I. I. Pyatetskii-Shapiro, *Representation theory and automorphic functions*, Academic Press, 1990

[Ha] K. Hashimoto, *Zeta functions of finite graphs and representations of p-adic groups*, Advanced Study in Pure Math. **15** (1989), 211–280.

[H1] D. A. Hejhal, *The Selberg trace formula for $PSL(2,\mathbb{R})$*, Vol. 1, Lecture Note in Math. **548**, Springer.

[H2] D. A. Hejhal, *The Selberg trace formula for $PSL(2,\mathbb{R})$*, Vol. 2. Lecture Note in Math. **1001**, Springer.

[H3] D. A. Hejhal, *The Selberg trace formula and the Riemann zeta function*, Duke Math. J. **43** (1976), 441–481.

[Hu] N. E. Hurt, *Mathematical Physics of Quantum Wires and Devices*, Kluwer Academic Publishers, 2000.

[Ih] Y. Ihara, *On discrete subgroups of the two by two projective linear group over p-adic fields*, J. Math. Soc. Japan **18** (1966), 219–235.

[In] A. E. Ingham, *The Distribution of Prime Numbers*, Cambridge University Press, 1990

[Ka] S. Katok, *Fuchsian Groups*, Chicago Lectures in Mathematics Series, 1992.

[KS] M. Kotani, T. Sunada, *Zeta functions of finite graphs*, J. Math. Sci. Univ. Tokyo **7** (2000), 7–25.

[La] J. C. Lagarias, *Number Theory Zeta Functions and Dynamical Zeta Functions*, Contemp. Math. **237** (1999), 45–86.

[L1] W. W. Li, *On modular functions in characteristic p*, Trans. Amer. Math. Soc. **246** (1978), 231–259.

[L2] W. W. Li, *A survey of Ramanujan graphs*, In: Arithmetic, Geometry and Coding Theory, Proc. International Conference held at Luminy, de Gruyter, 1996, 127–143.

[L3] W. W. Li, *Eigenvalues of Ramanujan graphs*, In: Emerging Applications of Number Theory, IMA Vol. Math. Appl. **109** (1999), 387–403

[Lu] A. Lubotzky, *Discrete Groups, Expanding Graphs and Invariant Measures*, Progress in Math. **125**, Birkhäuser, 1994.

[LPS] A. Lubotzky, R. Phillips, P. Sarnak, *Ramanujan graphs*, Combinatorica, **8** (1988), 261–277.

[LS] W. Luo, P. Sarnak, *Quantum ergodicity of eigenfunctions on* $PSL(2,\mathbb{Z})\backslash\mathbb{H}^2$, IHES Publ. Math. **81** (1995), 207–237.

[MS] H. Mizuno, I. Sato, *Zeta functions for images of graph coverings by some operations*, Interdiscip. Inform. Sci. **7** (2001), 53–60.

[Ma] G. Margulis, *Explicit group theoretic constructions of combinatorial schemes and their application to the design of expanders and concentrators*, J. Prob. of Info. Trans., **24** (1988), 39–46.

[M] D. Mayer, *The thermodynamic formalism approach to Selberg's zeta function for* $PSL(2,\mathbb{Z})$, Bull. Amer. Math. Soc. **25** (1991), 55–60.

[Mc] B. McKay, *The expected eigenvalue distribution of a large regular graph*, Lin. Alg. Appl. **40** (1981), 203–216.

[Me] M. L. Mehta, *Random Matrices*, Second Edition, Academic Press, 1991.

[M1] M. Morgenstern, *Ramanujan diagrams*, SIAM J. Discrete Math. **7** (1994), 560–570.

[M2] M. Morgenstern, *Natural bounded concentrators*, Combinatorica, **15** (1995), 111–122.

[M3] M. Morgenstern, *Existence and explicit constructions of* $q+1$ *regular Ramanujan graphs for every prime power* q, J. Comb. Theory, Series B, **62** (1994), 44–62.

[N1] H. Nagoshi, *Selberg zeta functions over function fields*, J. Number Theory, **90** (2001), 207–238

[N2] H. Nagoshi, *Spectra of arithmetic infinite graphs and their application*, Interdiscip. Inform. Sci. **7** (2001), 67–76.

[N3] H. Nagoshi, *The distribution of eigenvalues of arithmetic infinite graphs*, Forum Math. **14** (2002), 807–829.

[Ni] A. Nikitin, *The Ihara-Selberg zeta function of a finite graph and symbolic dynamics*, Algebra i Analiz **13** (2001), 134–149.

[No] S. Northshield, *Two proofs of Ihara's theorem*, In: Emerging Applications of Number Theory, IMA Vol. Math. Appl. **109** (1999), 469–478.

[Ru] D. Ruelle, *Generalized* ζ*-functions for axiom A basic sets*, Bull Amer. Math. Soc. **82** (1976), 156–153.

[Sa1] P. Sarnak, *Class numbers of indefinite binary quadratic forms*, J. Number Theory **15** (1982), 229–247.

[Sa2] P. Sarnak, *Some applications of modular forms*, Cambridge Tracts in Math. 99, Cambridge Univ. Press., 1990.

[Sc1] O. Scheja, *Zetafunktionen arithmetisch definierter Graphen*, Ph. D. Thesis, Saarbrücken, 1998

[Sc2] O. Scheja, *On zeta functions of arithmetically defined graphs*, Finite Fields Appl. **5** (1999), 314–343.

[S1] A. Selberg, *Harmonic analysis and discontinuous groups in weakly symmetric Riemannian spaces with applications to Dirichlet series*, J. Indian Math. Soc. **20** (1956), 47–87.

[S2] A. Selberg, *Harmonic analysis*, Collected papers I, 626–674, Springer-Verlag, 1989.

[Se] J. P. Serre, *Trees*, Springer-Verlag, 1980.

[ST1] H. Stark, A. Terras, *Zeta functions of finite graphs and coverings*, Adv. Math. **121** (1996), 124–165.

[ST2] H. Stark, A. Terras, *Zeta functions of finite graphs and coverings, Part II*, Adv. Math. **154** (2000), 132-195.

[S] T. Steger, *Local fields and building*, Contemp. Math. **206** (1997), 79–107.

[Su] T. Sunada, *L-functions in geometry and some applications*, Lecture Notes in Math. 1201, 266–284, Springer, 1986.

[T1] A. Terras, *Survey of Spectra of Laplacians on finite symmetric spaces*, Experimental Math. **5** (1996), 15–32.

[T2] A. Terras, *Fourier Analysis on Finite Groups and Applications*, London Math. Soc. Stud. Text **43**, Cambridge University Press, 1999.

[T3] A. Terras, *A survey of discrete trace formulas*, In: Emerging Applications of Number Theory, IMA Vol. Math. Appl. **109** (1999), 643–681.

[T4] A. Terras, *Statistics of graph spectra for some finite matrix groups: Finite quantum chaos*, In: Proceedings International Workshop on Special Functions, 1999, Hong Kong, World Scientific, Singapore, 2000, 351–374.

[T5] A. Terras, *Finite quantum chaos*, Amer. Math. Monthly, **109** (2002), 121-139.

[Ve] A. B. Venkov, *Spectral theory of automorphic functions, the Selberg zeta function and some problems of analytic number theory and mathematical physics*, Russian Math. Surveys **34** (1979), 79–153.

[VN] A. B. Venkov, A. M. Nikitin, *The Selberg trace formula, Ramanujan graphs and some problems in mathematical physics*, St. Petersburg Math. J. **5** (1993), 419–484.

[Zh] W. Zhang, *The abstract prime number theorem for algebraic function fields*, In: Analytic number theory, Progr. Math. 85, 529–558, Birkhäuser, 1990.

Department of Mathematics, Faculty of Science, Niigata University, Niigata 950-2181, Japan.

E-mail address: `nagoshih@ybb.ne.jp`

Contemporary Mathematics
Volume **347**, 2004

Radial Geometric Analysis on Groups

Amos Nevo

Abstract. Radial geometric analysis on a group G is the study of convolution operators on G which are naturally defined by a distance function, such as the ball, shell and sphere averages associated with an invariant admissible metric on G. Motivated by some recent results, we formulate several general principles that may prove to hold for all radial averages on all locally compact second countable groups. We will present the existing evidence for their validity, together with many open problems. The three main themes we will focus on are comparisons between the behaviour of radial (and some other geometric) averages on groups along the following dichotomies.

(1) Polynomial volume growth versus exponential volume growth.
(2) Amenable groups versus non-amenable groups.
(3) Isometry groups of Euclidean versus non-Euclidean symmetric spaces of higher rank.

We will survey some of the relevant motivating recent results, focusing primarily on those regarding geometric analysis on connected Lie groups. In particular, we will discuss the following facts.

1) The type of volume growth, namely polynomial or exponential, is detected in the behaviour of the maximal function associated with shell averages defined by an invariant admissible metric.

2) There exists an exponential-maximal inequality satisfied by the ball averages, much stronger than the classical Hardy-Littlewood-Wiener maximal inequality, which is connected with the existence of a spectral gap in the regular representation of a non-amenable group.

3) There exists natural product-type averages of very small support on non-Euclidean symmetric spaces of higher rank which behave very differently than their Euclidean analogs. This phenomenon has its root in the difference between polynomial and exponential volume growth, and leads to some striking differences in the behaviour of the associated maximal functions.

1. Group actions and averaging operators

1.1. Two basic problems in analysis. Let G be a locally compact second countable (lcsc) group, S a proper homogeneous space, namely an lcsc space on which G acts continuously and transitively as a group of homeomorphisms, with

1991 *Mathematics Subject Classification.* Primary 22D40, 22E30, 28D10; Secondary 43A10, 43A90.

Key words and phrases. Geometric Analysis, Maximal Inequalities, Symmetric Spaces, Radial Averages, Volume Growth, Amenable Groups, Convolution Inequalities, Spectral Gaps.

Supported by Israel Science Foundation ISF Grant #126-01.

compact stability group K, so that $S = G/K$. Let m be a G-invariant σ-finite Borel measure on S. The G-action on S gives rise to a representation π of G as a group of isometries of the Banach spaces $L^p(S,m)$, given by $\pi(g)f(x) = f(g^{-1}x)$, and for $1 \le p < \infty$ the representation $\pi : G \to Iso(L^p(S))$ is strongly continuous.

For any Borel probability measure on G, we can consider the averaging operator given by $\pi(\mu)f(x) = \int_G f(g^{-1}x)d\mu(g) = \int_G \pi(g)f(x)d\mu(g)$. Given a family of probability measures μ_t $(t \in \mathbb{R}_+)$ on G, the (a-priori) maximal function associated with the family is defined by :

$$M_\mu^* f(x) = \sup_{t\in\mathbb{R}_+} |\pi(\mu_t)f(x)| \ , \quad f \in C_c(S)\,.$$

Note that the maximal function is well-defined and measurable for $f \in C_c(S)$.

Similarly, given any strongly continuous unitary representation $\tau : G \to \mathcal{U}(\mathcal{H})$ to the unitary group $\mathcal{U}(\mathcal{H})$ of a Hilbert space $\mathcal{H}$, we can consider the operator $\tau(\mu) = \int_G \tau(g)d\mu(g) \in \text{End}(\mathcal{H})$, given by the convex average of the unitary operators $\tau(g)$, $g \in \operatorname{supp} \mu$.

Two very broad classical themes in analysis, which are fundamental problems with a diverse array of applications can be formulated as follows.

(1) Establish decay estimates for the operator norms $\|\pi(\mu_t)\|$ as a function of t, for unitary representations π of G.
(2) Establish maximal inequalities in $L^p(S,m)$, $1 \le p < \infty$ for the family μ_t.

We recall that a strong maximal inequality is an L^p-norm inequality for the maximal function, of the form $\left\|M_\mu^* f\right\|_p \le C_p \|f\|_p$, $\forall f \in L^p(S)$, where $1 < p \le \infty$. A weak-type maximal inequality is an estimate on the distribution function associated with the maximal function. Of particular interest is the weak-type $(1,1)$ maximal inequality

$$m\{x \in X\,;\, M_\mu^* f(x) > \delta\} \le \frac{C}{\delta} \|f\|_1 \quad , \quad \forall f \in L^1(S)\,.$$

We will discuss some aspects of these problems in more detail below, but first let us define the averages we will be interested in. To that end we start by briefly recalling some facts related to volume growth in groups.

1.2. Growth type of a group, admissible invariant metrics. Let G be an lcsc and compactly generated group, and assume also that G is unimodular for simplicity. If V is a compact set generating G, namely $\cup_{n\in\mathbb{N}} V^n = G$, we can define a distance function $|g|_V$ on G via $|g|_V = \min\{n\,;\, g \in V^n\}$, where $V^0 = \{e\}$ and $V^n = V \cdot V \cdots V$ is the set of n-fold products of elements of V. The distance function is inversion-invariant if and only if the set V is symmetric, namely $V = V^{-1}$. The latter condition is equivalent to the symmetry of the function $d_V(g,h) = \left|g^{-1}h\right|_V$. We will assume $V = V^{-1}$ from now on, and then the associated function d_V is a left-G-invariant metric on G. Since G is lcsc, for some $n \in \mathbb{N}$, V^n contains an open set and then $m_G(V^{n+k}) > 0$ for all $k \in \mathbb{N}$, where m_G is a Haar measure. By subadditivity, the limit $\lim_{n\to\infty} \frac{1}{n} \log m_G(V^n) = h_V$ exists and if $h_V > 0$ for some V, then $h_{V'} > 0$ for any other compact generating set V'. If $h_V > 0$, G is called a group of exponential volume growth, and if $h_V = 0$ for some, or equivalently, all compact generating sets V, G is called a group of subexponential growth. In that case we can consider the quantity $\limsup_{n\to\infty} \frac{\log m_G(V^n)}{\log n} = p_V$, and we recall the well known fact that if $p_V < \infty$ for one compact generating set V, then $p_{V'} < \infty$ for any other compact generating set V'. If $p_V < \infty$ for some compact generating set then G is called a group of polynomial volume growth, and if $h_V = 0$ and $p_V = \infty$ for some compact generating set V, then G is called a group of intermediate growth.

We have used the metrics of the form d_V described above to define the growth type of a group G. There are of course many other left-invariant metrics on a group, but since invariant metrics can be easily rescaled, not all of them will give the same growth type. For example, taking the metric $d(t,s) = \log(1 + |t-s|)$ on $G = \mathbb{R}$, we obtain an invariant metric whose balls centered at 0 (denoted B_t) satisfy $m_{\mathbb{R}}(B_t) = \exp(t) - 1$, namely have exponential growth. It is therefore natural to define a left-invariant metric d on G to be admissible, if the quantities $h_d = \lim_{t\to\infty} \frac{1}{t} \log m_G(B_t)$ and $p_d = \limsup_{t\to\infty} \frac{\log m_G(B_t)}{\log t}$ exhibit behaviour similar to that of the quantities h_V and p_V defined by the metrics d_V associated with compact generating sets V. In other words, $h_d > 0$ iff $h_V > 0$, and $p_d < \infty$ iff $p_V < \infty$. In the sequel, we will consider only admissible metrics on G.

Note that we allow the case where the metric $d(g,h)$ takes only a discrete set of values, an obvious example being the metrics d_V on any lcsc group, and in particular word metrics on a discrete group. Another example is the metric on a discrete subgroup $\Gamma \subset G$ inherited from some invariant metric on G, or G/K, K a compact subgroup. Discreteness of the possible values of d has no bearing on the discussion of ball averages, but when discussing sphere averages we will always assume that the discrete metric d takes values in the set of integers, and consider spheres with integer radii.

1.3. Geometric averaging operators. The averaging operators $\pi(\mu_t)$ defined above are most directly related to the distance geometry of the group when μ_t are the natural probability measures supported on balls (or spheres, shells, etc) defined by a left-invariant admissible metric. These will account for many of the examples we consider below, but it is useful first to generalize this set-up to include a larger family of geometric averaging operators which is of considerable interest and occurs naturally in the discussion of proper homogeneous spaces. Let G be a unimodular lcsc compactly generated group, K an arbitrary compact subgroup, and let $S = G/K$ be the associated homogenous space, namely the space of right cosets $G/K = \{gK\,;\, g \in G\}$. We fix a left-$G$-invariant admissible metric on G denoted D, and note that such a metric gives rise naturally to a G-invariant metric, denoted ℓ_D, on $S = G/K$, via $\ell_D(gK, hK) = \min_{k,k'\in K} D(gk, hk')$. We will call an arbitrary invariant metric d on G/K an admissible metric on G/K provided that $h_d > 0$ iff $h_D > 0$ and $p_d < \infty$ iff $p_D < \infty$. Clearly the metric ℓ_D is admissible on G/K iff the metric D is admissible on G. Note further that since K is compact, S carries a canonical G-invariant σ-finite Radon measure, unique up to a positive multiple, namely the projection of Haar measure on G.

Let us now consider one of our most important class of examples, namely the case where G is a connected Lie group, K a compact subgroup. In that case there exists a G-invariant Riemannian metric on S, and we denote its associated (admissible) distance function by d. There is also a G-invariant Riemannian volume form on G projecting to an invariant volume form on G/K.

Given a continuous function f on S, we can consider the averaging operators on S associated to balls and spheres, namely

$$\mathcal{B}_t f(x) = \frac{1}{vol(\tilde{B}_t)} \int_{y\in\tilde{B}_t(x)} f(y)dm(y)\ ,\ \mathcal{S}_t f(x) = \int_{y\in\tilde{S}_t(x)} f(y)d\tilde{\sigma}_t(y)\,.$$

where $\tilde{B}_t(x)$ (resp. $\tilde{S}_t(x)$) is the ball (resp. sphere) of radius t and center x in $S = G/K$, and $\tilde{\sigma}_t(x)$ the normalized volume induced on $S_t(x)$ by the restriction of the Riemannian structure.

Another family of averaging operators which appears here naturally and will be important in our discussion is the family of shell averages :

$$\mathcal{C}_t f(x) = \int_{y\in\tilde{C}_t(x)} f(y)d\tilde{\gamma}_t(y)$$

where $\tilde{C}_t(x) = \{gK \in G/K\,;\, t \le d(xK, gK) < t+1\}$ and $\tilde{\gamma}_t$ is the absolutely continuous probability measure supported on $\tilde{C}_t(x)$ given by $\tilde{\gamma}_t = \int_t^{t+1} \tilde{\sigma}_t dt$.

As is well-known, these operators (which act on $C_c(S)$) can be represented as (right) convolution operators on G itself. Indeed, first let us consider the following bi-K-invariant subsets of G, which we call K-balls, K-spheres and K-shells, defined by

$$B_t = \{g\,;\, d(gK,K) \le t\},\ S_t = \{g\,;\, d(gK,K) = t\},\ C_t = \{g\,;\, t \le d(gK,K) < t+1\}\,.$$

B_t (resp. S_t, C_t) is the pre-image in G of the ball $\tilde{B}_t$ (resp. sphere $\tilde{S}_t$, shell $\tilde{C}_t$) of radius t and center $K = o$ in $X = G/K$. We have the associated bi-K-invariant probability measures

$$\beta_t = \frac{1}{m_G(B_t)} \int_{g\in B_t} m_K * \delta_g * m_K dm_G(g)$$

and

$$\sigma_t = \int_{g\in S_t} m_K * \delta_g * m_K d\tilde{\sigma}_t(gK)\ ,\ \gamma_t = \int_t^{t+1} \sigma_t dt$$

where m_G is Haar measure on G, δ_g the Dirac δ-measure supported at g.

The averaging operators $\mathcal{B}_t$, $\mathcal{S}_t$ and $\mathcal{C}_t$ defined above can be identified with the operators of right convolution on G by the measures β_t, σ_t and γ_t, respectively. Indeed, let μ be any bi-K-invariant finite Borel measure on G, namely μ satisfies $m_K * \mu * m_K = \mu$, and define its reflection by $\mu^\vee(A) = \mu(A^{-1})$. Given $f \in C_c(G/K)$, the function $f * \mu^\vee(gK) = \int_G f(ghK)d\mu(h)$ is well defined. Since d is a G-invariant metric, $d(gK, ghK) = d(K, hK)$ so that if $\mu = \sigma_t = \sigma_t^\vee$ for example then $f * \sigma_t(gK)$ is the average of f over points at distance t from gK, namely $f * \sigma_t(gK) = \mathcal{S}_t f(gK)$. Similarly, the same holds for β_t, γ_t or any other bi-K-invariant measure. We conclude that the action of the radial averages $\mathcal{B}_t$, $\mathcal{S}_t$ and $\mathcal{C}_t$ on functions on the homogeneous space $S = G/K$ is equivalent to the action of β_t, σ_t and γ_t on functions on G via right translation.

More generally, let G be an lcsc (unimodular) group, and $N : G \to \mathbb{R}_+$ be a continuous and proper function satisfying $N(gh) \le C(N(g) + N(h))$, namely a quasi-norm on G. We can consider the sets $B_t^N = \{g \in G\,;\, N(g) \le t\}$, the normalized measures β_t^N with density $\chi_{B_t^N}/m_G(B_t^N)$, and the corresponding right convolution operators on $L^p(G)$. When G is a connected Lie group and N is smooth, (or when G is lcsc and the norm N takes integer values) it is also possible to define for every t the natural probability measure σ_t^N supported on $\partial B_t^N = S_t^N = \{g \in G\,;\, N(g) = t\}$. Our discussion below can be in principle extended to this more general context, but in the interest of simplicity we will restrict ourselves to a discussion of functions of the form $N(g) = d(e, g)$ where d is an invariant admissible metric (so that the constant C in the definition of a quasi-norm is equal to 1). We remark that one particularly interesting natural set of examples consist of a homogeneous invariant quasi-norm on a homogeneous connected nilpotent Lie group. Such quasi-norms give rise to a quasi-distance which is not necessarily a metric. but an equivalent homogeneous invariant norm can always be found. We refer to [**S2**](Ch. XIII, §5 and §7B) for more on this topic.

2. Examples of groups, invariant metrics and radial averages

2.1. Some classical radial averages on connected Lie groups. Let us note the following concrete and well-known examples of connected Lie groups and invariant metrics on them, accompanied by a brief mention of the corresponding results concerning the maximal functions associated with ball and sphere averages.

(1) $G = SO(n,\mathbb{R})\ltimes\mathbb{R}^n$ the Euclidean motion group, $K = SO(n,\mathbb{R})$ a maximal compact subgroup, d the Euclidean distance on $\mathbb{R}^n$ derived from the unique (up to scalar) G-invariant Riemannian metric on $\mathbb{R}^n = G/K$. Here we retrieve the usual theory of radial averages on $\mathbb{R}^n$. The fact that M_β^* is bounded in every L^p, $1 < p \leq \infty$ (and satisfies the weak-type $(1,1)$-maximal inequality) is classical[**HL**][**W**]. The spherical maximal operator M_σ^* is bounded in L^p iff $p > \frac{n}{n-1}$ ([**SW**] for $n > 2$, [**Bn**] for $n = 2$).

(2) $G = Iso^0(\mathbb{H}^n) \cong SO^0(n,1)$, the connected component of the isometry group of hyperbolic space $\mathbb{H}^n$, $K = SO(n,\mathbb{R})$ a maximal compact subgroup, d the non-Euclidean distance on $\mathbb{H}^n$ derived from the unique G-invariant Riemannian metric on $\mathbb{H}^n = G/K$ with constant sectional curvature -1. Here we retrieve the usual theory of radial averages on $\mathbb{H}^n$, and the fact that M_β^* is bounded in every L^p, $1 < p \leq \infty$ is due to [**CS**] (the weak-type $(1,1)$-maximal inequality is proved in [**Str**]). The spherical maximal operator M_σ^* is bounded in L^p iff $p > \frac{n}{n-1}$ ([**EK**] for $n > 2$, [**I**] for $n = 2$).

(3) $G = SU(n,\mathbb{C})\ltimes H^n$ the Heisenberg motion group ($H^n = \mathbb{C}^n \oplus \mathbb{R}$ the Heisenberg group), $K = SU(n,\mathbb{C})$ a maximal compact subgroup, d the metric on H^n derived from the homogeneous K-invariant norm $N(z,t) = (\|z\|^4 + |t|^2)^{1/4}$, $z \in \mathbb{C}^n$, $t \in \mathbb{R}$ (see [**S2**](p. 638)). The boundedness of M_β^* in every L^p, $1 < p \leq \infty$ (as well as the weak-type $(1,1)$-maximal inequality) is a consequence of a Hardy-Littlewood maximal inequality for a doubling measure given by a polynomial density on Euclidean space (see [**Ca**], and [**S2**](Ch. I) for a complete discussion). M_σ^* was shown to be bounded in L^p iff $p > \frac{2n+1}{2n} = \dim_{\mathbb{R}}(H^n)/(\dim_{\mathbb{R}}(H^n) - 1)$ in [**Co2**]. Results on the maximal operator associated with singular spherical averages supported on the spheres in $\mathbb{C}^n$ (defined by the *usual* Euclidean distance on $\mathbb{C}^n$) were proved in [**NT**], and recently improved in [**MS**].

2.2. Classes of connected Lie groups. More generally, we can consider the following classes of connected Lie groups, with the following choices of an invariant admissible metric. In the generality suggested here, the results established in the literature are very partial, and in certain cases non existent. We will comment very briefly on some open problems as we present the examples.

(1) **Cartan Motion groups, and amenable Gelfand pairs**. Let $G = K\ltimes\mathfrak{p}$, where K is a maximal compact subgroup of a non-compact semisimple Lie group L with finite center, and $\mathfrak{p}$ is the corresponding Cartan subspace, namely $\mathfrak{k} \oplus \mathfrak{p}$ are the complementary eigenspaces of a Cartan

involution of $\mathfrak{l}$. We take d to be the distance on $G/K = \mathfrak{p}$ derived from a K-invariant Riemannian metric on $\mathfrak{p}$. Note that this set of examples generalizes the cases of the K-radial Riemannian averages on $\mathbb{R}^n$ and H_n. The boundedness of M_β^* is a simple consequence of the corresponding results in the Euclidean case, but not much seems to be known on boundedness of M_σ^* on Cartan motion groups. A more general context is that of amenable Gelfand pairs, namely where the convolution algebra of bi-K-invariant finite Borel measures on G is commutative (see [**BJR1**]). Note that it is also possible to generalize this set-up and choose a proper subgroup K_0 of the isometry group K together with a K_0-invariant metric (see [**BJR2**]).

(2) **Homogeneous nilpotent groups**. Let $G = K \ltimes N$, where N is homogeneous nilpotent connected simply connected Lie group, K is a compact group of automorphisms of N, and d is the (quasi) on N distance derived from a K-invariant homogeneous invariant (quasi) norm on $N = G/K$. For the boundedness of M_β^* in every L^p, $1 < p \leq \infty$ (and the weak-type $(1,1)$ maximal inequality) for any admissible (quasi) metric on any simply connected Lie group of polynomial volume growth see [**S2**](Ch. I, §3). This set of examples also generalizes the averages discussed above on $\mathbb{R}^n$ and H^n. Note however that except for certain specific choices of admissible metrics (or quasi-norms) on H^n, not much seems to be known regarding the boundedness of M_σ^* in the case of a general Riemannian metric, for example.

(3) **Semisimple Lie groups**. Let G be a connected semisimple Lie group with finite center, K a maximal compact subgroup, $S = G/K$ a symmetric space of non-compact type without Euclidean factors. Taking d as the distance function on $S = G/K$ derived from the Riemannian metric associated with the Killing form on S, boundedness of M_β^* in every L^p, $1 < p \leq \infty$ is due to [**CS**] (the weak-type $(1,1)$-maximal inequality is proved in [**Str**]). For arbitrary G-invariant Riemannian (and many other) metrics on symmetric spaces S the boundedness of M_β^* in every L^p, $1 < p < \infty$ is established in [**N4**] (but the weak-type $(1,1)$-maximal inequality remains unresolved). M_σ^* (w.r.t. the Killing metric) was shown to be bounded in L^p iff $p > \dim_{\mathbb{R}}(G/K)/\dim_{\mathbb{R}}(G/K) - 1$ for every simple Lie group of real rank one in [**N1**][**N2**][**NS1**], except for the case of $SL_2(\mathbb{R})$ which was considered in [**I**] (see [**EK**] for hyperbolic space of dimension at least three). M_σ^* was shown to be bounded in L^2 for every complex semisimple group in [**CN**], but in the general case, it has not been established that M_σ^* is bounded in any L^p, $1 < p < \infty$.

(4) **Exponential solvable groups**. Let $G = A \ltimes N$ be a connected Lie group, where N is nilpotent, A an Abelian group of automorphisms of N, such that G has exponential volume growth. Let d be the distance on G derived from an invariant Riemannian metric, and K a subgroup of the isometry group of (G, d). This is a particularly interesting and difficult case, and it seems that the boundedness of M_σ^* in any L^p has not been established even for one Riemannian metric on even one solvable (unimodular) group G of exponential volume growth. The same seems to be the case even for the boundedness of M_β^*. An intriguing set of relevant examples to consider are the groups discussed in [**CM**].

(5) **Parabolic groups**. Another family that arise naturally and generalizes exponential solvable groups consists of groups of the form $Q = MAN$, M a non-compact reductive group, N nilpotent and normal, A a vector group commuting with M. This class includes the parabolic subgroups of semisimple Lie groups.

2.3. General groups. There is considerable interest in solving the problems posed in §1.1 also for radial averages on general lcsc compactly generated groups. Among the most important cases are groups acting isometrically on simplicial (and more generally, polyhedral) complexes, with an invariant metric taking integer values. Note that an integer valued metric does determine spheres and spherical measures, but of course here $\gamma_t = \sigma_t$. Examples include

(1) **Semisimple algebraic groups over local fields**. Let G be the group of F-points of a semisimple algebraic group defined over a locally compact totally disconnected field F. We let K be a compact open subgroup, X the Bruhat-Tits building (on which G acts with finitely many orbits of vertices), d an admissible invariant metric on X. As in the homogeneous space G/K discussed in §1.3 we can consider the ball and sphere averages on X, and their natural lifts to probability measures on G. When G has split F-rank one, M_σ^* and hence also M_β^* is bounded in every L^p, $1 < p \leq \infty$, as follows from the results established in [**N0**] and [**NS1**] (the weak-type $(1,1)$ maximal inequality follows from the results of [**CMS**]).
(2) **Automorphism groups of graphs.** Closed subgroups of the automorphism group of locally finite graphs form an extensive class of examples. Among them we can mention in particular closed subgroups of the automorphism group of a graph with infinitely many ends (in particular, a tree), and groups of automorphisms of cube complexes and of affine and other buildings.
(3) **Discrete groups**. Let G be any finitely generated discrete group acting on its Cayley graph, with the word metric associated with a finite symmetric set on generators. For the case of $\mathbb{Z}^n$ with a symmetric set of free generators, or $\mathbb{Z}^n$ with Euclidean distance (and other norms and admissible metrics) boundedness of M_β^* in every L^p, $1 < p \leq \infty$ (and the weak-type $(1,1)$-maximal inequality) is of course classical [**HL**][**W**](see [**S2**](Ch. I) for a discussion). For the free group $\mathbb{F}_r$ on $r > 1$ generators M_σ^* (and consequently also M_β^*) w.r.t. the metric defined by a symmetric set of free generators is bounded in every L^p, $1 < p \leq \infty$, as follows immediately from the exponential decay of the convolution norms and Riesz-Thorin interpolation (see [**H**] for the decay of the norms in ℓ^2, and [**N0**][**NS1**] for a more general result). The weak-type $(1,1)$-maximal inequality was recently proved in [**CMS**]. For word hyperbolic groups with an arbitrary finite symmetric set of generators , the boundedness of M_σ^* and M_β^* in L^p, $1 < p < \infty$ was shown (in a much stronger form, see §4 below) in [**N3**], based on the results regarding rapid decay [**J**].
(4) **Lattice subgroups**. Let $\Gamma \subset G$ where G is an arbitrary compactly generated non-compact lcsc group, and Γ an arbitrary discrete group. Here it is natural to consider the invariant metric on Γ inherited from an invariant metric on G or G/K (K compact). In the case Γ is a lattice

subgroup of a semisimple Lie group, the boundedness of M_β^* for the ball averages defined using the invariant Riemannian metric on the symmetric space G/K was established (in a much stronger form, see §4 below) in [**N3**].

Having defined the operators and the problems under consideration, we now turn in the three succeeding sections to describing some principles that may be useful in developing a general theory of radial averages on groups and proper homogeneous spaces. These principles are an attempt to formalize the influence that three basic dichotomies have on the averages under study, as follows.

(1) Polynomial versus exponential volume growth and its connection with the shell averages.
(2) Amenable versus non-amenable groups and the appearance of exponential-maximal inequalities.
(3) Euclidean versus non-Euclidean symmetric spaces of higher rank, and its relevance to general product structures.

3. Polynomial versus exponential volume growth

3.1. Volume growth and shell averages. The first phenomenon we would like to consider is demonstrated by the following result.

THEOREM 3.1 (**ball, shell and sphere averages**). *Let β_t, σ_t and γ_t denote the ball, sphere and shell averages on $G = Iso^0(S)$, where $S = G/K$ is a symmetric space without compact factors, M_β^*, M_σ^* and M_γ^* the associated maximal functions.*

(1) If S is Euclidean, namely $S \cong \mathbb{R}^n$, then M_γ^* and M_σ^* satisfy exactly the same strong maximal inequalities.
(2) If S has no Euclidean factors (namely G is semisimple) then M_γ^* and M_β^* satisfy exactly the same strong maximal inequalities.

PROOF. (1) Clearly, any strong maximal inequality satisfied by the sphere averages is also satisfied by the shell averages, which are convex combinations of the sphere averages. The other direction, namely that a strong maximal inequality holding for the shell averages also holds for the sphere averages, is a consequence of the following observations concerning convolution operators on $\mathbb{R}^n$. First, it suffices to prove that the desired maximal inequality for M_σ^* holds on the dense subspace of continuous functions with compact support. For a fixed $\varepsilon > 0$, let j_ε be the map $v \mapsto \varepsilon v$ on $\mathbb{R}^n$, and define $\gamma_u^\varepsilon = \int_u^{u+\varepsilon} \sigma_s ds/\varepsilon$. Then j_ε maps the probability measure γ_t to the probability measures $\gamma_{\varepsilon t}^\varepsilon = \int_{\varepsilon t}^{\varepsilon t+\varepsilon} \sigma_s ds/\varepsilon$. It is easily seen that $f * \gamma_{\varepsilon t}^\varepsilon(v) = j_\varepsilon(f) * \gamma_t(v/\varepsilon)$, where $j_\varepsilon(f)(w) = f(\varepsilon w)$. Hence $\sup_{t>0} |f * \gamma_{\varepsilon t}^\varepsilon(v)| = \sup_{t>0} |j_\varepsilon(f) * \gamma_t(v/\varepsilon)|$, and so

$$\left\|M_{\gamma^\varepsilon}^* f\right\|_p = \left\|j_{1/\varepsilon} \circ M_\gamma^* \circ j_\varepsilon(f)\right\|_p \leq \varepsilon^{-n} C_p \left\|j_\varepsilon(f)\right\|_p = C_p \left\|f\right\|_p$$

for every $\varepsilon > 0$. Now assuming f is continuous with compact support, clearly for every $x \in \mathbb{R}^n$ we have $\lim_{\varepsilon \to 0} f * \gamma_u^\varepsilon(x) = f * \sigma_u(x)$. If P is an arbitrary finite set of non-negative rational numbers, and f is non-negative, then

$$\sup_{p \in P} f * \sigma_p = \lim_{\varepsilon \to 0} \sup_{p \in P} f * \gamma_p^\varepsilon = \liminf_{\varepsilon \to 0} \sup_{p \in P} f * \gamma_p^\varepsilon \leq \liminf_{\varepsilon \to 0} M_{\gamma^\varepsilon}^* f \ .$$

Since for a continuous f the supremum of $f * \sigma_p$ over $p \in \mathbb{R}$ is equal to the supremum over $p \in \mathbb{Q}$, going over all the finite sets $P \subset \mathbb{Q}$, we conclude that $M_\sigma^* f \leq \liminf_{\varepsilon \to 0} M_{\gamma^\varepsilon}^* f$. By Fatou's Lemma, $\|M_\sigma^* f\|_p \leq \liminf_{\varepsilon \to 0} \left\|M_{\gamma^\varepsilon}^* f\right\|_p \leq C_p \|f\|_p$.
(2) The second assertion has been proved in [**NS2**] for simple Lie groups of real rank one. In general, for arbitrary higher rank groups we use the fact that there exist positive constants c_1, c_2 and D such that for $t \geq 1$:

$$c_1 \leq \frac{vol\tilde{S}_t}{t^D \exp(2t\,\|\rho\|)} \leq c_2 \ , \quad c_1 \leq \frac{vol\tilde{B}_t}{t^D \exp(2t\,\|\rho\|)} \leq c_2$$

where $\tilde{S}_t$ and $\tilde{B}_t$ are the sphere and the ball of radius t and center $[K]$ in G/K (ρ is half the sum of the positive roots, see §5.2 below). The same estimate holds of course also for the lifts S_t and B_t to G. Using this estimate of the volume of the ball and the sphere, for every non-negative function f and $t \geq 1$

$$\pi(\beta_t)f(x) \leq \frac{c_2}{c_1} t^{-D} e^{-2t\|\rho\|} \int_0^t s^D e^{2s\|\rho\|} \pi(\sigma_s) f(x) ds \leq$$

$$\leq \frac{c_2}{c_1} t^{-D} e^{-2t\|\rho\|} \left(\sum_{k=0}^{[t]-1} \int_{t-k-1}^{t-k} s^D e^{2s\|\rho\|} \pi(\sigma_s) f(x) ds + C_1 \pi(\gamma_{t-[t]}) f(x) \right)$$

$$\leq C_2 \sum_{k=0}^{[t]-1} \left(\frac{t-k}{t}\right)^D e^{-2k\|\rho\|} \pi(\gamma_{t-k-1}) f(x) + C_{3'} \pi(\gamma_1) f(x)$$

$$\leq C_4 \left(\sum_{k=0}^{\infty} e^{-2k\|\rho\|} \right) \sup_{0 \leq s \leq t} \pi(\gamma_s) f(x) + C_3 \pi(\gamma_1) f(x) \leq C_5(\|\rho\|, D) \sup_{s \geq 0} \pi(\gamma_s) f(x).$$

Taking the supremum over $t \geq 1$ on the left hand side, we conclude that the global part of the maximal operator (namely the supremum over $t \geq 1$) for balls is dominated by the maximal operator for the shells. A simple variant of the Hard-Littlewood-Wiener-Calderon covering argument (see e.g. [**N6**] §5, §6) shows that the operator $\sup_{0<t\leq 1} \pi(\beta_t)$ (as well as the operator) $\sup_{0<t\leq 1} \pi(\gamma_t)$) is bounded in every L^p, $1 < p \leq \infty$. It follows that any strong maximal inequality satisfied by M_γ^* is also satisfied by M_β^*.

For the converse inequality, note that if $t \geq 1$:

$$\beta_{t+1} \geq \frac{c_1}{c_2}(t+1)^{-D} e^{-2(t+1)\|\rho\|} \int_t^{t+1} s^D e^{2s\|\rho\|} \sigma_s ds \geq C_6(\|\rho\|, D)\gamma_t$$

as measures on G. Hence the global part of the maximal operator for shells is dominated by the global part of the maximal operator for balls, and using again the foregoing local argument, any strong maximal inequality satisfied by M_β^* is also satisfied by M_γ^*. This concludes the proof of part (2). □

Remarks on Theorem 3.1 and its proof.

(1) As noted in §2.1, M_σ^* on $\mathbb{R}^n$ is unbounded in $L^{n/(n-1)}(\mathbb{R}^n)$ (see the counter-example in [**SW**]). By Theorem 3.1 we conclude that M_γ^* on $\mathbb{R}^n$ does not satisfy the strong maximal inequality in L^p, $p \leq \frac{n}{n-1}$. On the other hand, on $\mathbb{H}^n$ M_γ^* satisfies the maximal inequality in every L^p, $p > 1$, and similarly on any non-compact symmetric space without Euclidean factors, again by Theorem 3.1 and [**CS**]. Hence the behaviour of

the shell averages detects the type of volume growth of the (Euclidean or non-Euclidean) symmetric space under discussion.

(2) The problem of the connection between volume growth and shell averages can be reformulated naturally so as to apply to an integer valued admissible invariant metric on arbitrary lcsc compactly generated groups (for example discrete groups). Here of course spheres and shells are the same, and the relevant question is whether exponential volume growth of the group is equivalent to $M_\sigma^* \cong M_\beta^*$, namely to the fact that M_σ^* and M_β^* satisfy exactly the same maximal inequalities. We remark that volume estimates on the growth of spheres and balls in a non-amenable group which are as tight as those which we used above for symmetric spaces easily imply that M_σ^* and M_β^* are indeed equivalent. However such estimates for discrete groups have not been established except in certain special cases, where the growth series on the group yields a rational function, for example word-hyperbolic groups [**Coo**], or Coxeter groups. For another application of such growth estimates to maximal inequalities we refer to [**FN**].

(3) The proof of Theorem 3.1, part (1) is valid for any norm on $\mathbb{R}^n$ and any homogeneous (quasi) norm on a homogeneous nilpotent group. Thus the equivalence $M_\gamma^* \cong M_\sigma^*$ is valid for an extensive class of (quasi) metrics on an extensive class of polynomially growing groups. For the ℓ^1 and ℓ^∞ norm on $\mathbb{R}^n$, the "spheres" (which are the boundaries of cubes) do not satisfy any strong (or weak) maximal inequality in L^p, $1 \le p < \infty$ (see [**Wa**]), and hence neither do the shells. It is unknown whether such a situation can arise for M_β^* and M_γ^* on a connected Lie group of exponential volume growth.

(4) A similar question presents itself also in the discrete case. For the free group $\mathbb{F}_r$ on $r > 1$ generators M_σ^* and M_β^* (w.r.t. metric defined by any symmetric set of generators) are clearly equivalent. For the case of $\mathbb{Z}^n$ with a symmetric set of free generators M_σ^* is unbounded in every L^p, $1 < p < \infty$, but M_β^* is bounded in every L^p, $1 < p \le \infty$ (and satisfies the weak-type $(1,1)$-maximal inequality). Again it is unknown if such a situation can arise for M_β^* and M_σ^* on a discrete group with exponential volume growth.

(5) Finally note also that in the Euclidean case of $\mathbb{R}^n$ (or $\mathbb{Z}^n$), it is interesting to determine the right condition on the (variable) size $t - \theta(t)$ of the shell which will guarantee that the averages $\gamma_t^\theta = \int_{t-\theta(t)}^t \sigma_s ds/\theta(t)$ satisfy the same strong maximal inequalities as the ball averages. Clearly the size $\theta(t) = \theta \cdot t$ (where $0 < \theta < 1$ is a constant) is sufficient, but $\theta(t) = 1$ is not, by Theorem 3.1.

3.2. Radial averages and volume growth : some open problems. We now list some of the most basic open problems that arise from the preceding discussion. First let G be a connected Lie group, d the distance associated with an invariant *Riemannian* metric.

(1) **Ball averaging problem.** Is M_β^* bounded in every L^p, $1 < p \le \infty$ and does it satisfy the weak-type $(1,1)$-maximal inequality ?

(2) **Equivalence problem for shell averages.** Determine whether $M^*_\gamma \cong M^*_\sigma$ is equivalent to polynomial volume growth on G. Determine whether $M^*_\gamma \cong M^*_\beta$ is equivalent with exponential volume growth.
(3) **Sphere averaging problem.** Determine whether there exists $q_d < \infty$ such that M^*_σ is bounded on L^p, $p > q_d$?

A particularly intriguing case in point for all three problems is that of exponential solvable Lie groups.

Now let G be an arbitrary lcsc compactly generated (unimodular) group of exponential volume growth, with d an arbitrary G-invariant admissible metric, and formulate:

(1) **Ball averaging problem.** Is M^*_β bounded in every L^p, $1 < p \leq \infty$ and does it satisfy the weak-type $(1,1)$-maximal inequality ?
(2) **Equivalence problem for sphere averages.** If the metric is integer valued, do M^*_β and M^*_σ satisfy exactly the same maximal inequalities ?

We remark that the boundedness of M^*_β in every L^p, $1 < p \leq \infty$ (as well as the weak-type $(1,1)$ maximal inequality) has been established for groups of polynomial volume growth (see the discussion of [**N6**] or [**S2**, Ch I]).

We now turn to discuss the dichotomy between amenable and non-amenable groups. The spectral aspects of this dichotomy lead us to expect that M_β satisfies a maximal inequality much stronger than the one discussed so far, namely an exponential maximal inequality, which is the subject of the succeeding section.

4. Amenable versus non-amenable groups

4.1. Spectral gaps and operator norm estimates. There is another distinctly non-Euclidean phenomenon appearing in the discussion of radial averages on semisimple Lie groups (as well as many other groups, e.g. lattice subgroups, semisimple algebraic groups, hyperbolic groups, and others). This phenomenon is the existence of a spectral gap in the regular unitary representation of G on itself. The existence of a spectral gap bears on both of the questions we raised in §1.1, namely the norm of the radial averaging operators, and the behaviour of the associated maximal functions.

Let us first note briefly some facts regarding unitary representations with a spectral gap of semisimple Lie groups (we refer to [**N3**], [**MNS**] for a more detailed discussion). The basis of this discussion is the following result, which is due to M. Cowling.

THEOREM 4.1 (**Estimate of spherical functions [Co1]**). *Assume that G is a connected non-compact simple Lie group with finite center, K a maximal compact subgroup. Let $(\tau, \mathcal{H}_\tau)$ denote a strongly continuous unitary representation of G. Then the following two conditions are equivalent.*

(1) *$\mathcal{H}_\tau$ does not contain a sequence of unit vectors asymptotically invariant under $\tau(G)$, i.e. a sequence satisfying $\|\tau(g)v_k - v_k\| \to 0$ uniformly on compact subsets of G.*
(2) *There exist $\eta_\tau > 0$ such that the following holds for every $g \in G$, any two K-invariant vectors $v, w \in \mathcal{H}_\tau$, and every $\eta < \eta_\tau$:*

$$|\langle \tau(g)v, w\rangle| \leq C_\eta e^{-\eta|g|} \|v\| \|w\| ,$$

where $|g| = d(gK, K)$, *d the Riemannian metric associated with Killing form on* G/K.

The representation $(\tau, \mathcal{H}_\tau)$ is said to have a spectral gap if these equivalent conditions hold. The maximum value of η_τ that satisfies the above condition can be taken to parameterize the size of the spectral gap.

We recall that by definition, a group G satisfies property T if every unitary representation which does not contain invariant unit vectors does not contain an asymptotically invariant sequence, i.e. has a spectral gap. In the semisimple case where $G = \prod_{i=1}^N G_i$ is a product group, and G_i are connected simple with finite center, a representation $(\tau, \mathcal{H}_\tau)$ of G is said to have a strong spectral gap if the restriction of τ to every non-compact simple component G_i has no $\tau(G_i)$ asymptotically invariant sequence.

If τ has a spectral gap with parameter η_τ, then for a family of bi-K-invariant probability measures ν_t there exists a constant $\theta > 0$ satisfying : $\|\tau(\nu_t)\| \leq B\exp(-\theta t)$, if the following integral condition holds : for some $\eta < \eta_\tau$

$$\int_{\overline{\mathfrak{a}_+}} \exp(-\eta\,|\exp(H)|)d\nu_t^{rad}(H) \leq B\exp(-t\theta) ,$$

where $G = K\overline{A_+}K$ is the Cartan polar coordinates decomposition of G, $g = k_1\exp(H)k_2$, and ν_t^{rad} the radial part of ν_t (see §5.2 below).

Clearly, this is the case for all three of the bi-K-invariant averaging families β_t, γ_t and σ_t (w.r.t. the Killing form), namely these averages have exponentially decaying operator norm in every representation τ of G with a strong spectral gap. In particular, such an estimate holds for the regular representation π of a connected non-compact semisimple Lie group with finite center, which does indeed have a strong spectral gap in the sense above.

More generally, a group G is called non-amenable if the right regular representation satisfies the first condition given by Theorem 4.1, i.e. it admits no asymptotically invariant sequence. Equivalently, a group G is called amenable if it admits an asymptotically translation-invariant sequence of sets F_n of positive finite measure, namely a sequence with the following property. If χ_{F_n} is the characteristic function of F_n, and $f_n = \chi_{F_n}/\sqrt{m_G(F_n)}$ is the corresponding unit vector in $L^2(G, m_G)$ (m_G Haar measure), then $\lim_{n\to\infty} \|\pi(g)f_n - f_n\|_{L^2(G,m_G)} = 0$ for every $g \in G$ uniformly on compact sets in G (here $\pi(g)$ the right regular representation). Such a sequence is also called a (right) Følner sequence, and its existence implies immediately that *every* probability measure μ on G has the property that the operator norm $\|\pi(\mu)\| = 1$ in every $L^p(G)$, $1 \leq p \leq \infty$.

Referring to the first basic question posed in §1.1, namely the norm decay of the averaging operators, we note that for semisimple Lie groups, Theorem 4.1 gives exponential decay of the operator norms $\pi(\beta_t)$, $\pi(\gamma_t)$ and $\pi(\sigma_t)$ acting in $L^2(G)$, and hence by Riesz-Thorin interpolation in every $L^p(G)$, $1 < p < \infty$. Thus the answer in the semisimple case is very different from the answer for Abelian, or more generally, amenable groups, where the phenomenon of decay (with any rate at all) in the operator norm of the averaging operators $\pi(\mu_t)$ (π the regular representation) does not appear.

We note that the assumption of exponential decay of the operator norms $\|\pi(\beta_t)\|$ is satisfied by the representations π associated with many natural actions of semisimple groups (say without compact factors). In particular, it holds for the regular representation of G on $L^2(G)$, the representation on the symmetric space $L^2(G/K)$, the representation on $L_0^2(G/\Gamma)$ where Γ is a lattice subgroup, and L_0^2 the space of functions with zero integral. It holds also in the representation in $L^2(G/H)$, where H is a proper algebraic unimodular subgroup, and in $L^2(G/\Lambda)$ where Λ is discrete and not Zariski dense (see [**G**] or [**IN**]). Furthermore, if G is a simple non-compact Lie group and has property T, then of course *any* representation without invariant unit vectors has a spectral gap.

4.2. Exponential maximal inequality. As to the second basic question posed in §1.1 regarding the associated maximal functions, we note that for certain semisimple groups it is possible to obtain a much stronger result than the strong maximal inequality for $\pi(\beta_t)$ and $\pi(\gamma_t)$ discussed in §2.1. This result is formulated as follows. As usual, we assume that G is a connected non-compact semisimple Lie group with finite center, K a maximal compact subgroup, and π the representation of G by right translations on itself. We consider the bi-K-invariant averages on G, defined using the Riemannian metric on G/K given by the Killing form.

THEOREM 4.2 (**Exponential-maximal inequality for ball averages**[**N5**]). *Let G be a connected semisimple Lie group with finite center. Then the bi-K-invariant ball averages β_t satisfy an $(L^p \to L^p)$ exponential-maximal convolution inequality, namely there exist positive constants $\theta_p = \theta_p(G)$, satisfying, for all $f \in L^p(G)$, $1 < p < \infty$:*

$$\left\|\sup_{t\geq 1} \exp(t\theta_p)\left|\pi(\beta_t)f(x)\right|\right\|_{L^p(G)} \leq B_p \left\|f\right\|_{L^p(G)} .$$

A similar statement holds for the family γ_t of shell averages.

We note also the following result.

THEOREM 4.3 (**Exponential maximal inequality for sphere averages in groups of real-rank one** [**I**]). *Let G be a connected simple Lie group of real-rank one and finite center. Then the sphere averages satisfy an $(L^p \to L^p)$ exponential-maximal convolution inequality, namely there exist positive constants $\eta_p = \eta_p(G)$, satisfying, for all $f \in L^p(G)$, $\frac{\dim_{\mathbb{R}}(G/K)}{\dim_{\mathbb{R}}(G/K)-1} < p < \infty$:*

$$\left\|\sup_{t\geq 1} \exp(t\eta_p)\left|\pi(\sigma_t)f(x)\right|\right\|_{L^p(G)} \leq B_p \left\|f\right\|_{L^p(G)} .$$

Remarks regarding the exponential-maximal inequalities.

(1) The norm inequality of Theorem 4.2 implies of course a pointwise bound holding almost everywhere, of the form

$$\left|\pi(\beta_t)f(x)\right| \leq C_p(f,x)\exp(-t\theta_p)$$

for every $f \in L^p(G)$, where the estimator satisfies the norm inequality $\left\|C_p(f,\cdot)\right\|_{L^p(G)} \leq B\left\|f\right\|_{L^p(G)}$.

(2) The sphere averages σ_t satisfy the exponential-maximal inequality in L^p only for $p > n/(n-1)$. Indeed, as noted in §2.1, on $\mathbb{H}^n$ the sphere averages do not even satisfy the strong maximal inequality in L^p, $p \leq \frac{n}{n-1}$. Nevertheless, the operator norm of $\pi(\sigma_t)_{L^p \to L^p}$ does decay exponentially in every L^p, $1 < p < \infty$. Thus the exponential-maximal inequality in L^p, $1 < p < \infty$ formulated in Theorem 4.2 is not simply a consequence of

exponential decay of the operator norm, but requires additional regularity conditions on the averages involved. For a discussion of such regularity conditions, which are not very stringent and are satisfied by many families of measures μ_t (including non radial families, and some singular ones) we refer to [**N3**][**N4**].

(3) Note also the simple fact that for the spheres $\sigma_{n\delta}$, $n \in \mathbb{N}$ ($\delta > 0$ fixed) the maximal operator $\sup_{n\in\mathbb{N}} \exp(\eta_p n\delta) |\pi(\sigma_{n\delta})f(x)|$ is bounded in L^p, where $1 < p < \infty$, and this fact is indeed an obvious corollary of the exponential decay of the operator norms $\|\pi(\sigma_{n\delta})\|$. Thus the problem of the boundedness of the maximal operator arises only for continuous families μ_t.

(4) It is possible to prove exponential maximal inequalities much more general than those of Theorem 4.2 and Theorem 4.3. Such results apply to general measure-preserving actions of the group, not only its action by convolutions, and require additional spectral and other considerations. We refer to [**N5**] for the details.

4.3. Non-amenable groups : some open problems. In view of the discussion of amenability in §4.1 and Theorem 4.2, it is natural to formulate the following problems. Let G be an lcsc compactly generated group, d an admissible metric, and β_t the corresponding ball averages. Assume G is a non-amenable unimodular group, and let π be the regular representation.

(1) **Norm decay problem.**
Is the condition $\|\pi(\beta_t)\|_{L^2(G)} < 1$, for all $t \geq t_0$, equivalent to the condition $\lim_{t\to\infty} \|\pi(\beta_t)\|_{L^2(G)} = 0$?

(2) **Exponential decay of convolution norms.**
For $1 < p < \infty$ does there exist $\theta_p > 0$, satisfying for all $f \in L^p(G)$

$$\|\pi(\beta_t)\|_{L^p(G)} \leq C_p \exp(-\theta_p t) \|f\|_{L^p(G)} \text{ ?}$$

(3) **Exponential maximal inequality.**
For $1 < p < \infty$ does there exist $\theta_p > 0$, satisfying for all $f \in L^p(G)$

$$\left\| \sup_{t\geq 0} \exp(\theta_p t) |f * \beta_t| \right\|_{L^p(G)} \leq C_p \|f\|_{L^p(G)} \text{ ?}$$

We remark that when the metric is integer valued, a positive solution to the second problem immediately implies a positive solution to the third problem (see e.g. remark (3) above or [**N3**]). As evidence for the plausibility of a positive solution to these problems we can cite (in addition to Theorem 4.2) a number of cases where a positive solution to the exponential norm decay problem has been obtained. These include the case of sphere and ball averages w.r.t. an arbitrary word metric on a hyperbolic group (see [**N3**]), and uniform lattices in $SL_3(\mathbb{R})$ and $SL_3(\mathbb{C})$. These results are based on the property of rapid decay established in the first case in [**J**] and in the second by [**La**]. Similar results hold for one-relator groups, for groups acting simply transitively on $\tilde{A}_2$ buildings (where the balls are w.r.t. a well chosen set of generators [**RRS**]) and for lattices in semisimple Lie groups (where the balls are defined by the Riemannian metric on the symmetric space [**N3**]).

We note that the remarkable phenomenon of rapid decay (discussed e.g. in [**J**],[**La**]) is very useful and has many important applications, but it cannot be regarded as a general principle that one might hope to establish for non-amenable groups. In fact, even for the non-uniform lattice $SL_3(\mathbb{Z})$ rapid decay fails, but one would still expect that exponential decay of the convolution norm of the balls holds.

Finally, we remark that an interesting case to examine the problem of exponential decay of convolution norms is that of non-amenable Baumslag-Solitar groups, which satisfy only a weak (logarithmic) form of an isoperimetric inequality.

5. Euclidean and non-Euclidean maximal inequalities in higher rank

We now turn to discuss higher rank phenomena, and the contrast between the behaviour of some natural averaging operators on Euclidean versus non-Euclidean symmetric spaces. This discussion is interesting in its own right, but also sheds some light on phenomena that might be expected to arise in general product groups, and general semisimple algebraic groups of higher rank.

We will use exponential volume growth on semisimple Lie groups together with convolution inequalities related to a quantitative form of the wave front Lemma, in order to establish some general maximal inequalities which are in rather striking contrast to the Euclidean case. To explain our direction, we first consider a very simple special case, which hopefully will serve to illustrate some of the ideas involved.

5.1. Averages on product groups. Consider the simplest real-rank two case, namely when G is a product $G = L_1 \cdot L_2$ of two groups of real rank one, say $L_1 = L_2 \cong PSL_2(\mathbb{R})$ for definiteness. The ball averages on the factor groups L_i are given in this case by

$$\beta_t = \frac{\int_0^t \sinh s \; \sigma_s \; ds}{\int_0^t \sinh s \; ds} \, .$$

Let us define $D_t = \{(u, v) \, ; \, u^2 + v^2 \leq t^2\}$. Then the ball averages on $G = L_1 \cdot L_2$ satisfy :

$$\beta_t^G = \frac{\int_{D_t} |\sinh u \sinh v| \, \sigma_{|u|}^{(1)} \sigma_{|v|}^{(2)} du dv}{\int_{D_t} |\sinh u \sinh v| \; du dv} \leq B \frac{\sum_{D_t \cap \mathbb{N}^2} \exp n \exp m \; \gamma_n^{(1)} \gamma_m^{(2)}}{\sum_{D_t \cap \mathbb{N}^2} \exp n \exp m} \, ,$$

where $\gamma_t^{(i)}$ (resp. $\sigma_t^{(i)}$) are the unit shell (resp. sphere) averages on the real-rank one group L_i. On the l.h.s. we have the integral w.r.t. the G-invariant Riemannian volume, which in geodesic polar coordinates is given by the integral w.r.t. the density $|\sinh u \sinh v|$ on a disc of radius t in $\mathbb{R}^2$. Using the Weyl group symmetry, this integral is equal to four times its value on $D_t \cap \mathbb{R}_+^2$. On the r.h.s. the integral was then further replaced by the weighted sum of integrals (w.r.t. the uniform Euclidean measure) on unit squares whose lower left hand corner is a lattice point of norm at most t. The weight attached to each unit square is the obvious one : we estimate the function $\sinh u \sinh v$ on the square whose lower left hand corner is (n, m) by C $\exp n$ $\exp m$. Since the shell averages on L_i satisfy the strong maximal inequality in L^p, $1 < p \leq \infty$ by [**NS2**], it follows that the same holds for the operators $\beta_t^G = \beta^{L_1 \times L_2}$

The support of $\gamma_t^{(1)} \gamma_s^{(2)}$, $t, s \geq 0$ has radial coordinates which constitute (the W-orbit of) a square of unit side length, whose lower left hand corner is the element (tH_1, sH_2). Let us now note the fact that already the family of unit square averages $\gamma_t^{(1)} \gamma_s^{(2)}$ (for arbitrary non-negative s and t) satisfies a strong maximal inequality in L^p, $p > 1$. Indeed, each square average is the product of two one-dimensional shell (or "interval", in this context) averages on real rank one groups. As a result, any strong maximal inequality which holds for the unit square averages can be established for any family of sets which admit a reasonable covering by unit squares, not only for discs (which correspond to Riemannian balls in G/K). To take a very simple example, consider the dilations $U_t = tU$ of a fixed convex open neighbourhood U of the origin in $\mathfrak{a} \cong \mathbb{R}^2$. In this case we can compare the bi-K-invariant average on G whose radial components lie in U_t, to the weighted average formed by the corresponding convex combination of square averages (with sufficiently small side-length) as was done above for the disc D_t. Thus as in the case of the ball averages w.r.t. the Killing metric, the foregoing comparison implies a strong maximal inequality in a similar fashion for the normalized averages with radial coordinates in U_t. In particular, the shell (and ball) averages associated with any G-invariant Riemannian metric on the symmetric space G/K satisfy a strong maximal inequality in L^p, $1 < p \leq \infty$, and the same holds for any admissible invariant bi-K-invariant metric whose balls admit a covering as described above. It is possible to greatly expand the scope and generality of results of this kind, and we proceed to describe some of the results for higher rank semisimple groups. More details on the comparison to the Euclidean case will be found in below, together with some remarks and open problems.

5.2. Polar coordinates on symmetric spaces. We devote this section to recalling some of the terminology related to semisimple groups and symmetric spaces. Let G denote a connected semi-simple Lie group with finite center, $\mathfrak{g}$ its Lie algebra, $\mathfrak{g} = \oplus_{i=1}^N \mathfrak{g}_i$, where $\mathfrak{g}_i$ are the simple factors of $\mathfrak{g}$. Let θ denote a Cartan involution on G and $\mathfrak{g}$, and let $\mathfrak{g} = \mathfrak{k} \oplus \mathfrak{p}$ be the corresponding Cartan decomposition, so that $\mathfrak{k}$ is the Lie subalgebra corresponding to a connected maximal compact subgroup K.

Let $\mathfrak{a} \subset \mathfrak{p}$ denote a maximal Abelian subalgebra, and let $\Phi(\mathfrak{a}, \mathfrak{g}) = \Phi \subset \mathfrak{a}^*$ denote the (real) root system of $\mathfrak{a}$ in $\mathfrak{g}$. Let $\mathfrak{g}_\alpha$ denote the root space corresponding to $\alpha \in \Phi$, and $\mathfrak{g} = \mathfrak{m} + \mathfrak{a} + \sum_{\alpha \in \Phi} \mathfrak{g}_\alpha$ the root space decomposition. Fix a system of simple roots $\Delta \subset \Phi$, the corresponding ordering of $\mathfrak{a}^* = \hom(\mathfrak{a}, \mathbb{R})$ and the system of positive roots Φ_+, and let ρ denote half the sum of the positive roots. Let $W = W(\mathfrak{a}, \mathfrak{g})$ denote the Weyl group of the root system, $\mathfrak{a}_+$ the positive Weyl chamber, and $\overline{\mathfrak{a}_+}$ its closure. Let $A = \exp \mathfrak{a}$ denote the Lie subgroup corresponding to $\mathfrak{a}$, $A_+ = \exp \mathfrak{a}_+$ and $\overline{A_+}$ its closure. The Cartan (or polar coordinates) decomposition in G is given by $G = K\overline{A_+}K$ and $g = k_1 e^{H(g)} k_2$, where $H(g)$ is an $\overline{\mathfrak{a}_+}$ component of g. We recall that $H(g)$ is uniquely determined for $g \in G$. Let $\langle , \rangle$ denote the Killing form on $\mathfrak{g}$, and let d denote the induced Riemannian metric on the symmetric space G/K. Then the restriction of $\langle , \rangle$ to $\mathfrak{a}$ is an inner product, and we have $d(\exp(H)o, o) = \sqrt{\langle H, H \rangle}$ for all $H \in \mathfrak{a}$, where $o = [K]$ denotes our choice of origin in G/K. Let $\hat{\Delta} = \{\hat{H}_1, \ldots, \hat{H}_r\} \subset \mathfrak{a}$ denote the basis of $\mathfrak{a}$ dual to the chosen basis $\Delta \subset \mathfrak{a}^*$ w.r.t. the natural pairing. We recall that the Cartan polar coordinates decomposition yield the following well-known integration formula for Haar measure on G

$$\int_G f(g)dm_G(g) = \int_K \int_{\overline{\mathfrak{a}_+}} \int_K f(ke^H k')\,\xi\,(H)\,dm_K\,(k)dH\,dm_K(k')\,.$$

Here $\xi(H) = \prod_{\alpha\in\Phi_+}(\sinh\alpha(H))^{m_\alpha}$, $H \in \overline{\mathfrak{a}_+}$, $m_\alpha = \dim_{\mathbb{R}}\mathfrak{g}_\alpha$, and m_G, m_K denote Haar measures on G and K, dH denotes Lebesgue measure on $\mathfrak{a}$. Note that $|\xi|$ coincides with the W-invariant extension of $\xi(H)$ from $\overline{\mathfrak{a}_+}$ to all of $\mathfrak{a}$, W the Weyl group of $\mathfrak{a}$ in $\mathfrak{g}$.

5.3. Some averages of product type on higher rank groups. We now introduce some natural bi-K-invariant probability measures on G (See for more details [**MNS**],[**CN**],[**N4**]). First note that any bi-K-invariant measure ν on G determines uniquely a measure ν^{rad} on $\mathfrak{a}$ which is W-invariant. We will usually describe a bi-K-invariant measure on G by specifying its associated W-invariant radial measure on $\mathfrak{a}$. On occasion, just the part in a fundamental domain of W in $\mathfrak{a}$ (namely in a Weyl chamber) will be specified.

DEFINITION 5.1. (1) **Directional sphere averages.** Fix $H \in \mathfrak{a}$, and let

$$\sigma^H = m_K * \delta_{\exp H} * m_K \ , \ H \in \mathfrak{a}\,.$$

(2) **Directional d-shell averages.** Fix $H \in \mathfrak{a}$, and let $\gamma_d^{t,H} = \int_t^{t+d}\sigma^{sH}ds$.
(3) d**-cube in** $\mathfrak{a}$. Given any basis $H_1,\cdots,H_r$ of $\mathfrak{a}$, the d-cube $R_d = R_d(0)$ is defined by

$$R_d = \{\sum_{i=1}^r s_iH_i \mid 0 \le s_i \le d\}\,,\ \text{and}\ R_d(H) = R_d + H\,.$$

(4) d**-cube averages.** Let $vol_{\mathfrak{a}}$ denote the Euclidean volume on $\mathfrak{a}$, normalized by taking $vol_{\mathfrak{a}}(R_1) = 1$, and define

$$\gamma_d^H = \frac{1}{vol_{\mathfrak{a}}(R_d)}\int_{H'\in R_{\vec{d}}}\sigma^{H+H'}dH' = \frac{1}{d^r}\int_0^d\cdots\int_0^d\sigma^{H+\sum_{i=1}^r s_iH_i}ds_1\cdots ds_r\,.$$

By definition, γ_d^H is a bi-K-invariant measure whose associated radial measure $(\gamma_d^H)^{rad}$ on $\mathfrak{a}$ is the average over the W-orbit of the uniform measure on a Euclidean d-cell (determined by a given basis).

We denote the measure on $\mathfrak{a}$ with density $|\xi(H)|\,dH$ by vol_G, to distinguish it from the Euclidean volume of $\mathfrak{a}$, denoted by $vol_{\mathfrak{a}}$. A ball (w.r.t. the norm associated with the Killing form) with center H and radius d in $\mathfrak{a}$ will be denoted by $b_d(H)$.

We now turn to generalizing the considerations discussed in §5.1. To formulate the generalization, we first introduce :

DEFINITION 5.2. (1) **Ample sets**. Fix constants $d > 0$ and $D > 1$. For a given set $E \subset \mathfrak{a}$, we denote by $U_d(E)$ the d-neighbourhood of E, namely the union of all balls of radius d centered at all points of E.

A set $E \subset \mathfrak{a}$ (of positive finite measure) will be called (d, D)-ample, if it satisfies $vol_G(U_d(E)) \le D vol_G(E)$. (Note that for any W-invariant set $E \subset \mathfrak{a}$, $\frac{1}{|W|}vol_G(E) = m_G(K\exp(E)K)$, where m_G is Haar measure on G.)

(2) **Condition** $A_{(c,C)}$. A set $E \subset \mathfrak{a}$ (of finite measure) satisfies condition $A_{(c,C)}$ if for every $H \in E$, there exists $H' \in E$, satisfying the conditions $\|H - H'\| \le C$, and $b_c(H') \subset E$.

Examples of ample sets are provided by the following

EXAMPLE 5.3. The following subsets of $\mathfrak{a}$ are (d, D)-ample for some $d > 0$ and $D > 1$.

(1) Any set satisfying condition $A_{(c,C)}$ is (d, D)-ample for some (d, D) depending on (c, C).
(2) Any union of sets satisfying condition $A_{(c,C)}$ still satisfies condition $A_{(c,C)}$, and is hence (d, D)-ample.
(3) All balls in $\mathfrak{a}$ of radius at least a given positive constant.
(4) All d'-cubes in $\mathfrak{a}$ w.r.t. to a given basis of $\mathfrak{a}$ ($d' > 0$), and therefore also all cubes with side length at least d'.

Given a set $E \subset \overline{\mathfrak{a}_+}$ of positive finite measure, we consider the probability measure :

$$\nu_E = \frac{\int_E \sigma^H \, |\xi(H)| \, dH}{\int_E |\xi(H)| \, dH} \, .$$

The average ν_E is supported on the bi-K-invariant set $K \exp(E) K \subset G/K$, whose radial coordinates are the W-orbit of E. The probability measure ν_E is normalized by the G-invariant Riemannian volume of its support, since $vol(K \exp(E) K) = vol_G(E) = vol_G(W \cdot E)/ |W|$.

We can now formulate the following general maximal inequality :

THEOREM 5.4 (**Maximal inequality for ample sets [N4]**). *Let G be a non-compact semisimple Lie group with finite center, and π the representation of G itself by right translations. Assume $\mathfrak{a} \subset \mathfrak{g}$ has a strongly admissible basis (see definition 5.8 below). Fix positive constants $D > 1$, $d > 0$, and let*

$$\mathcal{A}^* f(x) = \sup \left\{ |\pi(\nu_E) f(x)| \; : \; E \subset \mathfrak{a}, \text{ and } E \text{ is } (d, D) \text{ ample} \right\} \, .$$

Then $\mathcal{A}^* f$ satisfies the strong L^p-maximal inequality :

$$\|\mathcal{A}^* f\|_p \le B_p(G, D, d) \, \|f\|_p$$

for all $f \in L^p(G)$, $1 < p \le \infty$.

A basic ingredient in the proof of Theorem 5.4 is a maximal inequality for averaging on small neighbourhoods in $\mathfrak{a}$, but with respect to the density $|\xi(H)| \, dH$. The cube averages γ_d^H are defined above w.r.t. the Euclidean measure dH. However, these averages give rise to equivalent maximal operators, namely, maximal operators which satisfy exactly the same maximal inequalities, according to the following

LEMMA 5.5. *Consider the d-ball averages, given by :*

$$\nu_{b_d(H)} = \frac{\int_{H' \in b_d(H)} \sigma^{H'} \, |\xi(H')| \, dH'}{\int_{H' \in b_d(H)} |\xi(H')| \, dH'}$$

and the associated the maximal operator, $M_{b_d}^* f(x) = \sup_{H \in \mathfrak{a}} \left| \pi(\nu_{b_d(H)}) f(x) \right|$. Then, in any measure-preserving action of G, $M_{b_d}^*$ is equivalent to the maximal operator

$$M_{\gamma_{d'}}^* f(x) = \sup_{H \in \mathfrak{a}} \left| \pi(\gamma_{d'}^H) f(x) \right|$$

associated with the d'-cube averages. The equivalence holds for any choice of basis of $\mathfrak{a}$, where d and d' are arbitrary positive constants.

We now introduce another natural family of averages, which will also provide an interesting comparison to the Euclidean case, as follows.

DEFINITION 5.6. (1) **Rectangles.** Let H_i, $1 \le i \le r$ be a basis of $\mathfrak{a}$. Given $\overrightarrow{t} = (t_1, \dots t_r) \in \mathbb{R}^r_+$ with $\min_{1\le i\le r} t_i > 0$, rectangles in $\mathfrak{a}$ are defined

$$R_{\overrightarrow{t}} = \left\{ H \in \mathfrak{a} \mid H = s_1 H_1 + \ldots + s_r H_r, 0 \le s_i \le t_i \right\} \quad , \quad R_{\overrightarrow{t}}(H_0) = H_0 + R_{\overrightarrow{t}}$$

(2) **Rectangle Averages.** The following probability measure on G :

$$\nu_{R_{\overrightarrow{t}}} \quad = \quad \frac{\int_{R_{\overrightarrow{t}}} |\xi(H)| \, \sigma^H \, dH}{\int_{R_{\overrightarrow{t}}} \xi(H) \, dH}$$

will be called rectangle averages.

Rectangles of minimum side length at least $c_0 > 0$ (but with arbitrary orientation of the axes) clearly satisfy condition $A_{(c,C)}$ for some $c, C > 0$, and therefore are ample sets. Hence Theorem 5.4 implies that the associated maximal operator satisfies the strong L^p, $1 < p \le \infty$ maximal inequality. This fact is in marked contrast to the Euclidean case, and we refer to §5.5 for more on this comparison.

In the next section we turn to explaining more about the main ingredients of the proof of Theorem 5.4. The method of proof will also allow us to formulate a stronger version of Theorem 5.4, namely a weak-type maximal inequality for averaging on ample sets.

5.4. Convolution inequalities : A quantitative wave front lemma. We now describe a geometric approach to proving maximal inequalities on symmetric spaces, for bi-K-invariant averages. This approach utilizes direct geometric comparison arguments based on structure theory, and yields certain convolution inequality formulated in Theorem 5.7 below. The convolution inequality can be viewed as a quantitative version of the wave front lemma on symmetric spaces, and the gist of it is as follows. Fix any basis $H_1, \cdots, H_r$ of $\mathfrak{a}$ where $H_i \in \overline{\mathfrak{a}_+}$, $1 \le i \le r$. Then the d-cube average γ_d^H, where $H = \sum_{i=1}^r t_i H_i$, $t_i \ge 0$, can be dominated by the convolution of the r directional d-shell averages $\gamma_d^{t_i, H_i}$, $1 \le i \le r$, each smoothed by convolution with β_ε, for any $\varepsilon > 0$. More precisely, the key convolution inequality is the following :

THEOREM 5.7 (**Quantitative wave-front Lemma** [N4]). *Let G be a connected non-compact semisimple Lie group with finite center. Fix $\varepsilon > 0$, $\delta > 0$, and $d > 0$. Assume $H_i \in \overline{\mathfrak{a}_+}$, $1 \le i \le r$ form a basis of $\mathfrak{a}$. Then there exists $C_G(\delta, \varepsilon, d)$ such that for any $t_i \ge 0$, and $H = t_1 H_1 + \cdots + t_t H_r$*

$$\nu_{b_{\varepsilon/32}(H)} \le C_G(\delta, \varepsilon, d) \, (\gamma_d^{t_1, H_1} * \beta_\varepsilon) * (\gamma_d^{t_2, H_2} * \beta_\varepsilon) * \cdots * (\gamma_d^{t_r, H_r} * \beta_\varepsilon) \ .$$

The utility of this inequality is particularly evident when $\mathfrak{a} \subset \mathfrak{g}$ has a strongly admissible basis, which we define as follows.

DEFINITION 5.8. (1) A basis $H_1, \cdots, H_r$ of $\mathfrak{a}$ will be called an admissible basis if $H_i \in \overline{\mathfrak{a}_+}$ for each $1 \le i \le r$, and there exists a real-rank-one subalgebra $\mathfrak{l}_i \subset \mathfrak{g}$, with $\mathbb{R} \cdot H_i \subset \mathfrak{l}_i$.

(2) If in addition the basis spans a closed Weyl chamber $\mathcal{C}$ as a cone (namely $\mathcal{C} = \mathbb{R}_+ \cdot H_1 + \cdots + \mathbb{R}_+ \cdot H_r$) then it will be called strongly admissible.

Let us therefore assume that each $\mathbb{R} \cdot H_i$ lies in a real-rank-one subalgebra $\mathfrak{l}_i$. Then the maximal function associated with the directional d-shell averages $\gamma_d^{t_i,H_i}$ can be dominated by the maximal function associated with the shell averages of width d supported on the group supported on the associated real-rank-one subgroup L_i. The latter maximal functions satisfy the strong L^p-maximal inequalities in a general measure-preserving action, as shown in [**NS2**], for $p > 1$. Assume now that we can find a basis $H_1, \cdots, H_r$ of such vectors which also span the positive Weyl chamber as a cone, namely a strongly admissible basis. Then the convolution inequality 5.7 can be applied and implies the strong maximal inequality for the d-cube averages γ_d^H, $H \in \overline{\mathfrak{a}_+}$. Using Lemma 5.5, we obtain finally a proof of Theorem 5.4.

Now note that the shell averages on a real-rank-one group L actually satisfy the weak-type $(1,1)$ maximal inequality in their action by convolution on the associated symmetric space L/K_L, as follows from [**Str**] together with [**NS2**](§3). This implies, using again the convolution inequality 5.7, that the higher rank cube averages γ_d^H, $H \in \overline{\mathfrak{a}_+}$ are dominated by a product of $2r+1$ averages each satisfying a weak-type $(1,1)$ inequality. Using a well-known result on the composition of $2r+1$ weak-type $(1,1)$ operators [**Fa**], we obtain a weak-type result (formulated below) for the maximal operator defined by convolution with cube averages on the symmetric space G/K, in the higher rank case. This weak type maximal inequality can then be extended to the maximal function $\mathcal{A}^*$ associated with averaging on all (d,D)-ample sets (for appropriate constants d and D).

In order to state the weak-type result alluded to above, we must first introduce the following function space, for any given σ-finite measure space (Y,m) (for a detailed discussion we refer to [**Fa**]).

$$L^{(k)}(Y) = \left\{ f : Y \to \mathbb{C}\,;\ \int_{|f|>t} \frac{|f|}{t}\left(\log\frac{|f|}{t}\right)^k dm < \infty\ ,\ \forall t > 0 \right\}.$$

$L^{(k)}(Y)$, $k \geq 1$ is a linear subspace of $L(\log L)^k(Y)$, and these spaces coincide if and only if $m(Y) < \infty$. On the other hand, $L^{(k)}$, $k \geq 1$ strictly contains $\cup_{p>1} L^p(Y)$ (see [**Fa**]).

We can now formulate the following weak-type result for the maximal operator associated with averaging on all ample sets.

THEOREM 5.9 (**Weak-type maximal inequality for ample sets** [**N4**]). *Let G be a connected semisimple Lie group with finite center and real rank r. Assume the basis $H_i \in \overline{\mathfrak{a}_+}$, $1 \leq i \leq r$ is strongly admissible. Then for every $\varepsilon > 0$,*

(1) *The maximal operator $M_{b_\varepsilon}^*$ associated with the averages $\nu_{b_\varepsilon(H)}$, $H \in \mathfrak{a}$ satisfies a weak-type maximal inequality in $L^{(2r)}(G)$, when acting by right convolution on G (or the symmetric space G/K). More explicitly, $\forall t > 0$*

$$m_G\left\{x \in G \mid M_{b_\varepsilon}^* f(x) > t\right\} \leq C_\varepsilon \int_{|f|>t} \frac{|f|}{t}\left(\log\frac{|f|}{t}\right)^{2r} dm_G\,,$$

where C is independent of f (and t). In particular, the maximal function $M_{b_\varepsilon}^* f$ is integrable on any set of finite measure, provided $f \in L^{(2r)}$.

(2) *The maximal operator $\mathcal{A}^*$ defined by averaging on all (d, D)-ample sets satisfies the maximal inequality in $L^{(2r)}(G)$ stated in* (1). In particular, the inequality holds for the maximal operators associated with d-cube averages γ_d^H defined using any given basis of $\mathfrak{a}$, and rectangle averages $\nu_{R_{\vec{t}}}$ (both of side length at least some fixed positive constant).

5.5. Comparison with the Euclidean case. We note the following regarding the comparison between the results stated in the previous section and some classical Euclidean averages which may be considered analogously.

(1) **Averaging on cubes of fixed size**. Theorem 5.9 namely the maximal inequality for d-cube averages, is in a striking contrast to the Euclidean case. The cubes $R_d(H)$ and the cube averages can be defined for any given basis of $\mathbb{R}^n$, but the maximal function that these averages define (i.e. taking the supremum over all $H \in \mathbb{R}^n$) is unbounded in *every* L^p, $1 \leq p < \infty$. Indeed, if f is a continuous non-negative function of unit L^1-norm, supported in the d-cube $R_d(0)$, then

$$M^*_{\gamma_d} f(H) \geq f * \gamma_d^H(H) = \int_{R_d(0)} f(v)dv = 1$$

so that the maximal function has no strong (or weak) L^p-bound, $1 \leq p < \infty$. Of course, it follows that the same holds for the (larger) maximal operator associated with averaging on all (d, D)-ample sets in $\mathbb{R}^n$ (with the obvious definition).

(2) **Rectangles averages**. We can also consider the Euclidean analogs of the maximal operator associated with averaging on all the rectangles $R_{\vec{t}}$ (with sides parallel to the directions determined by a given basis) with $\min \vec{t} \geq c_0 > 0$. We observe that on $\mathbb{R}^n$ any strong maximal inequality satisfied by this operator is also satisfied by the maximal operator $\mathcal{S}^*$ of averaging on all rectangles with sides parallel to the basis vectors, which are centered at a given point (namely replacing the condition $t \geq c_0$ by the condition $t > 0$). This fact follows from a dilation argument similar to the one given in the proof of Theorem 3.1.

The operator $\mathcal{S}^*$ and many of its variants were considered in detail in the classical Abelian theory (see e.g. [**S2**](Ch. X)). In particular, $\mathcal{S}^* f = \sup_{\min \vec{t} > 0} |\nu_{R_{\vec{t}}} * f|$ is bounded in $L^p(\mathbb{R}^n)$, $1 < p \leq \infty$. This is an immediate consequence of an n-fold iteration of the classical Hardy-Littlewood maximal inequality. In fact, the Jessen-Marcinkiewicz-Zygmund maximal inequality (see e.g. [**S2**][**Fa**]) implies that $\mathcal{S}^*$ is also bounded on $L^{(n-1)}(\mathbb{R}^n)$ (see §5.4 for the definition of the function space). On the other hand, if we add to this collection also the rectangle averages defined w.r.t. an arbitrary orthogonal rotation of the axes, the corresponding maximal operator is unbounded in every L^p, $1 \leq p < \infty$ (see [**S2**](Ch. X, §2.2)).

By way of comparison, Theorem 5.9 gives a bound for the maximal operator associated with rectangles in $\mathfrak{a}$ (satisfying $\min \vec{t} \geq c_0 > 0$), together with those obtained from them by applying an arbitrary orthogonal rotation of the axes. This follows since all of the rectangles in question satisfy the same $A_{(c,C)}$-condition and hence are all (d, D)-ample with the same constants. This is again in striking contrast to the Euclidean case.

(3) **Poly-radial averages on $\mathbb{R}^n$.** An interesting class of averages on $\mathbb{R}^n$ which may be compared to the bi-K-invariant product-type averages such as cube and rectangles averages discussed above is the class of poly-radial averages on $\mathbb{R}^n$. For example, we can consider the probability measures on $\mathbb{R}^4$ invariant under the natural diagonal action of $SO(2,\mathbb{R}) \times SO(2,\mathbb{R})$. These measures have poly-radial coordinates parametrized by two real parameters, and so we can form the "poly-radial cubes" analogous to those discussed on $\mathbb{H}^2 \times \mathbb{H}^2$ in §5.1. Clearly, the poly-radial cubes satisfy a maximal inequality in L^p, $p > 2$, since the shell averages on $\mathbb{R}^2$ satisfy a maximal inequality in L^p, $p > 2$. Similarly, sets in the parameter plane which are well-covered by squares will satisfy every maximal inequality that the squares satisfy.

5.6. Averages on higher rank groups : some open problems. We note some natural open problems regarding averages on higher rank groups which the discussion above raises. We assume here that G is a connected semisimple Lie group with finite center and real-rank at least two, acting by right convolutions on itself.

(1) **L^1-problem for the cube averages.** Is there a weak-type $(1,1)$ maximal inequality averages for the cube averages γ_d^H on a symmetric space?
(2) **Maximal inequality for unrestricted rectangles.** Is the strong L^p, $p > 1$ maximal inequality (or possibly the weak-type $(1,1)$) valid for the maximal operator associated with all rectangle averages with sides parallel to a given basis vectors, without the restriction $\min \overrightarrow{t} \geq c_0 > 0$? Are these maximal inequalities valid for the (larger) maximal operator associated with the set of all rectangle averages obtained also from all possible rotations of a basis ?

6. Proper actions versus ergodic actions

The regular action of G on itself by right translations, and more generally on homogeneous spaces $X = G/K$ with K compact are examples of proper actions. Such actions have a particularly simple orbit structure, and their study can easily be reduced to the case of the regular action. The results presented above for the regular action all hold without change for general proper actions, and the assumption of homogeneity is not necessary.

At the other extreme, we can consider the case where $m(X) < \infty$ and the action is ergodic, namely every G-invariant set has full measure or zero measure. Such actions have an extremely complicated orbit structure and bear no obvious relation to the regular action of G on itself, unless the group G is compact, which we assume is not the case.

When the group G is an amenable group, there exists a fundamental transfer principle, which allows the reduction of the proof of maximal inequalities that arise in a general ergodic action to the corresponding ones for the regular action. This is an indispensable tool in the ergodic theory of amenable groups, but which has as yet no analog for non-amenable group. We refer to [**N6**] for a comprehensive discussion of the transfer principle for amenable groups.

We note that for those non-amenable groups where the classification of irreducible unitary representations is known (for example for the groups $SL_2(\mathbb{R})$ and $SL_2(\mathbb{Q}_p)$), the regular representation turns out to be supported on a small subset of the unitary dual. Thus the regular representation of a non-amenable group is not representative of general unitary representations, which arise in general ergodic actions. One significant exception is the case of actions with a spectral gap of semisimple Lie groups. Then there exists a spectral transfer principle which allows the reduction of the norm decay problem for such an action to the same problem for the regular representation. In particular, this principle makes it possible to establish exponential-maximal inequalities for ball averages (as well as many other radial and non-radial averages) on a semisimple Lie group, in an arbitrary probability measure-preserving action with a spectral gap. We refer to [**MNS**] for the case of ball averages, and to [**N3**],[**N4**] for other more general averages, including the cube averages γ_d^H. In particular, these results show that the convergence of the averaging operators to the ergodic mean is at an exponential rate, for almost every starting point. Thus these results establish, in the setting of semisimple Lie groups, a much stronger version of the maximal and pointwise ergodic theorems for ball averages in ergodic actions of $\mathbb{R}^n$ due to N. Wiener [**W**] (at least for the case where $p > 1$). No rate of convergence result can hold for the case of properly ergodic actions of Abelian, or more generally, amenable groups, due to to the existence of asymptotically invariant sets. Thus our discussion of exponential-maximal inequalities in the regular representation of non-amenable groups in §4 can be generalized to the set-up of probability measure-preserving ergodic actions of semisimple groups. We refer to [**N6**] for a survey on these and other maximal and ergodic theorems for actions of connected Lie groups.

References

[Bn] J. Bourgain, *Averages in the plane over convex curves and maximal operators.* J. Anal. Math.**47** (1983), pp. 69–85.

[BJR1] C. Benson, J. Jenkins and G. Ratcliff, *On Gelfand pairs associated with solvable Lie groups.* Trans. Amer. math. Soc. **321** (1990) pp. 85-116.

[BJR2] C. Benson, J. Jenkins and G. Ratcliff, *The orbit method and Gelfand pairs associated with nilpotent groups.* J. Geom. Anal. **9** (1999), pp. 569-582.

[Ca] A. Calderon, *A general ergodic theorem.* Ann. of Math. **58** (1953), pp. 182-191.

[CM] M. Christ and D. Müller, *On L^p-spectral multipliers for a solvable Lie group.* Geom. & Func. Anal. **6** (1996), pp. 860-876.

[CS] J-L. Clerc and E. M. Stein, *L^p-multipliers for noncompact symmetric spaces.* Proc. Nat. Acad. Sci. U.S.A. **71**(1974), pp. 3911–3912.

[Coo] M. Coornaert, *Mesures de Patterson-Sullivan sur le bord d'un espace hyperbolique au sens de Gromov.* Pacific J. Math., **159** (1993), pp. 241-270.

[Co1] M. Cowling, *Sur les coefficients des représentations unitaires des groupes de Lie simples.* In : Analyse harmonique sur les groupes de Lie. Séminaire Nancy–Strasbourg 1975–77, pp. 132–178. Lecture Notes in Mathematics, vol. 739, Springer Verlag, 1979.

[Co2] M. Cowling, *On Littlewood–Paley–Stein theory.* Suppl. Rend. Circ. Mat. Palermo **1** (1981), pp. 21-55.

[CMS] M. Cowling, S. Meda and A. G. Setti, *An overview of harmonic analysis on the group of isometries of a homogeneous tree.* Exposition. Math. **16** (1998), pp. 385-423.

[CN] M. Cowling and A. Nevo, *Uniform estimates for spherical functions on complex semisimple Lie groups.* GAFA, Geom. Fuct. Anal. **11** (2001), pp. 900-932.

[EK] A. El-Kohen, *Maximal operators on hyperboloids.* J. Operator Theory **3** (1980), pp. 41-56.

[Fa] N. A. Fava, *Weak-type inequalities for product operators.* Studia Math. **XLII** (1972), pp. 271-288.

[FN] K. Fujiwara and A. Nevo, *Maximal and pointwise ergodic theorems for word-hyperbolic groups.* Erg. Th. & Dyn. Sys. **18** (1998), pp. 1-16.

[G] Y. Guivarc'h, *Sur la loi de grands nombres et la rayon spectrale d'une march aléatoire.* Conference on random walks, Astérisue **74**, pp. 47-98, Soc. Math. FRance (1980).

[H] U. Haagerup, *An example of a non-nuclear C^*-algebra which has the metric approximation property.* Inven. Math. **50** (1979), pp. 279-293.

[HL] G. H. Hardy and J. E. Littlewood, *A maximal theorem with function theoretic applications.* Acta Math. **54** (1930), pp. 81-116.

[I] A. Ionescu, *Fourier integral operators on non-compact symmetric spaces of real rank one.* J. Funct. Anal. **174** (2001), pp. 274-300.

[IN] A. Iozzi and A. Nevo, *Algebraic hulls and the Følner property.* Geom. & Funct. Anal. **6** (1996), pp. 666-688.

[J] P. Jolissaint, *Rapidly decreasing functions in reduced C^*-algebras of groups.* Trans. Amer. Math. Soc. **317** (1990), pp. 167-196.

[Ka] D. A. Kazhdan, *On a connection between the dual space of a group and the structure of its closed subgroups.* Funct. Anal. Appl. **1**(1967), pp.63–65.

[La] V. Lafforgue, *A proof of property (RD) for cocompact lattices of $SL(3,\mathbb{R})$ and $SL(3,\mathbb{C})$.* J. Lie Theory **10** (2000), pp. 255-267.

[MS] D. Müller and A. Seeger, *Singular spherical maximal operators on a class of step two nilpotent Lie groups.* Preprint, 2003.

[MNS] G. A. Margulis, A. Nevo and E. M. Stein, *Analogs of Wiener's ergodic theorems for semi-simple Lie groups II.* Duke Math. J.,**103**, (2000), pp.233–259.

[N0] A. Nevo, *Harmonic analysis and pointwise ergodic theorems for non-commuting transformations.* J. Amer. Math. Soc. **7** (1994), pp. 875-902.

[N1] A. Nevo, *Pointwise ergodic theorems for radial averages on simple Lie groups I.* Duke Math. J., **76** (1994), pp.113–140.

[N2] A. Nevo, *Pointwise ergodic theorems for radial averages on simple Lie groups II.* Duke Math. J., **86** (1997), pp. 239–259.

[N3] A. Nevo, *Spectral transfer and pointwise ergodic theorems for semi-simple Kazhdan Lie groups.* Mathematical Research Letters, **5** (1998), pp. 305–325.

[N4] A. Nevo, *Exponential volume growth, maximal functions on symmetric spaces, and ergodic theorems for semisimple Lie groups.* Submitted (2003).

[N5] A. Nevo, *Best possible speed of convergence in pointwise ergodic theorems on semisimple groups.* Preprint.

[N6] A. Nevo, *Pointwise ergodic theorems for actions of connected Lie groups.* Handbook of Dynamical Systems, Eds. B. Hasselblatt and A. Katok, to appear.

[NS1] A. Nevo and E. M. Stein, *A generalization of Birkhoff's pointwise ergodic theorem.* Acta Math. **173** (1994), pp. 135-154.

[NS2] A. Nevo and E. M. Stein, *Analogs of Wiener's ergodic theorems for semi-simple Lie groups I.* Ann. of Math., **145** (1997), pp. 565–595.

[NT] A. Nevo, and S. Thangavelu, *Pointwise ergodic theorems for radial averages on the Heisenberg group.* Adv. in Math., **127** (1997), pp. 307–334.

[RRS] J. Ramagge, G. Robertson and T. Steger, *A Haagerup inequality for $\tilde{A}_1 \times \tilde{A}_1$ and $\tilde{A}_2$ buildings.* Geom. & Funct. Anal. **8** (1998), pp. 702-731.

[S1] E. M. Stein, *Topics in Harmonic Analysis Related to the Littlewood–Paley Theory.* Annals of Math. Studies, **63**, Princeton Univ. Press, 1970.

[S2] E. M. Stein, *Harmonic Analysis,* Princeton Mathematical Series, **43**, Princeton Univ. Press, 1993.

[SW] E. M. Stein, and S. Wainger, *Problems in harmonic analysis related to curvature.* Bull. Amer. Math. Soc. **84** (1978), pp. 1239–1295.

[Str] J-O. Stromberg, *Weak type L^1 estimates for maximal functions on noncompact symmetric spaces.* Ann. of Math. **114** (1981), pp. 115–126.

[Wa] S. Wainger, *Averages and singular integrals over lower dimensional sets.* Beijing Lectures in Harmonic Analysis, Annals of Math. Studies **112** (1986), Princeton Univ. Press.

[W] N. Wiener, *The ergodic theorem.* Duke Math. J. **5** (1939), pp. 1–18.

Department of Mathematics, Technion IIT, Haifa 32000, Israel

E-mail address: `anevo@tx.technion.ac.il`

Contemporary Mathematics
Volume **347**, 2004

The heat kernel and the Green kernel of an infinite graph

Hajime Urakawa

ABSTRACT. This paper gives a new approach to estimate the Cheeger constant, the heat kernel and the Green kernel of the combinatorial Laplacian for an infinite graph.

1. Introduction

Random walks on infinite graphs are related to various fields of mathematics including probability and potential theory, harmonic analysis, geometry, topology, and discrete mathematics. Graph theory treats a completely discrete object, but surprisingly it has a closed relationship to Riemannian geometry. Many ideas in differential geometry have been brought into graph theory. This paper is one of them. In this paper, we want to determine and estimate the n-step transition probability of a simple random walk, i.e., the heat kernel, the Green kernel and the Cheeger constant, quite explicitly.

In this section, we first explain the Cheeger constant, the heat kernel and the Green kernel for an infinite graph. An infinite graph $\mathcal{G} = (V, E)$ is a collection of a set V of infinite number of vertices and a set E of edges connecting two vertices. We always assume that $\mathcal{G}$ is connected and *locally finite*, i.e.. for each $x \in V$, the degree $m(x) = \#\{y \in V;\, y \sim x\}$ is finite, where $y \sim x$ means that x and y are connected by an edge.

For an infinite graph $\mathcal{G} = (V, E)$, the *Cheeger constant* $h(\mathcal{G})$ is defined (cf. [**A**], [**DK**], [**DKa**], [**F1**], [**Mo**], [**MW**], [**T**], [**U3**]) by

$$h(\mathcal{G}) = \inf\left\{\frac{\mathrm{vol}(\partial S)}{\mathrm{vol}(S)};\, \emptyset \neq S \subset V, \#S < \infty\right\},$$

where $\mathrm{vol}(S) = \sum_{x\in S} m(x)$, and $\mathrm{vol}(\partial S) = \#(\partial S)$, $\partial S = \{e = (x,y) \in E;\, x \in S, y \notin S\}$. The spectrum $\sigma(\mathcal{G})$ of the discrete Laplacian Δ (cf. (2.2)) is estimated

1991 *Mathematics Subject Classification.* Primary 58J65, 05C50; Secondary 60C05.

Key words and phrases. infinite graph, heat kernel, Green kernel, Cheeger constant, semi-regular graph.

Supported by the Grant-in-Aid for Scientific Research, No.13440051, The Ministry of Education, Science, Sport and Culture.

by $h(\mathcal{G})$ as follows:

$$\sigma(\mathcal{G}) \subset \left[1 - \sqrt{1 - h(\mathcal{G})^2}, 1 + \sqrt{1 - h(\mathcal{G})^2}\right] \subset [0, 2].$$

Let $u_1(x,y) = p(x,y)$ be the one-step transition probability of the simple random walk on $\mathcal{G}$ starting at x to move to y, i.e., $p(x,y) = \frac{1}{m(x)}$ $(y \sim x)$, and $p(x,y) = 0$ (otherwise). Let $u_t(x,y)$ be the t-step transition probability from x to y for a non-negative integer t also called the *heat kernel*, where $u_0(x,y)$ is the Dirac measure at y. The *Green kernel* $G(x,y)$, x, $y \in V$, is defined by $G(x,y) = \sum_{t=0}^{\infty} u_t(x,y)$, and the graph $\mathcal{G} = (V,E)$ is called *transient* (resp. *recurrent*) if $G(x,y) < \infty$ (resp. $= \infty$) for all x and y in V. Stochastically, $G(x,y)$ is the expected value of the number of visits of the random walk starting at x to y.

Here, we give a brief survey on the study of the heat kernel and the Green kernel for an infinite graph. It is not difficult to see that the Green kernel of the homogeneous infinite tree T_d $(d \geqq 3)$ is given by

$$G(x,y) = \frac{(d-1)^{1-d(x,y)}}{d-2}$$

for all vertices x and y of T_d, where $d(x,y)$ is the distance between them. For upper and lower estimations of the Green kernel for a general infinite graph and many examples are given in [**U1**]. See also [**W2**]. Furthermore, in [**U4**], the notion of a harmonic morphism φ from a graph $G_1 = (V_1, E_1)$ onto $G_2 = (V_2, E_2)$ is introduced. That is, the composition $f \circ \varphi$ is harmonic at $x \in V_1$ whenever f is harmonic at $\varphi(x) \in V_2$. Then, the Green kernels satisfy that

$$G_1(x,y) \leqq \frac{m(y)}{m_h(y)} G_2(\varphi(x), \varphi(y)), \quad x, y \in V_1,$$

where $m_h(y) := \#\{z \in V_1;\ z \sim y,\ z \notin \varphi^{-1}(\varphi(y))\}$.

On the other hand, there are some studies of explicit determination of the heat kernel for some trees or the integer lattice, and the asymptotic behavior for abelian covering graphs of finite graphs (see also [**CG**], [**KSS**], [**H**]).

In this paper, we determine the Cheeger constant $h(\mathcal{G})$, the heat kernel $u_t(x,y)$ and the Green kernel $G(x,y)$ for a semi-regular infinite graph explicitly (cf. Theorem 5.7) and show lower estimate of the heat kernel for every infinite graph (cf. Theorem 7.1 and Corollary 7.3) and upper estimate under some condition (Theorem 6.1 and Corollary 6.7). We also illustrate by examples to show how our estimates are sharp in Sect. 8.

2. Preliminaries

The curvature of a Riemannian manifold plays an essential role in Riemannian geometry and analysis. We first want to define the notion of curvature in graph theory.

We consider a connected, locally finite graph $\mathcal{G} = (V,E)$. Recall that $\mathcal{G} = (V,E)$ is *connected* if for all x, y in V, there exists a path connecting them. The *graph distance* $d(x,y)$, x, $y \in V$, is the minimum of length of a path connecting x and y.

We fix arbitrarily a vertex $x_0 \in V$. Let us denote by $r(x) = d(x, x_0)$, $x \in V$ and we consider (cf. [**DKa**]) the function K_{x_0} on V, called the *curvature* (*polar at*

x_0), defined by

$$K_{x_0}(x) = \Delta r(x), \quad x \in V, \tag{2.1}$$

where the Laplacian Δ on the space $C(V)$ of all real valued functions on V is defined by

$$\Delta f(x) = f(x) - \frac{1}{m(x)} \sum_{y \sim x} f(y), \quad x \in V, \ f \in C(V). \tag{2.2}$$

To calculate the curvature K_{x_0}, let us define for each $x \in V$ with $x \neq x_0$,

$$\begin{aligned} m_+(x) &= \#\{z \in V;\ z \sim x, \text{ and } r(z) = r(x) + 1\}, \\ m_-(x) &= \#\{z \in V;\ z \sim x, \text{ and } r(z) = r(x) - 1\}, \\ m_0(x) &= \#\{z \in V;\ z \sim x, \text{ and } r(z) = r(x)\}. \end{aligned}$$

We have $m(x) = m_+(x) + m_0(x) + m_-(x)$. Furthermore, we have immediately

LEMMA 2.3. (*cf.* [**DKa**]) • *We have for $x \neq x_0$,*

$$K_{x_0}(x) = -\frac{m_+(x)}{m(x)} + \frac{m_-(x)}{m(x)},$$

and $K_{x_0}(x_0) = -1$.

• *For all $x \in V$ with $x \neq x_0$,*

$$-1 < \frac{2 - m(x)}{m(x)} \leqq K_{x_0}(x) \leqq 1.$$

• *In particular, $\mathcal{G}$ is a tree if and only if the equality $K_{x_0}(x) = \frac{2-m(x)}{m(x)}$ holds for all $x \in V$ with $x \neq x_0$, and x is extremal, i.e., $m_+(x) = m_0(x) = 0$, if and only if $K_{x_0}(x) = 1$.*

We give an example of homogeneous regular tree T_d ($d \geqq 2$) which is a regular tree with $m(x) = d$ for all vertices x. In this case, $K_{x_0}(x) = \frac{2-d}{d}$ for all x, that is, T_d is a constant negative curvature. The positivity of curvature $K_{x_0}(x) > 0$ means that $m_-(x) > m_+(x)$ by definition, so that $K_{x_0}(x) > 0$ for all $x \neq x_0$ implies that $G = (V, E)$ is a finite graph.

Upper and lower bounds of the curvature are particularly important. The upper and lower bounds of $K_{x_0}(x)$ are given by

$$-C + D \leqq K_{x_0}(x) \leqq -A + B \quad (x \neq x_0),$$

where

$$A = \inf\left\{\frac{m_+(x)}{m(x)};\ x \in V - \{x_0\}\right\}, \quad B = \sup\left\{\frac{m_-(x)}{m(x)};\ x \in V - \{x_0\}\right\},$$

$$C = \sup\left\{\frac{m_+(x)}{m(x)};\ x \in V - \{x_0\}\right\}, \quad D = \inf\left\{\frac{m_-(x)}{m(x)};\ x \in V - \{x_0\}\right\}.$$

Here $0 \leqq A < 1$, $0 < B \leqq 1$, $0 < C \leqq 1$, $0 \leqq D < 1$, $A + B \leqq 1$, and $C + D \leqq 1$.

3. The Cheeger constant

We have the following estimate of the Cheeger constant. For a proof, see [**U7**] or [**DKa**]. Let $\mathcal{G} = (V, E)$ be any infinite graph, and fix $x_0 \in V$. Let $r(x)$ be the distance function in V from x_0. For a finite subset $S \subset V$, put $r_1(S) = \min\{r(x); x \in S\}$, and $r_2(S) = \max\{r(x); x \in S\}$, and define $r_1^-(S) := \max\{0, r_1(S) - 1\}$, and denote them also by r_1, r_2, and r_1^-, for simplicity.

LEMMA 3.1. *Let S be any finite subset in V. Then we have,*

$$\mathrm{vol}(S) \leqq \mathrm{vol}(\partial S)\, \kappa(r_1(S), r_2(S)),$$

where

$$\kappa(r_1, r_2) = \begin{cases} \dfrac{1}{A-B}\left(1 - \left(\dfrac{B}{A}\right)^{r_2 - r_1^-}\right) + \max\{0, 1 - r_1\}\left(\dfrac{B}{A}\right)^{r_2}, & \text{if } A \neq B, \\[2ex] \dfrac{r_2 - r_1^-}{B} + \max\{0, 1 - r_1\}, & \text{if } A = B. \end{cases}$$

Due to Lemma 3.1, we have immediately

THEOREM 3.2. *For all infinite graph $\mathcal{G} = (V, E)$ and all non-empty finite subset S of V, we have*

$$\mathrm{vol}(S) \leqq \mathrm{vol}(\partial S)\left\{\frac{1}{A-B}\left(1 - \left(\frac{B}{A}\right)^{\mathrm{diam}(S)+1}\right) + \left(\frac{B}{A}\right)^{\mathrm{diam}(S)}\right\} \quad (A \neq B),$$

$$\mathrm{vol}(S) \leqq \mathrm{vol}(\partial S)\left\{\frac{\mathrm{diam}(S) + 1}{B} + 1\right\} \quad (A = B),$$

*where diam(S) is the diameter of S. Thus, we have $h(\mathcal{G}) \geqq A - B$ (cf. [**DKa**]).*
In particular, if $A > B$, then

$$\sigma(\mathcal{G}) \subset \left[1 - \sqrt{1 - (A-B)^2}, 1 + \sqrt{1 - (A-B)^2}\right] \subset [0, 2].$$

Remark that Theorem 3.2 is a discrete analogue of the following theorem in differential geometry (cf. [**Y**], [**K1**],[**K2**],[**K3**]):

THEOREM 3.3. *Let M be an n-dimensional simply connected complete Riemannian manifold whose sectional curvature is bounded above by a negative constant $-a$. Then the Cheeger constant $h(M)$ of M satisfies that $h(M) \geqq (n-1)a$. Then, the infimum of the spectrum of the Laplacian Δ of M satisfies that* $\inf \mathrm{Spec}(\Delta) \geqq \frac{1}{4}h(M)^2 \geqq \frac{(n-1)^2}{4}a$.

4. Semi-regular infinite graphs

In Riemannian geometry, the model space is particularly important. We adopt the following semi-regular graph as the model space like the space of constant curvature space or rotationally symmetric space in Riemannian geometry. In this section, we give examples of infinite graphs satisfying the condition $A > B$ of Theorem 3.2, which are also useful in the estimation of the heat kernel.

DEFINITION 4.1. For integers $e \geqq 1$, $d \geqq 2$, $d-1 \geqq a \geqq 1$, and a fixed vertex $x_0 \in V$, an infinite graph $\mathcal{G} = (V, E)$ is *semi-regular* (*centered at* x_0), denoted by $\mathcal{G} = \mathcal{G}_e^{d,a}$, if $m(x_0) = e$, and

$$m(x) = d,\ m_+(x) = a \text{ and } m_-(x) = 1 \quad (\text{for all } x \in V - \{x_0\}).$$

For every semi-regular infinite graph, $\frac{m_+(x)}{m(x)} = \frac{a}{d}$, $\frac{m_-(x)}{m(x)} = \frac{1}{d}$, for all $x \in V - \{x_0\}$, and $m_0(x) = d - 1 - a$, say c.

We have $h(\mathcal{G}_e^{d,a}) = A - B = \frac{a-1}{d}$, and $G(x, x_0) = \frac{1}{a-1} a^{1-r(x)}$, $x \in V$.

We call a semi-regular infinite graph $\mathcal{G}_e^{d,a}$ to be *rotationally symmetric at* x_0, if, for each $k = 1, 2, \cdots, \partial B_k(x_0)$ is rotationally homogeneous, i.e.. for all two elements x and y in $\partial B_k(x_0)$, there exists an isometry φ of $\mathcal{G}_e^{d,a}$ such that $\varphi(x_0) = x_0$ and $y = \varphi(x)$.

EXAMPLE 4.2. We can construct a semi-regular infinite graph which is rotationally symmetric at x_0, denoted $\widetilde{\mathcal{G}_e^{d,a}}$, in the following procedure:

(I) For each positive integer $k = 1, 2, \cdots$, let $K_{ea^{k-1}}$ be the complete graph of order ea^{k-1}. We put all vertices of $K_{ea^{k-1}}$ on a circle and label them counterclockwise by $x_1^k, \cdots, x_{ea^{k-1}}^k$. At each vertex x_j^k, we eliminate $ea^{k-1} - 1 - c$ edges which are connected to x_j^k and can be seen symmetrically left and right and seen to stand at the same position, looking at x_j^k. (Possibly we eliminate one edge doubly). We denote the resulting graph by $\partial B_k(x_0)$.

(II) We connect $\partial B_{k-1}(x_0)$ and $\partial B_k(x_0)$ in the following way:

(i) For $k = 1$, we connect by an edge, each vertex x_j^1 $(j = 1, \cdots, e)$ to x_0.

(ii) For $k = 2, 3, \cdots$, we connect by an edge, x_j^k $(j = (i-1)a + 1, \cdots, ia)$ to x_i^{k-1}, for each $i = 1, 2, \cdots, ea^{k-2}$.

Then, $\widetilde{\mathcal{G}_e^{d,a}}$ is rotationally symmetric at x_0.

PROPOSITION 4.3. *Let $\mathcal{G}_e^{d,a} = (V, E)$ be an arbitrary semi-regular infinite graph.*

(1) *If $c = 0$, $\mathcal{G}_e^{d,a}$ is always a tree.*

(2) *For each $x \in \partial B_k(x_0)$, where $k = 1, 2, \cdots$, let us define the cone $C_k(x)$ at the vertex x by the set of all vertices $y \in V$ such that there exists a chain $c = (y_0, y_1, \cdots, y_i)$ for some positive integer i such that $y = y_i \sim y_{i-1} \sim \cdots \sim y_1 \sim y_0 = x$ and $r(y_j) = k + j$ $(j = 0, 1, \cdots, i)$ Then, the complement $V - B_k(x_0)$ of $B_k(x_0)$ in V is the disjoint union of $C_k(x)$, where x runs over $\partial B_k(x_0)$, i.e.,*

$$V - B_k(x_0) = \bigsqcup_{x \in \partial B_k(x_0)} C_k(x).$$

For a proof, see [**U7**].

5. The heat kernel of a semi-regular graph

In this section, we first note the properties of the heat kernel of a semi-regular infinite graph with center x_0.

Let $\mathcal{G}_e^{d,a}$ be a semi-regular infinite graph with center x_0. Then the heat kernel $u_t(x, y)$ of $\mathcal{G}_e^{d,a}$, is of the form:

$$u_t(x, x_0) = H_t^{d,a}(r(x)), \tag{5.1}$$

where $H_t^{d,a}(\bullet)$, $t = 0, 1, \cdots$ are functions only of $r = 0, 1, \cdots$, and $r(x)$ is the graph distance from x_0. We sometimes denote it by $H_t(r)$ simply.

The function $H_t(\bullet)$, $t = 0, 1, \cdots$ satisfies the following:

$$(5.2)\qquad \begin{cases} H_{t+1}(r) = \dfrac{1}{d}\,H_t(r-1) + \dfrac{c}{d}\,H_t(r) + \dfrac{a}{d}\,H_t(r+1), & t,\, r = 1, 2, \cdots, \\ H_{t+1}(0) = H_t(1), \quad t = 1, 2, \cdots, \\ \quad H_t(r) = 0, \quad r \geq t+1, \end{cases}$$

and for $r = 0, 1, 2, \cdots$,

$$(5.3)\qquad H_1(r) = \begin{cases} \dfrac{1}{d} & (r = 1), \\ 0 & (r > 1), \end{cases} \qquad H_0(r) = \begin{cases} 1 & (r = 0), \\ 0 & (r > 0). \end{cases}$$

Furthermore, we obtain

THEOREM 5.4. *Let $\mathcal{G}_e^{d,a}$ be a semi-regular infinite graph with center x_0, and $k := \frac{d}{a+1}$. Define the function $Q_t^{d,a}(\bullet)$ in $r = 0, 1, 2, \cdots$, by*

$$(5.5)\qquad Q_t^{d,a}(\bullet) := H_t^{d,a}(\bullet) + kH_{t+1}^{d,a}(\bullet).$$

We also denote it by $Q_t(\bullet)$ simply. Then, we have, for all $t = 0, 1, 2, \cdots$,

$$(5.6)\qquad \begin{cases} Q_{t+1}(0) = Q_t(1), \\ \quad Q_t(r) \geqq Q_t(r+1), \quad r = 0, 1, 2, \cdots. \end{cases}$$

For a proof, see [**U7**].

Furthermore, we determine the heat kernel of any semi-regular graph $\mathcal{G}_e^{d,a}$.

THEOREM 5.7. *Let $\mathcal{G}_e^{d,a}$ be a semi-regular infinite graph. Let $u_t(x, x_0)$ be the heat kernel, and put $Q_t(r(x)) = u_t(x, x_0) + ku_{t+1}(x, x_0)$, where $k = \frac{d}{a+1}$ and $r(x)$ is the graph distance between x and x_0. Then $u_t(x, x_0)$ and $Q_t(r(x))$ are given as follows:*

(I) *In the case that $t + r$ is an even integer,*

$$u_t(x, x_0) = \sum_{p=0}^{(t-r)/2} \binom{t}{(t-r-2p)/2,\, 2p,\, (t+r-2p)/2} S^{(t-r-2p)/2} T^{2p} U^{(t+r-2p)/2},$$

$$\begin{aligned} Q_t(r(x)) = \sum_{p=0}^{(t-r)/2} &\binom{t}{(t-r-2p)/2,\, 2p,\, (t+r-2p)/2} S^{(t-r-2p)/2} T^{2p} U^{(t+r-2p)/2} \\ &+ \frac{U}{S+U} \sum_{q=0}^{(t-r)/2} \binom{t}{(t-r-2q)/2,\, 2q+1,\, (t+r-2q-2)/2} \times \\ &\qquad \times S^{(t-r-2q)/2} T^{2q+1} U^{(t+r-2q-2)/2}. \end{aligned}$$

(II) *In the case that $t+r$ is an odd integer,*

$$u_t(x,x_0) = \sum_{p=0}^{(t-r-1)/2} \binom{t}{(t-r-2p-1)/2,\, 2p+1,\, (t+r-2p-1)/2} \times$$
$$\times\, S^{(t-r-2p-1)/2} T^{2p+1} U^{(t+r-2p-1)/2},$$

$$Q_t(r(x)) = \sum_{p=0}^{(t-r-1)/2} \binom{t}{(t-r-2p-1)/2,\, 2p+1,\, (t+r-2p-1)/2} \times$$
$$\times\, S^{(t-r-2p-1)/2} T^{2p+1} U^{(t+r-2p-1)/2}$$
$$+ \frac{U}{S+U} \sum_{q=0}^{(t-r+1)/2} \binom{t}{(t-r+1-2q)/2,\, 2q,\, (t+r-1-2q)/2} \times$$
$$\times\, S^{(t-r+1-2q)/2} T^{2q} U^{(t+r-1-2q)/2},$$

where $S = \frac{a}{d}$, $T = \frac{c}{d}$, and $U = \frac{1}{d}$. Here $\binom{n}{\alpha,\beta,\gamma} = \frac{n!}{\alpha!\beta!\gamma!}$ with $n = \alpha+\beta+\gamma$, and $\alpha,\ \beta,\ \gamma \geqq 0$.

In particular, in the case $c = 0$, i.e., $T = 0$, the heat kernel $u_t(x,x_0)$ is given by

$$u_t(x,x_0) = \begin{cases} \binom{t}{(t+r)/2} \dfrac{a^{(t-r)/2}}{d^t} & \text{if } t+r \text{ is even,} \\ 0 & \text{if } t+r \text{ is odd.} \end{cases}$$

For a semi-regular infinite graph $\mathcal{G}_e^{d,a}$ which is rotationally symmetric at a fixed point x_0, and $k := \frac{d}{a+1}$, define the function ${Q^{d,a}}_t(\bullet)$ in $r = 0,\ 1,\ 2,\ \cdots$, by (5.5) and denote by $Q_t(\bullet)$ simply. Then $H_t(r)$ and $Q_t(r)$ satisfy the same equation in $R_t(r)$,

$$R_{t+1}(r) = SR_t(r+1) + TR_t(r) + UR_t(r-1), \tag{5.8}$$

and since (5.3),

$$Q_0(r) = \begin{cases} 1 & (r=0), \\ \dfrac{U}{S+U} & (r=1), \\ 0 & (r \geqq 2). \end{cases} \tag{5.9}$$

For a proof of Theorem 5.7, we first observe that the solution $R_t(r)$ of (5.8) satisfies that

$$\begin{aligned} R_{t+1}(r) &= SR_t(r+1) + TR_t(r) + UR_t(r-1) \\ &= S^2 R_{t-1}(r+2) + 2STR_{t-1}(r+1) + 2SUR_{t-1}(r) + \\ &\quad + T^2 R_{t-1}(r) + 2TUR_{t-1}(r-2) \\ &= \cdots\cdots\cdots\cdots \\ &= \sum_{\substack{\alpha+\beta+\gamma=m \\ \alpha,\beta,\gamma \geqq 0}} \binom{m}{\alpha,\beta,\gamma} S^\alpha T^\beta U^\gamma R_{t+1-m}(r+\alpha-\gamma). \end{aligned}$$

Taking $m = t+1$, and replacing $t+1$ by t, we have

$$R_t(r) = \sum_{\substack{\alpha+\beta+\gamma=t\\ \alpha,\beta,\gamma\geqq 0}} \binom{t}{\alpha,\beta,\gamma} S^{\alpha}T^{\beta}U^{\gamma}R_0(r+\alpha-\gamma).$$

For $R_t(r) = Q_t(r)$, the initial condition is $Q_0(r) = 1\,(r=0)$, $\frac{U}{S+U}\,(r=1)$, and $0\,(r \geqq 2)$, since (5.9). Thus, we obtain the formulas of $Q_t(r)$ in Theorem 5.4. Since the initial condition for $R_t(r) = H_t(r)$ is (5.3), we have the formulas of $H_t(r)$.

Conversely, we can see by a direct computation, that the functions $H_t(r)$ (resp. $Q_t(r)$) of the form in Theorem 5.4 satisfies (5.2) and (5.3) (resp. (5.8) and (5.9)), making use of the binomial formula

$$\binom{m}{\alpha_1,\alpha_2,\cdots,\alpha_r} = \sum_{i=1}^{r}\binom{m-1}{\alpha_1,\cdots,\alpha_i-1,\cdots,\alpha_r}$$

for $m = \alpha_1+\cdots+\alpha_r$ with non-negative integers α_i $(i=1,\cdots,r)$. □

We have also the following bounds for $Q_t(r)$.

PROPOSITION 5.10. (I) *In the case that $t+r$ is an even integer, we have*

$$\binom{t}{(t+r)/2}\frac{a^{(t-r)/2}}{d^t} \leqq Q_t(r) \leqq g_t(r),$$

where

$$g_t(r) := \binom{t}{(t+r)/2}\frac{a^{(t-r)/2}}{d^t}\frac{d}{1+a}\binom{(t-r)/2}{[(t-r)/4]}\binom{(t+r)/2}{[(t+r)/4]}\sum_{p=0}^{(t-r)/2}\frac{c^{2p}}{a^p}.$$

(II) *In the case that $t+r$ is an odd integer, we have*

$$\binom{t}{(t+r-1)/2}\frac{a^{(t-r+1)/2}}{d^t}\frac{1}{1+a} \leqq Q_t(r) \leqq h_t(r),$$

where

$$h_t(r) := \binom{t}{(t+r)/2}\frac{a^{(t-r)/2}}{d^t}\left\{\left(\frac{1}{1+a}+\frac{c}{a}\right)\sum_{p=0}^{(t-r-1)/2}\frac{c^{2p}}{a^p} + \frac{1}{1+a}\frac{c^{t-r+1}}{a^{(t-r+1)/2}}\right\}$$
$$\times \binom{(t-r+1)/2}{[(t-r+1)/4]}\binom{(t+r-1)/2}{[(t+r-1)/4]}.$$

6. An upper estimate of the heat kernel

In this section, we give an upper estimate of the heat kernel.

THEOREM 6.1. *Let $\mathcal{G} = (V,E)$ be an infinite graph, and $u_t(x,y)$, the heat kernel of $\mathcal{G}$, where $t = 0, 1, \cdots$ and $x, y \in V$. We fix an arbitrary vertex x_0 in V. Let*

$$A := \inf\left\{\frac{m_+(x)}{m(x)};\, x \in V-\{x_0\}\right\},\ \textit{and}\ B := \sup\left\{\frac{m_-(x)}{m(x)};\, x\in V-\{x_0\}\right\}.$$

Assume that

$$A > 0 \quad \text{and} \quad A\lfloor \frac{1}{B} \rfloor \geqq 1. \tag{6.2}$$

Let $d = \lfloor \frac{1}{B} \rfloor \geqq 1$*, and* $a = \lfloor Ad \rfloor \geqq 1$*. Since* $d > a$*, we can write* $d = 1 + c + a$ *for some non negative integer* c*. For these integers* d*,* a*, and* c*, let us take* $\mathcal{G}_e^{d,a}$ *with center* $\widetilde{x_0}$ *with some positive integer* e *and define* $H_t^{d,a}(r)$ *by (5.1). Then, we have*

$$u_t(x, x_0) + ku_{t+1}(x, x_0) \leqq H_t^{d,a}(r(x)) + kH_{t+1}^{d,a}(r(x)), \quad t = 0, 1, 2, \cdots ; x \in V,$$

where $r(x)$ *is the graph distance from* x_0*, and* $k = \frac{d}{a+1}$.

We remark that $1 \leqq d$, $a = \lfloor Ad \rfloor \leqq Ad < d$, $d \leqq m(x)$ $(x \in \partial B_1(x_0))$, $B \leqq \frac{1}{d}$, and $\frac{a}{d} \leqq A$.

Before going to the proof, we prepare the following Lemma.

LEMMA 6.3. *Let us define the function* $v_t(x)$ *in* (t, x) *by*

$$v_t(x) = u_t(x, x_0) + ku_{t+1}(x, x_0) - Q_t(r(x)),$$

where $Q_t(r(x)) = H_t(r(x)) + kH_{t+1}(r(x))$*. Then we have*

$$v_{t+1}(x) - v_t(x) + \Delta_x v_t(x) \leqq 0, \quad t = 0, 1, 2, \cdots : x \in V. \tag{6.4}$$

PROOF. Taking $t = 0, 1, 2, \cdots ; x \in V$, the left hand side of (6.4) coincides with

$$\begin{aligned} u_{t+1}(x, x_0) + ku_{t+2}(x, x_0) &- Q_{t+1}(r(x)) - u_t(x, x_0) - ku_{t+1}(x, x_0) + Q_t(r(x)) \\ &+ \Delta_x u_t(x, x_0) + k\Delta_x u_{t+1}(x, x_0) - \Delta_x Q_t(r(x)) \\ &= -Q_{t+1}(r(x)) + Q_t(r(x)) - \Delta_x Q_t(r(x)). \end{aligned} \tag{6.5}$$

Here, if $x \neq x_0$,

$$\begin{aligned} \Delta_x Q_t(r(x)) &= -\frac{m_+(x)}{m(x)}\{Q_t(r(x) + 1) - Q_t(r(x))\} \\ &\quad + \frac{m_-(x)}{m(x)}\{Q_t(r(x)) - Q_t(r(x) - 1)\} \\ &\geqq -\frac{a}{d}\{Q_t(r(x) + 1) - Q_t(r(x))\} \\ &\quad + \frac{1}{d}\{Q_t(r(x)) - Q_t(r(x) - 1)\} \\ &= (\Delta_e^{d,a})_{\widetilde{x}} Q_t(r(\widetilde{x})), \end{aligned}$$

by definitions of A and B, and Remark below Theorem 6.1. Here $\Delta_e^{d,a}$ is the discrete Laplacian for the graph $\mathcal{G}_e^{d,a}$, and $\widetilde{x}$ is the vertex of $\mathcal{G}_e^{d,a}$ which the graph distance $r(\widetilde{x})$ from the fixed vertex $\widetilde{x_0}$ satisfies that $r(\widetilde{x}) = r(x)$.

If $x = x_0$,

$$\Delta_x Q_t(r(x)) = Q_t(0) - \frac{1}{m(x_0)} \sum_{r(y)=1} Q_t(1) = Q_t(0) - Q_t(1) = (\Delta_e^{d,a})_{\widetilde{x}} Q_t(r(\widetilde{x})).$$

Therefore, the right hand side of (6.5) is smaller than or equal to

$$-Q_{t+1}(r(\widetilde{x})) + Q_t(r(\widetilde{x})) - (\Delta_e^{d,a})_{\widetilde{x}} Q_t(r(\widetilde{x})) = 0. \qquad \square$$

PROOF OF THEOREM 6.1. Let us recall first the following maximum principle:

LEMMA 6.6. *Let $\mathcal{G} = (V, E)$ be any locally finite graph, and $D \subset V$, a finite subset of V which is connected, i.e., for two vertices x and y in D, there exist vertices $\{v_i\}_{i=0}^n$ in D satisfying that $x = v_0 \sim v_1 \sim \cdots \sim v_{n-1} \sim v_n = y$. Assume that a real valued function $u(t, x)$ on $\{0, 1, \cdots, m\} \times \overline{D}$ satisfies that*

$$u(t+1, x) - u(t, x) + \Delta_x u(t, x) \leqq 0, \quad t = 0, 1, \cdots, m-1; \ x \in D.$$

Then u attains its maximum at $S_1 \cup S_2$, where $S_1 = \{0, 1, \cdots, m\} \times \partial D$, $S_2 = \{0\} \times D$. Here $\overline{D} = D \cup \partial D$, $\partial D = \{y \in V - D;\ y \sim x \text{ for some } x \in D\}$.

For the proof, see p.213, Lemma 3.1 in [**U1**].

By Lemma 6.3, we apply Lemma 6.6 for the function $v_t(x)$ in (t, x), so we obtain:

$$\begin{aligned}
&\max\{v_t(x);\ t = 0, 1, \cdots, m,\ x \in B_{m+2}(x_0)\} \\
&= \max\{\max\{v_0(x);\ x \in B_{m+2}(x_0)\}, \\
&\qquad \max\{v_t(x);\ t = 0, 1, \cdots, m,\ x \in \partial B_{m+2}(x_0)\}\}.
\end{aligned}$$

We see that v_0 is non positive on V, so that $\max\{v_0(x);\ x \in B_{m+2}(x_0)\} \leqq 0$. Indeed, by definition of $v_0(x)$,

$$v_0(x) = u_0(x, x_0) + k u_1(x, x_0) - Q_0(r(x)).$$

By (5.2) and (5.3), we have

$$Q_0(r(x)) = \begin{cases} 1 & (r(x) = 0), \\ \dfrac{k}{d} & (r(x) = 1), \\ 0 & (r(x) \geqq 2), \end{cases}$$

and by definition of $u_t(x, x_0)$,

$$u_0(x, x_0) + k u_1(x, x_0) = \begin{cases} 1 & (r(x) = 0), \\ \dfrac{k}{m(x)} & (r(x) = 1), \\ 0 & (r(x) \geqq 2). \end{cases}$$

By Remark below Theorem 6.1, $\frac{1}{m(x)} \leqq \frac{1}{d}$ for all $x \in \partial B_1(x_0)$, hence $v_0(x) \leqq 0$ everywhere on V.

On the other hand, by definition of $v_t(x)$ and (5.2), we have

$$v_t(x) = u_t(x, x_0) + k u_{t+1}(x, x_0) - Q_t(r(x)) = 0, \ t = 0, 1, \cdots, m, \ x \in \partial B_{m+2}(x_0).$$

Thus, we obtain

$$\max\{v_t(x);\ t = 0, 1, \cdots, m,\ x \in B_{m+2}(x_0)\} \leqq 0.$$

As $m \to \infty$, $v_t(x) \leqq 0$ $(t = 0, 1, 2, \cdots, x \in V)$, which is the desired. □

We have immediately

COROLLARY 6.7. *Under the same assumption of Theorem 6.1, the heat kernel and the Green kernel of $\mathcal{G} = (V, E)$ can be estimated from above as follows:*

$$u_t(x, x_0) + k u_{t+1}(x, x_0) \leqq Q_t^{d,a}(r(x)) \quad (t = 0,\, 1,\, 2,\, \cdots;\ x \in V),$$
$$G(x, x_0) \leqq \frac{1}{a-1} a^{1-r(x)}, \quad x \in V.$$

We remark that an upper bound for $Q_t^{d,a}(r)$ is given by Proposition 5.10.

7. A lower estimate of the heat kernel

In this section, we give a lower estimate of the heat kernel.

THEOREM 7.1. *Let $\mathcal{G} = (V, E)$ be an infinite graph, and $u_t(x, y)$, the heat kernel of $\mathcal{G}$. We fix an arbitrary vertex x_0. Let*

$$C := \sup\left\{\frac{m_+(x)}{m(x)};\ x \in V - \{x_0\}\right\}, \text{ and } D := \inf\left\{\frac{m_-(x)}{m(x)};\ x \in V - \{x_0\}\right\}.$$

Let $\widetilde{d} = \lceil \frac{1}{D} \rceil$, and $\widetilde{a} = \lceil C\widetilde{d}\, \rceil$. Since $\widetilde{a} < \widetilde{d}$ (cf. Remark below), we can write $\widetilde{d} = 1 + \widetilde{c} + \widetilde{a}$ for some non negative integer $\widetilde{c}$. For these integers $\widetilde{d}$, $\widetilde{a}$, and $\widetilde{c}$, taking $\mathcal{G}_{\widetilde{e}}^{\widetilde{d},\widetilde{a}}$ with center $\widetilde{x_0}$ for some positive integer $\widetilde{e}$. Let us denote the heat kernel of $\mathcal{G}_{\widetilde{e}}^{\widetilde{d},\widetilde{a}}$ by $\widetilde{H}_t(r)$ (cf. (5.1)). Then we have

$$\widetilde{H}_t(r) + \widetilde{k}\widetilde{H}_{t+1}(r) \leqq u_t(x, x_0) + \widetilde{k} u_{t+1}(x, x_0), \quad t = 0,\, 1,\, 2,\, \cdots;\ x \in V,$$

where $r(x)$ is the graph distance from x_0 and $\widetilde{k} = \frac{\widetilde{d}}{\widetilde{a}+1}$.

Here, we remark that $2 \leqq \widetilde{d}$, $1 \leqq \widetilde{a} < \widetilde{d}$, $m(x) \leqq \widetilde{d}$ for all $x \in \partial B_1(x_0)$, $\frac{1}{\widetilde{d}} \leqq D$, and $C \leqq \frac{\widetilde{a}}{\widetilde{d}}$.

For a proof of Theorem 7.1, see [**U7**].

We have immediately

COROLLARY 7.2. *Under the same situation of Theorem 7.1, the heat kernel and the Green kernel of $\mathcal{G} = (V, E)$ can be estimated from below as follows:*

$$u_t(x, x_0) + \widetilde{k} u_{t+1}(x, x_0) \geqq Q_t^{\widetilde{d},\widetilde{a}}(r(x)) \quad (t = 0,\, 1,\, 2,\, \cdots;\ x \in V),$$
$$G(x, x_0) \geqq \frac{1}{\widetilde{a}-1} \widetilde{a}^{1-r(x)}, \quad x \in V.$$

Combining the above and Proposition 5.10, we obtain

COROLLARY 7.3. *For an infinite graph $\mathcal{G} = (V, E)$ and any fixed vertex x_0 in V, let us consider the constants C and D and take $\widetilde{d} = \lceil \frac{1}{D} \rceil$, and $\widetilde{a} = \lceil C\widetilde{d}\, \rceil$. Then,*

(I) *In the case that $t + r$ is an even integer,*

$$u_t(x, x_0) + \widetilde{k} u_{t+1}(x, x_0) \geqq \binom{t}{(t+r)/2} \frac{\widetilde{a}^{(t-r)/2}}{\widetilde{d}^t}.$$

(II) *In the case that $t + r$ is an odd integer, we have*

$$u_t(x, x_0) + \widetilde{k} u_{t+1}(x, x_0) \geqq \binom{t}{(t+r-1)/2} \frac{\widetilde{a}^{(t-r+1)/2}}{\widetilde{d}^t} \frac{1}{1+\widetilde{a}}.$$

8. Examples

In this section, we give examples and see [**U7**] for the other examples.

EXAMPLE 1. We can give an example $\mathcal{G} = \mathcal{G}_e^{5,2}$ with $e \geqq 2$ (see Figure 8.1). We have that (1) $h(\mathcal{G}) = \frac{1}{5}$, (2) $u_t(x, x_0) = H_t^{5,2}(r(x))$, and (3) $G(x, x_0) = 2^{1-r(x)}$.

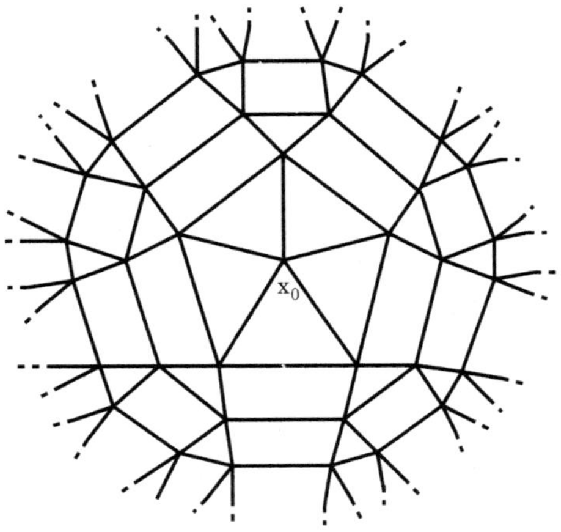

(Figure 8.1)

EXAMPLE 2. Let $\mathcal{G} = (V, E)$ be the distance regular graph $D_{m,s}$, m, $s \geq 2$, i.e., all of which vertices are the intersection of m copies of the complete graph K_s. $\mathcal{G}$ is $(s-1)m$-regular, and $m_+(x) = (s-1)(m-1)$, $m_-(x) = 1$, and $m_0(x) = s-2$ for all $x \in V - \{x_0\}$. $\mathcal{G}$ is a rotationally symmetric, semi-regular graph $\mathcal{G}_{(s-1)m}^{(s-1)m,(s-1)(m-1)}$.

$$h(\mathcal{G}) = \frac{(s-1)(m-1)-1}{(s-1)m}, \tag{1}$$

$$u_t(x, x_0) = H_t^{(s-1)m,(s-1)(m-1)}(r(x)), \tag{2}$$

$$G(x, x_0) = \frac{1}{(s-1)(m-1)-1}\Big((s-1)(m-1)\Big)^{1-r(x)}. \tag{3}$$

EXAMPLE 3. Let $\mathcal{G} = (V, E)$ be the free product $K_r * K_s$ $(r \geqq s \geqq 2)$. It is $(r+s-2)$-regular, and $m_+(x) = s-1$, or $r-1$, $m_-(x) = 1$, and $m_0(x) = r-2$ or $s-2$. We have $A = \frac{s-1}{r+s-2}$, $B = \frac{1}{r+s-2}$, $C = \frac{r-1}{r+s-2}$, and $D = \frac{1}{r+s-2}$, $d = r+s-2$,

$a = s-1$, $\widetilde{d} = r+s-2$, and $\widetilde{a} = r-1$. Hence,

$$(1) \qquad \frac{s-2}{r+s-2} \leqq h(\mathcal{G}),$$

$$(2) \qquad u_t(x,x_0) + \frac{r+s-2}{s} u_{t+1}(x,x_0) \leqq H_t^{r+s-2,s-1}(r(x)) + \frac{r+s-2}{s} H_{t+1}^{r+s-2,s-1}(r(x)),$$

$$H_t^{r+s-2,r-1}(r(x)) + \frac{r+s-2}{r} H_{t+1}^{r+s-2,r-1}(r(x)) \leqq u_t(x,x_0) + \frac{r+s-2}{r} u_{t+1}(x,x_0),$$

$$(3) \qquad \frac{1}{r-2}(r-1)^{1-r(x)} \leqq G(x,x_0) \leqq \frac{1}{s-2}(s-1)^{1-r(x)}.$$

If $r = s \geqq 2$, $\mathcal{G}$ is a rotationally symmetric, semi-regular graph $\mathcal{G}_{2r-2}^{2r-2,r-1}$, (1) $h(\mathcal{G}) = \frac{r-2}{2r-2}$, (2) $u_t(x,x_0) = H_t^{2r-2,r-1}(r(x))$, and (3) $G(x,x_0) = \frac{1}{r-2}(r-1)^{1-r(x)}$.

References

[A] N. Alon, *Eigenvalues and expanders*, Combinatorica **6** (1986), 83–96.

[BS] I. Benjamini and O. Schramm, *Every graph with a positive Cheeger constant contains a tree with a positive Cheeger constant*, Geom. Funct. Anal. **7** (1997), 403–419.

[CG] T. Coulhon and A. Grigoryan, *Random walks on graphs with regular volume growth*, GAFA **8** (1998), 656–701.

[DKa] J. Dodziuk and L. Karp, *Spectral and function theory for combinatorial Laplacians*, Contemp. Math. **73** (1988), 25–40.

[DK] ——— and W. S. Kendall, *Combinatorial Laplacians and isoperimetric inequality*, From Local Times to Global Geometry, Control and Physics (K. D. Elworthy, ed.), Longman Scientific & Technical, 1986, pp. 68–75.

[F1] K. Fujiwara, *On the bottom of the spectrum of the Laplacian on graphs*, Geometry and Its Applications (T. Nagano et al., eds.), World Scientific, Singapore, 1993, pp. 21–27.

[F2] ———, *Growth and the spectrum of the Laplacian for an infinite graph*, Tohoku Math. J. **48** (1996), 293–302.

[H] Yu. Higuchi, *Boundary area growth and the spectrum of discrete Laplacian*, Ann. Global Anal. Geom. **24** (2003), 201–230.

[K1] A. Kasue, *A Laplacian comparison theorem and function theoretic properties of a complete Riemannian manifold*, Japan. J. Math. **8** (1982), 309–341.

[K2] ———, *Ricci curvature, geodesics and some geometric properties of Riemannian manifolds with boundary*, J. Math. Soc. Japan **35** (1983), 117–131.

[K3] ———, *Applications of Laplacian and Hessian comparison theorems*, Advanced Studies Pure Math. **3** (1984), 333–386.

[KSS] M. Kotani, T. Shirai and T. Sunada, *Asymptotic behavior of the transition probability of a random walk on an infinite graph*, J. Funct. Anal. **159** (1998), 664–689.

[Mo] B. Mohar, *Isoperimetric inequalities, growth, and the spectrum of graphs*, Linear Alg. Appl. **103** (1988), 119–131.

[MW] ——— and W. Woess, *A survey on spectra of infinite graph*, Bull. London Math. Soc. **21** (1989), 209–234.

[T] J. Tan, *On Cheeger inequalities of a graph, thesis* (2000), Tohoku University.

[U1] H. Urakawa, *Heat kernel and Green kernel comparison theorems for infinite graphs*, J. Funct. Anal. **146** (1997), 206–235.

[U2] ———, *The eigenvalue comparison theorems for the discrete Laplacians for a graph*, Geometriae Dedicata **74** (1999), 95–112.

[U3] ———, *The spectrum of an infinite graph*, Canadian J. Math. **52** (2000), 1057–1084.

[U4] ———, *A discrete analogue of the harmonic morphism and Green kernel comparison theorems*, Glasgow Math. J. **42** (2000), 319–344.

[U5] ———, *A discrete analogue of the harmonic morphism*, Harmonic morphism, harmonic maps, and related topics (Brest, 1997), Res. Notes Math., vol. 413, Chapman & Hall/CRC, 2000, pp. 97–108.

[U6] ———, *Laplacian and Networks*, Shokabo, Tokyo, 1996. (Japanese)

[U7] ———, *The Cheeger constant, the heat kernel, and the Green kernel of an infinite graph*, Monatshefte für Math. **138** (2003), 225–237.

[W1] W. Woess, *Random walks on infinite graphs and groups – A survey on selected topics jour Bull. London Math. Soc.* **26** (1994), 1–60.

[W2] ———, *Random Walks on Infinite Graphs and Groups*, Cambridge Univ. Press, Cambridge, New York, Melbourne, Madrid, 2000.

[Y] S. T. Yau, *Isoperimetric constants and the first eigenvalue of a compact Riemannian manifold*, Ann. Scient. Éc. Norm. Sup. **8** (1975), 487–507.

Division of Mathematics, Graduate School of Information Sciences, Tohoku University, Aoba 09, Sendai, 980-8579, Japan

E-mail address: urakawa@math.is.tohoku.ac.jp

Titles in This Series